Advanced Textile Testing Techniques

Advanced Textile Testing Techniques

Edited by
Sheraz Ahmad, Abher Rasheed, Ali Afzal,
and Faheem Ahmad

CRC Press is an imprint of the
Taylor & Francis Group, an **informa** business

CRC Press
Taylor & Francis Group
6000 Broken Sound Parkway NW, Suite 300
Boca Raton, FL 33487-2742

Printed on acid-free paper

International Standard Book Number-13: 978-1-1387-4633-6 (Paperback)

Visit the Taylor & Francis Web site at
http://www.taylorandfrancis.com

and the CRC Press Web site at
http://www.crcpress.com

Contents

Preface

Textiles are being used in every walk of life from clothing to space stations. Their application depends upon their performance properties. Textile substrates are widely diversified, and modifications are being made to better their performance and prolong their life. Advanced and engineered materials are also included in textiles along with conventional textile materials to obtain particular properties in required applications. The intensive research in textiles opens new application fields that require high technical specifications. Their performance depends upon their quality with respect to the specifications of the application.

The global increase in the share of technical textiles increased the pressure on their quality of production and performance. The testing of products plays a vital role in judging and designating quality. The properties of technical textiles are highly diversified due to which a number of testing prototypes are developed according to the intended application. This book has been written keeping in view the basic knowledge of conventional and technical textile testing techniques. It provides key parameters required for particular products with their testing methods and techniques.

This book is divided into two main sections: one discusses testing of conventional textiles and the other explains testing of technical textiles. The first three chapters have been written jointly for both conventional and technical textiles. Chapter 1 aims to develop an understanding of the importance of textile testing and its influence on the quality of a product. Chapters 2 and 3 discuss the characterization of textile polymers considering conventional as well as advanced technical polymers. This section provides basic knowledge on different polymeric properties and their basic and advanced characterization methods.

Section I of this book explains the conventional textile testing techniques starting from fiber to final product. Chapter 4 has been written on textile fibers, their physical and chemical identification methods, and their properties. Chapter 5 deals with testing of conventional yarns and their properties. The testing of greige fabrics in both woven and knitted structures is introduced in Chapter 6. Fabric inspection and quality assurance are also discussed in this chapter. The quality of finished and dyed fabrics is discussed under testing procedures in Chapter 7. This chapter also includes the testing of finishing effects, which are commonly used in clothing and finished products. Chapter 8 is focused on testing procedures for products used in apparel and home textile applications.

Section II is focused on testing of technical textiles. Chapter 9 discusses testing of textile composites. It includes their physical, mechanical, and non-destructive testing techniques. Chapter 10 deals with nonwoven fabrics, their technology, construction, characterization, and destructive and non-destructive testing methods. Medical textiles, a widely recognized field, is covered in Chapter 11. The electronic and SMART textiles product testing techniques and their possible characterization methods has been discussed in Chapter 12.

The contributing authors and editors have put in their best efforts to summarize the basic and advanced knowledge on textiles in this book. To-the-point information has been reported, considering the constraint on the length of the book. The editors acknowledge the efforts made by all the contributing authors in their respective chapters. We hope that the reader will benefit from and enjoy reading this book.

Sheraz Ahmad
Abher Rasheed
Ali Afzal
Faheem Ahmad

Acknowledgments

We set our unfeigned and meek thanks before the Almighty, who created the universe and bestowed humankind with knowledge and wisdom to search for its secret and favored and invigorated us with the fortitude and capability to aptly complete this work and contribute a drop to the existing ocean of scientific knowledge. We extend our deepest gratitude to the chapter authors for their precious time toward completing this book. We are also obliged to all those who provided support for the improvement of the manuscript. We offer special thanks to Dr. Yasir Nawab and Khubab Shaker who provided guidance throughout the writing of this book. We also express our gratitude to CRC Press/Taylor & Francis Group for providing us the opportunity to contribute toward scientific knowledge.

Editors

Dr. Sheraz Ahmad holds a master's degree in fiber technology from the University of Agriculture and a PhD in textile engineering with a focus on textile materials from Université de Haute Alsace, France. He is working as an assistant professor at the National Textile University, Pakistan, since October 2012 as chairman of the materials and testing department; he teaches undergraduate, MS, and PhD level students; is doing research on textile fibers, conductive yarn, and recycled materials; and has authored more than 13 peer-reviewed journal articles, 2 books, and 8 conference communications. Dr. Ahmad is fluent in English, French, Urdu, and Punjabi. He has experience in developing course curricula as well as executing field trips, laboratory exercises, and other activities beyond traditional lectures.

Dr. Abher Rasheed is an assistant professor at the National Textile University, Pakistan. He is a textile engineer. Further, he has completed his masters in textile materials and processes and in total quality management. In addition to that, he received his PhD from Université de Haute Alsace, France, in the domain of SMART textiles. His research areas are e-textiles, clothing manufacturing, and quality management. He has filed one international patent and published 12 peer-reviewed publications. Furthermore, he has contributed a book chapter as well.

Ali Afzal is an assistant professor of materials science at the National Textile University, Pakistan. He did his doctorate from Université de Haute Alsace, France, in SMART textiles. He completed his master's degree from the National Textile University with distinction in 2012. He has published a number of research papers in high-impact-factor international journals. His research and teaching experience spans more than 5 years after having 1 year of industrial experience. His main research areas include melt extrusion, SMART textiles, and clothing comfort.

Faheem Ahmad is currently doing his PhD in materials science and engineering from Koç University, Turkey. His primary area of expertise is textile materials. He is currently working to develop nonwoven composites for heat insulation applications. He is also a teaching assistant at Koç University, Turkey. He has worked as a lecturer at the National Textile University, Pakistan, for more than 5 years. He has nine international research publications to his credit. Ahmad also has experience working in textile industries in Pakistan.

Contributors

Ali Afzal
Faculty of Engineering and Technology
National Textile University
Faisalabad, Pakistan

Faheem Ahmad
Faculty of Engineering and Technology
National Textile University
Faisalabad, Pakistan

Sheraz Ahmad
Faculty of Engineering and Technology
National Textile University
Faisalabad, Pakistan

Khurram Shehzad Akhtar
Faculty of Engineering and Technology
National Textile University
Faisalabad, Pakistan

Nauman Ali
Faculty of Engineering and Technology
National Textile University
Faisalabad, Pakistan

Alvira Ayoub Arbab
Department of Textile Engineering
Mehran University of Engineering and Technology
Jamshoro, Pakistan

Fiaz Hussain
Department of Polymer Science and Engineering
Sungkyunkwan University
Suwon, South Korea

Madeha Jabbar
Faculty of Engineering and Technology
National Textile University
Faisalabad, Pakistan

Awais Khatri
Department of Textile Engineering
Mehran University of Engineering and Technology
Jamshoro, Pakistan

Muhammad Mohsin
Department of Textile Engineering
University of Engineering and Technology Lahore, Faisalabad Campus
Faisalabad, Pakistan

Yasir Nawab
Faculty of Engineering and Technology
National Textile University
Faisalabad, Pakistan

Muhammad Umar Nazir
Faculty of Engineering and Technology
National Textile University
Faisalabad, Pakistan

Abher Rasheed
Faculty of Engineering and Technology
National Textile University
Faisalabad, Pakistan

Ateeq ur Rehman
Faculty of Engineering and Technology
National Textile University
Faisalabad, Pakistan

Iftikhar Ali Sahito
Department of Textile Engineering
Mehran University of Engineering and Technology
Jamshoro, Pakistan

Khubab Shaker
Faculty of Engineering and Technology
National Textile University
Faisalabad, Pakistan

Azeem Ullah
Faculty of Engineering and Technology
National Textile University
Faisalabad, Pakistan

Muhammd Umair
Faculty of Engineering and Technology
National Textile University
Faisalabad, Pakistan

Usman Zubair
Department of Applied Science and Technology
Politecnico di Torino
Turin, Italy

1

Introduction to Textile Testing

Sheraz Ahmad

CONTENTS

1.1 What Is Textile Testing?

The quality of a product or process is checked before it is put into large-scale usage. The quality of the product, its performance, and its reliability are the key factors while testing is performed. Testing can be defined as the methods or protocols adopted to verify/determine the properties of a product. It can be divided primarily into two types: regular process testing and quality assurance testing. Routine testing helps to streamline the daily process. Quality assurance testing helps the process or product in the long run to establish credibility. Testing can also be defined as the procedures adopted to determine a product's suitability and quality [1,3].

Textile testing is a vital basic tool during the processing of a textile raw material into the product. It also helps the distributors and consumer to determine the end product's quality. So, textile testing refers to the procedures adopted to determine quality throughout the textile product chain. It can be summarized as the application of engineering facts and science to determine the quality and properties of a textile product [2].

1.2 Objectives of Textile Testing

The objectives are

- For researchers, testing results aid the development of new products or new processes, which can save money and resources before production starts on an industrial scale. They also help in the choosing of the best possible route to achieve the end product.
- Testing helps in the selection of the best possible raw materials. "Raw material" is a relative term; for example, fiber is the raw material for spinning, and yarn is the raw material for weaving.
- Testing helps in the process control through the use of advanced textile process-control techniques.
- Testing ensures the right product is shipped to the consumer or customer and that the product meets the customer specifications.

Testing in general, and textile testing in particular, is affected by the following factors: Atmospheric conditions affect test results as textile products are greatly influenced by moisture and humidity. The test method adopted will also cause variation in test results. The testing instrument is also a vital part and, if not properly calibrated, can cause serious variation. Human error is another source of variation [3,5].

Textile testing starts with textile fibers and goes all the way through to the final product. The fiber test includes the length, strength, elongation, fineness, and maturity. The yarn test includes linear density, single yarn strength, yarn evenness, and yarn hairiness. The fabric test includes aerial density, weave type, and air permeability.

In order to carry out the testing of the textile products, a well-equipped laboratory with a wide range of testing equipment is needed. Well-trained operators are also a prerequisite for the running of the laboratory. The cost of establishing and running the lab is nonproductive and is added to the cost of the final product. These nonproductive costs increase the cost of the final product and therefore make it an expensive trade. Therefore, it is vital that testing is not performed without accumulation of some payback to the end product. Testing is carried out at a number of points in a production cycle to improve the quality of the product [2,4].

1.3 Types of Textile Testing

Textile testing can be classified according to the basic technique used and on the basis of the data obtained. The former can be divided into destructive and nondestructive testing, the latter being defined as the application of noninvasive methods to reach a conclusion as to the quality of a material, process, or product. In other words, it is inspection or measurement without doing damage to the test specimen. Examples include drape testing and assessment based on the Kawabata evaluation system. Destructive testing is performed to test for failure of the sample. This type of test is much easier to perform and yield precise information and is more simple to understand than nondestructive testing. Examples of destructive testing are tensile testing and tear testing [5–7].

On the basis of data obtained, testing can be classified into objective and subjective. The former can be defined as the testing that gives us quantitative data, which can be easily further processed and interpreted. Subjective testing can be defined as that which gives us qualitative data, which is difficult to interpret and is greatly influenced by operator bias [6].

1.4 Importance of Textile Testing

Testing importantly supports the personnel involved in the textile supply chain, from the textile fiber to the end product. Persons involved in textiles should have knowledge of production as well as statistics. This helps them to interpret data efficiently.

The testing of textile products aids persons involved in the running of the production line. During testing, the discrepancy of the product, for example, its strength, maturity, waste percentage (for fibers), aerial density, and weave design (for fabrics), is properly measured. Thus the selection of the proper raw material is an important factor. Standards of control should be maintained to reduce waste, minimize price, and so on. Faulty machine parts or improper maintenance of the machines can be easily detected with the help of textile testing. Improved, less costly, and faster protocols can be developed by researchers with the aid of testing. The efficiency and quality of the product can also be enhanced with the help of regular and periodic testing. Customer satisfaction and loyalty can also be won by producing according to customer specification in good time. In short, testing is an essential pivot to the whole textile product supply chain [4,8].

The cycle of testing starts with the arrival of raw material and continues up to delivery of the final product. The production of the required end quality is impossible if the raw material is incorrect. The textile product supply chain comprises different processes, which include the raw material (natural or man-made fiber), yarn manufacturing, fabric manufacturing, textile processing, and apparel and home furnishing manufacturing. It also includes some industrial products, like ropes, cords, and conveyer belts. All the aforementioned processes are performed in separate units or in a single unit if the establishment is a vertical production unit. So the raw material for a spinning unit is fiber, for a weaving unit is yarn, for a textile processing unit is greige fabric, and so on. "Raw material" is a relative term that depends upon the further process for which it is used. Its testing is an important step, as improper raw material or low grade raw material will not yield the required quality of the end product. The testing of raw material is also performed to verify whether the incoming material accords with the trade agreement. Its consignment is therefore accepted or rejected on the basis of test results. The agreed specifications should be realistic so that the incoming raw material properties can meet the required level easily [1,3].

Production monitoring involves the testing of production line samples, which is termed "quality control." Its purpose is to sustain certain definite properties of the end product within acceptable tolerance limits as per the agreement between the producer and the consumer. A product that does not meet the already agreed specification or the required quality will be termed a "fail."

The proper testing protocol as well as proper monitoring are also required. The sampling techniques in use should also be selected properly, since the wrong selection could lead to serious problems. At the same time, the statistical tool employed is also an important factor. The collection of data is one thing, but its proper evaluation and interpretation, and the action taken on the basis of it, is an important factor in quality control. In this process, before the delivery of the product to the customer, the whole consignment is examined to check whether the product meets the specification or not. At this stage, product specifications cannot be changed. Depending upon the quantity of the product, a sample check or even a 100% check is carried out. The results thus obtained are utilized to rectify any possible faults or faulty products. For example, some of the faults observed in fabric are considered to be mendable faults and are rectified by skilled workers. This is the normal process and faults rectified in this way are shipped as a fine quality consignment. Faults that are normally detected at the final stage of production, through end product inspection or at the customer end, are raised to the producer as a complaint. It is vital to identify the particular cause of that fault so that it can be avoided in future consignments. It will also help to rectify the running process and enable it to run more smoothly so that the final product accords with the customer's requirement. It also helps to isolate the faulty part or machine so as to resolve any dispute between the supplier and the producer. If there is a dispute, third-party audit or testing is preferred so that the dispute can be settled amicably. Third-party results are acceptable to all concerned as they provide unbiased results or observations [3].

Due to research and development in this sector, textile science and technology is an ever changing domain, and introduction of new and improved raw materials and optimized innovative production methods are being incorporated. New methods need to be verified before products reach the consumer, to see whether there is some improvement or degradation due to the new faster method. The process and the usage of different raw materials can be optimized by the use of testing. A good and well reputed organization will have a properly set up and well equipped research and development section to carry out new product development and the optimization of existing processes and products. It will help the production department with quality control. However, for small units, the testing staff are responsible for all these activities apart from their routine testing [2].

1.5 Importance of Standards

The tested textile materials should satisfy certain specifications. Some of these requirements are implicit and others are explicit. The latter are those that indicate a material's performance in service or whether it will meet its specifications or not. The implicit requirement is that the test is repeatable, that is the textile material will give the same results if it is tested again after some time by another technician or at some other place or the customer's laboratory. In other words the test can measure the correct value of the property being assessed. There is no use in testing if it is not reproducible, as it will then count for nothing [6]. A lack of reproducibility of results can be attributed to the following.

Textile materials have natural variation, for example, fibers obtained from a natural source have variation among their properties. In the material process from fiber to yarn to fabric, the variations in properties smooth out during the assembly of small variable units into large units. The problem of material variation can be rectified with the help of the proper selection of raw material and the use of appropriate statistical tools while analyzing and interpreting the data thus obtained. It is important to minimize the variations caused by the test method [5].

The possible reasons for variations caused by the test method are

1. The technician has significant influence on the result. This is attributed to human error, human negligence, and not following the proper testing protocols. The preparation of the test specimen, the use of the proper instrument, the placement of the specimen on the testing machine, the noting of the value, and the adjusting of the scale properly are all sources of these types of variations.
2. An improper specimen size will also give an inaccurate result. For example, the length of the specimen in the case of tensile testing will

affect the strength value of the specimen. So a change in specimen size will cause variation in the test results.

3. Atmospheric conditions are a very important factor when checking natural fibers. Fibers like cotton, viscose, and wool are greatly affected by changes in temperature and relative humidity. The results will show variation if conditions are changed while the test is being performed. The proper conditioning of the test specimen and the laboratory needs to adhere to specific parameters.
4. The use of proper test protocols is necessary to minimize variation. Pilling can be checked by a pilling box as well by the "Martindale abrasion tester." The results obtained from the two types of testing equipment will not be comparable, as the methods involved are different, resulting in a variation of results.
5. The parameters used to perform tests, such as the speed of the machine or the pressure applied, will affect the final results. When these change, the results will also be changed.

It is therefore necessary to lay down the conditions of a test and the specific dimensions of the specimen, and also to define a test procedure that minimizes operator variability, even within a single organization [1,3].

In the case of the selling and buying of the product, it is important that both parties will get the same results when they test the same material. Disputes may arise due to the improper testing of the end product, which can lead to severe legal action or the cancelation of the sale agreement. The test protocols employed by more than one industry should be carefully monitored and identified so that reproducibility of results can be obtained; for example, the atmospheric conditions of the testing laboratory should be properly specified. Procedures should be clearly defined and explained in detail so that there is no ambiguity; when a test is performed in different places it should yield the same results. From this, the adoption of standard test methods arises, which should be well written up for carrying out different types of tests. These standards should specify the dimensions of the test specimen, the different test speeds required to perform the test, and the atmospheric conditions needed so that the results are reproducible. So, large-scale organizations such as Levis® and Marks & Spencer® have developed their own testing protocols. The producers have to satisfy the requirement of these protocols if they want to sell their products to these organizations.

Most countries have their own standards organizations, for example, BS (Britain), ASTM (United States), and DIN (Germany). National standards are required to assist worldwide trade, hence the existence of International Organization for Standardization test methods and, within the European Union, the drive to European standards [3].

1.6 Sampling Techniques

1.6.1 Sampling

The total raw material bought or the total end production is not 100% tested due to cost effectiveness and time constraints. The destructive type of testing will also increase the wastage in the process, which will ultimately increase the cost of testing, resulting in overall profit being decreased. Therefore sampling techniques are employed and representative samples of the whole material are tested.

For the testing of cotton fiber, a 20 mg weight of sample is taken from a 250 kg bale. The sample only represents a portion of the bulk, but the quality of the whole population will be evaluated on the basis of it. The sample from the bale is taken in such a way that each group's fiber length has an equal opportunity to be selected. The main objective is to get an unbiased test sample that represents the whole population; each and every part of the possible length group is represented in the sample [1].

1.6.2 Sampling Techniques

1.6.2.1 Zoning Technique

Zoning is a popular technique for fiber as it is used for selecting samples from raw material such as cotton or wool or other loose natural fibers. The properties of these natural fibers may vary significantly from place to place. A small tuft of fibers is taken at random from each of at least 40 widely spaced places (zones) throughout the bulk of the consignment (Figure 1.1) [1].

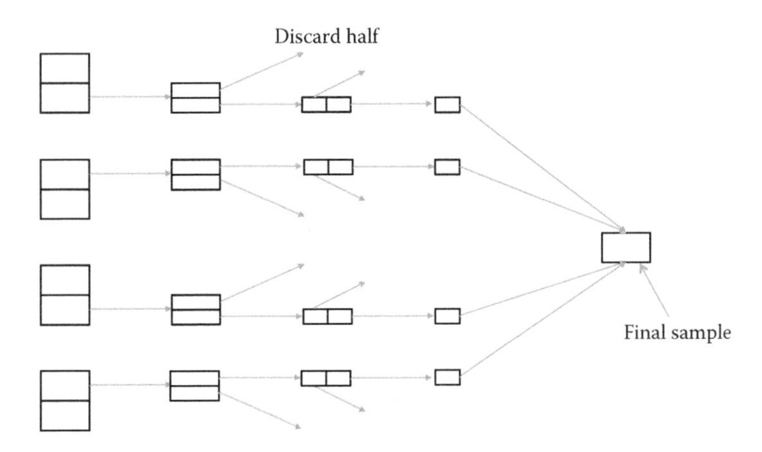

FIGURE 1.1
Sampling by zoning.

1.6.2.2 Core Sampling

The core sampling technique is used for assessing the proportion of foreign matter, the waste percentage, and the moisture content in the compressed unopened bales of cotton or wool. A tube with a sharpened tip is forced into the bale and a core of wool or cotton is withdrawn. The technique was first used as core boring in which the tube was rotated by a transportable electric drill. It was then developed further to facilitate the cores to be cut by pressing the tube into the bale by hand. This enables samples to be taken in areas distant from sources of power (Figure 1.2) [3].

1.6.2.3 Random Sampling

The random sampling technique is the most widely used technique. The steps involved are as follows: determine the size of the population; determine the sample size; prepare a random numbers table; determine the number of each item in a sample; collect the sample. The following types of random sampling are used in the industry:

Stratified random sampling. This is done by dividing the population into several mutually exclusive regions.

Cluster sampling. This is done by subdividing the population into groups or clusters and taking a sample from each.

Selected sampling. In this type, the samples are collected from one part of the population.

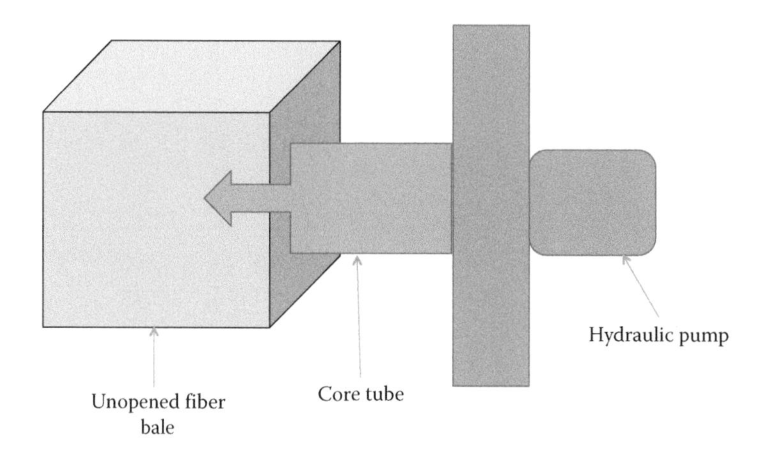

FIGURE 1.2
Core sampling.

Systematic sampling. This is performed systematically at a regular interval.

Acceptance sampling. This is used to accept the incoming raw material or for quality assurance of outgoing consignments [3].

1.6.2.4 Sampling Errors

The following types of errors might occur during sampling:

- Sampling only from the surface of a liquid at rest
- Sampling from edge of sheet
- Sampling from one segment of a lot
- Instrument calibration
- Improper reading
- Lack of accuracy of sample

1.6.2.5 Sources of Error

Errors can arise from a number of different sources. The properly calibrated instruments can still be a source of error if the test is not performed according to the specified protocol. Defects in machine parts, such as play due to wear and tear, slackness while mounting the specimen on the instrument, and vibrations (mechanical and electrical), can produce variation in test results.

The majority of tests involve technicians using testing instruments and then noting the values carefully. This can lead to human bias, as an operator who is well trained will manipulate the instrument more accurately as compared to one who is not. A well trained operator will prepare the sample carefully, mount it on the machine, adjust the machine, and then take the reading carefully. However, an untrained operator may cause variation due to the certain reason involved in the specimen testing. Further, the number of repetitions may be a cause of variation. So it is ideal if proposed protocols involve a minimum amount of human interaction so that results can be obtained with minimum variation.

A basic parameter to lay down is the accuracy of measurement by the testing equipment scale, which is primarily subdivided at fixed intervals. This shows that there is the possibility of error with respect to one-half of the fixed scale division. The accuracy of the digital scale is shown by the last digit of the display as it has by its nature to be a whole digit. The final digit implies that it is plus or minus half of what would be the next digit. However, digital scales usually read to more figures than the equivalent analog scale.

Other sources of variation may be due to some external factor like a sudden voltage fluctuation in the power supply, vibration due to wear and tear in mechanical parts, or fluctuation in the temperature and relative humidity. All these types of variation arise from outside of the instrument involved in the actual testing of the end product. Similarly, in areas with bright sunshine, factors such as intensity of sunlight may also be a cause of variation [1].

1.7 Statistical Terms

The majority of textile testing involves repetitions of a certain test for a number of times. Statistically, the number of repetitions should be a minimum of five. It is necessary to analyze statistically the data thus obtained. This shows the average of the measured values as well as their deviation. Commonly used statistical terms are defined briefly as follows:

- *Arithmetic mean or average.* In statistics, the term "average" refers to any of the measures of central tendency. The arithmetic mean is defined as being equal to the sum of the numerical values of each and every observation, divided by the total number of observations [9].
- *Standard deviation.* This is a numerical value used to indicate how widely individuals in a group vary [9].
- *Coefficient of variation.* This is also known as relative standard deviation and is a standardized measure of dispersion of a probability distribution or frequency distribution. It is often expressed as a percentage, and is defined as the ratio of the standard deviation to the mean [9].
- *Standard error of the mean.* This estimates the variability between sample means that you would obtain if you took multiple samples from the same population. The standard error of the mean estimates the variability between samples, whereas the standard deviation measures the variability within a single sample [9].

References

1. B. P. Saville, *Physical Testing of Textiles*, 1st edn. Cambridge, U.K.: CRC Press, 1999.
2. J. E. Booth, *Principle of Textile Testing*, 4th edn. London, U.K.: Chemical Publishing, 1968.
3. Y. Nawab, ed., *Textile Engineering: An Introduction*. Berlin, Germany: De Gruyter, 2016.

4. K. Slater, Subjective textile testing, *Journal of the Textile Institute*, 88(2), 79–91, 1997.
5. S. Ahmad, F. Ahmad, A. Afzal, A. Rasheed, M. Mohsin, and N. Ahmad, Effect of weave structure on thermo-physiological properties of cotton fabrics, *Autex Research Journal*, 15(1), 30–34, 2014.
6. A. Afzal, T. Hussain, M. Mohsin, A. Rasheed, and S. Ahmad, Statistical models for predicting the thermal resistance of polyester/cotton blended interlock knitted fabrics, *International Journal of Thermal Sciences*, 85, 40–46, 2014.
7. S. Ahmad, A. Sinoimeri, and S. Nowrouzieh, The effect of the sliver fibre configuration on the cotton inter-fibre frictional forces, *Journal of the Engineered Fibers and Fabrics*, 7, 87–93, 2012.
8. H. M. Reid, *Introduction to Statistics*. Thousand Oaks, CA: SAGE Publications, 2013.
9. E. B. Grover, *Handbook of Textile Testing and Quality Control*. New York: Textile Book Publishers, 1960.

2

Polymer Testing Methods for Conventional and Technical Textiles

Usman Zubair

CONTENTS

2.1 Introduction

Textile structures, in the form of fiber, yarn, fabric, or composite, are essentially constituted of polymers. Both natural and synthetic polymers have come to be used in conventional as well as advanced technical textiles. The production volume of textile fibers rose to 96 million tonnes in 2014, of which more than 67% is manufactured directly from thermoplastic polymers (The Fiber Year Consulting 2015). On the other hand, thermoset polymers appeared in the fiber-reinforced composites market to around 8.8 million tonnes in the same year. The requirement for carbon-fiber-reinforced plastics increased to 91,000 tonnes in 2015 globally. Glass-fiber- and natural-fiber-reinforced composites reached 92,000 tonnes in 2012 and 2.3 million tonnes in 2014 in Europe (Witten 2014). Polymer testing ensures quality control during manufacturing processes and provides a foundation for the development of new products and processes. Strictly speaking, polymer testing involves a broad range of characterizations from structure analysis of the raw material to product response to the environment.

Polymers are characterized for their structural, thermal, and rheological properties. Structural analysis fundamentally involves the assessment of the chemical composition, identification of functional groups, determination of average molecular mass, molecular weight distribution, and amount of crystallinity. Thermal analysis identifies first- and second-order thermal transitions and describes the thermal response of polymers under different heating and cooling profiles. Rheological characterization depicts the melt flow behavior of polymers under thermal and shear stresses. These analyses ultimately provide the essential combination of characteristics that determine the morphology, processability, and properties of polymeric materials. This chapter will provide the necessary details to carry out these structural, thermal, and rheological analyses of textile polymers.

2.2 Characterization Approaches

Characterization techniques of polymers continue to advance with progress in instrumentation and computer-enhanced data analysis. Structurally, polymers have to be investigated for their chemical composition and functionality, average molecular weight, polydispersity, and morphology. Compositional assay of materials could be carried out through a number of spectroscopic and chromatographic techniques, like Fourier transform infrared spectroscopy (FTIR), attenuated total reflection, Raman spectroscopy, atomic absorption spectroscopy, thermogravimetric analysis—gas chromatography, and energy dispersive X-ray spectroscopy. The molecular level configuration of polymers, such as the average molecular weight and polydispersity, is determined through chromatographic techniques like gel permeation chromatography (GPC) and high performance liquid chromatography. Polymer morphology at the nanolevel is usually observed through a number of microscopic techniques like scanning electron microscopy, atomic force microscopy, and X-ray diffraction. For the sake of brevity, structural analysis techniques except GPC will be dealt with in Chapter 3, as these are equally important for analyzing the structures of materials other than polymeric ones. The thermal characterization of polymers mainly relies on thermogravimetric analysis (TGA), differential thermal analysis (DTA), differential scanning calorimetry (DSC), and dynamic thermomechanical analysis (DTMA/DMA). TGA, DSC, and DMA will be discussed in the coming sections to assess the processing properties of polymers, which exhibit different viscoelastic behavior on heating and shearing, behavior that monitors process parameters during melting, spinning, and processing. For rheological characterization, polymers have to be investigated at a low shear rate for structural changes through parallel plate rheometery as well as at a high shear rate to simulate melt processing parameters via capillary rheometery.

2.3 Morphological Analysis

Morphological characterization involves molecular and structural analysis. Molecular level changes are related to changes in the molecular weight and distribution of intermolecular forces, while structural changes are related to crystalline and amorphous phase separation and phase mixing upon heating and shearing. Therefore, in order to obtain the desired product properties, a control over the morphology is essential. A profound knowledge of morphology is thus vital to understand the relationships between structure, processing, and property.

Morphological changes lie somewhere at the borders of pure physical and chemical changes. These changes could be purely due to the mixing of two separated phases, the reshuffling of hydrogen bonding in-between two phases, or to some sort of chemical polymerization between end groups that were deliberately left alone to make the melt processing of these polymers easy and feasible.

Textile and composite grade polymers have to be characterized morphologically for their average molar masses, crystallinity, and chemical structures through GPC, X-ray diffraction analysis, and infrared spectroscopy, respectively.

2.3.1 Gel Permeation Chromatography

GPC is a standard liquid chromatography technique for the determination of average molar masses and molar mass distributions (polydispersity) of polymers, also referred to as size exclusion chromatography (SEC) or gel filtration chromatography.

Polymers are composed of various chain lengths that determine the mix of properties (solubility, melt and solution viscosity, moldability, tensile strength) of polymers. Figure 2.1 depicts the influence of average molar mass on the properties of polymers (Gu et al. 2015). The molar mass of polymers is taken as an average because of the different chain lengths of individual polymer molecules. The average molar masses of polymers are assessed by different approaches as follows.

Number average molar mass

$$\langle M_n \rangle = \sum_{i=1}^{\infty} X_i M_i \quad \text{where} \quad X_i = \frac{N_i}{\sum_{i=1}^{\infty} N_i}$$

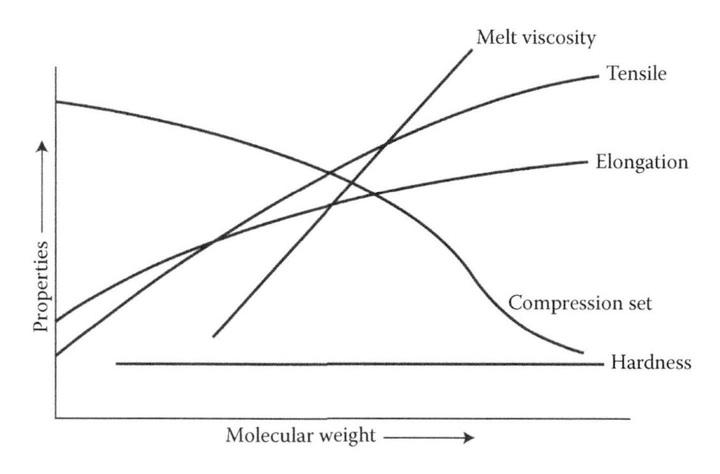

FIGURE 2.1
Effect of average molar mass on the processing and performance properties of polymers.

Weight average molar mass

$$\langle M_\omega \rangle = \sum_{i=1}^{\infty} \omega_i M_i \quad \text{where} \quad \omega_i = \frac{N_i M_i}{\sum_{i=1}^{\infty} N_i M_i}$$

Z-average molar mass

$$\langle M_z \rangle = \frac{\sum_{i=1}^{\infty} N_i M_i^3}{\sum_{i=1}^{\infty} N_i M_i^2}$$

Viscosity average molar mass

$$\langle M_z \rangle = \left[\frac{\sum_{i=1}^{\infty} N_i M_i^{1+\alpha}}{\sum_{i=1}^{\infty} N_i M_i} \right]^{1/\alpha}$$

where N_i is number of polymer molecules and M_i is the mass of each molecule.

Moreover, the different chain lengths impart a distribution of molar masses over a range called the molar mass distribution. The latter, which can only be determined by GPC, is an important parameter in polymer science for controlling the reaction kinetics of polymerization and the physical properties of polymers.

2.3.1.1 Theory and Principle of Measurement

Chromatography is essentially based on interactions between two phases (i.e., a stationary phase and a mobile phase). The stationary phase is a fixed platform where analytes (solutes) interact. The mobile phase is one that contains analytes that flow through the stationary phase. The interactions of different solute particles with a stationary medium must be differentially based on some intrinsic property of the analytes. The common modes of interaction are adsorption, partition, bonding, ion exchange, affinity, and size exclusion. The particles that interact with a stationary medium lag behind those that have little or no interaction. The mobile phase (solvent) takes out (elutes) the solute particles that do not interact with a stationary medium.

In 1964, J. C. Moore devised a standard GPC technique by separating synthetic polymers on cross-linked polystyrene gel in organic mobile phases (Moore 1964). The principle involved is the separation of polymer molecules based on their hydrodynamic volumes. Polymers are entangled masses of chains in solid form, but in solution form each chain acts as a single entity and curls up into a sphere like moieties due to entropic effects. The size of each sphere depends on the molecular mass of the polymer chain, and the corresponding volume of a sphere is labeled the "hydrodynamic volume." High molar mass chains curl up into big spheres and shorter chains form small spherical structures that are the basis of GPC. Polymer molecules with a higher hydrodynamic volume have limited accessibility to the porous packing of a stationary column. Therefore, large molecules exit first, followed by smaller ones; this is why GPC is also termed restricted diffusion chromatography. In principle, the polymer molecules with a lower dimension than the porous volume of a stationary phase can reside in pores of polymer gel (stationary phase). Smaller polymer molecules have to cover a longer path, thus they require a longer retention time to cross a column of the same length. So, the residence time depends on the size of the polymer and can be correlated to molar masses. The lower the molar mass of the polymer molecule, the more time it will consume to come across a stationary phase. The molar masses and concentrations of each fraction are determined either by molar mass sensitive detectors or by calibration (Lafita 2011, Trathnigg 2006). The graph obtained as a result of GPC is a chromatogram.

2.3.1.2 Experimental Protocol

The protocol for conducting a typical GPC analysis involved in the dissolution of polymers in a suitable solvent within a concentration range of 0.007%–0.200% (w/v) depends on the molar mass. The dissolved polymer is filtered before injection. The molecules in the solution are injected through a column of glass beads or styrene gel with various pore sizes to act as a molecular filtration system. Polymer molecules in a dilute solution are probed for their relative molecular weight by separating molecules of different sizes. The eluent is assessed by various detectors. Larger molecules are less prone to entering pores and so pass through the columns more quickly. Smaller molecules can fit into small pores and will be retained there for longer (Peter 2011).

2.3.1.3 Components of GPC Instrumentation

Main components of a typical GPC system are the mobile phase, injection pump, column, detectors, and data acquisition and processing, as illustrated in Figure 2.2. The quality of GPC analysis is dependent on every individual component.

- *Mobile phase.* The mobile phase in GPC must be a very good solvent of the polymer to avoid any nonexclusion effects (Trathnigg 2006). Sometimes to dissolve or to collapse aggregations of a polymer, an electrolyte or high temperature (140°C–150°C) has to be added as well as the solvent. The solvents required to dissolve various polymers are discussed in the following section.

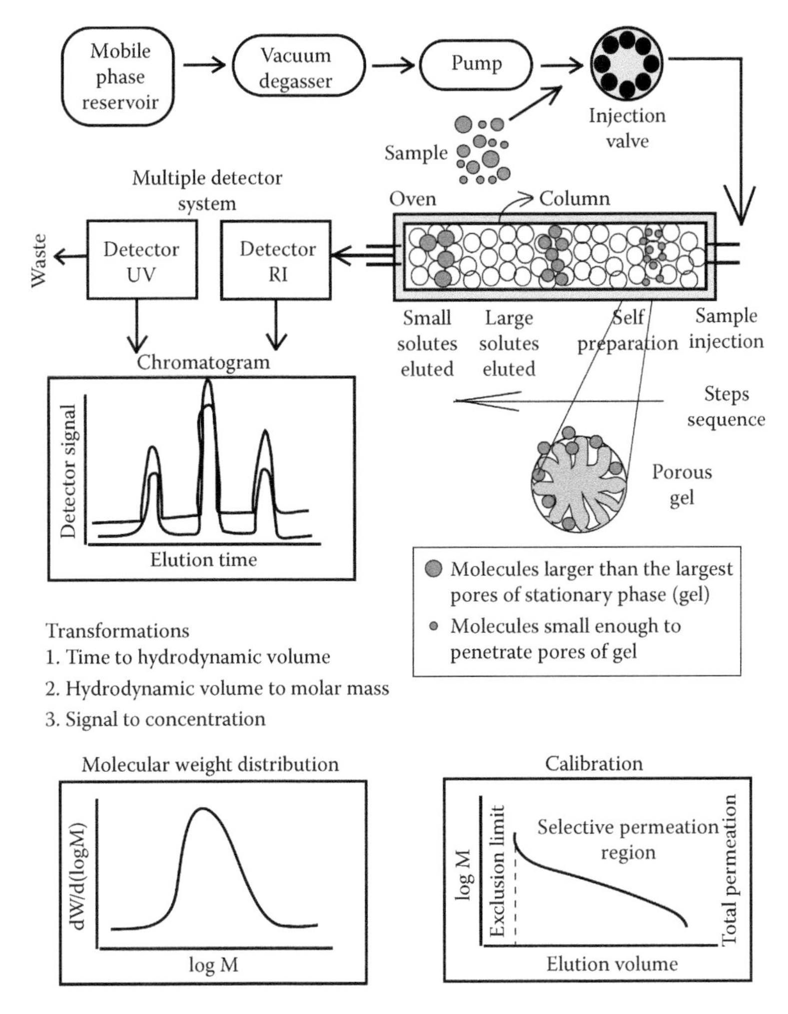

FIGURE 2.2
Schematic representation of GPC.

- *Pump.* A highly constant, accurate, and reproducible flow rate is required to construct a correct chromatogram. The flow rate is essentially related to a pump, so the selection of the pump and the designing of the pumping system should be precise to avoid any fluctuations. Because of the logarithmic relation between molar mass and elution volume, a change of 0.1% in the flow rate can cause a 10th-order error in the molar mass (Letot et al. 1980). Pressure delivered by the pump must be even to avoid any pulsation effect in the flow.
- *Oven.* GPC is usually carried out at room temperature. However, some instruments are available with thermostatically heated ovens

for higher viscosity solvents like trichlorobenzene or chloronaphthalene and for polymers like polyetheretherketone and polyolefins that are insoluble in solvents at room temperature (Holding 2011).

- *Stationary phase (columns).* In GPC, separation efficiency is directly linked with the performance of the stationary phase, so the right choice of column for a given polymer is crucial. Columns are hollow narrow tubes tightly packed with extremely small insoluble porous beads. The pores occupy 40% of the total volume of the column where separation takes place. The selection of pore size for the column packing material depends on the molecular weight of the polymer to be analyzed. In order to achieve high separation efficiencies in GPC, long columns or sets of several columns are requisite. Commercially available columns range from 2 to 25 mm in diameter and from 5 to 60 cm in length. The packaging material of the GPC columns is mostly made of porous silica or cross-linked polystyrene and divinylbenzene copolymer (Agilent Technologies 2015).
- *Detectors.* Different physical and chemical detectors have been developed to detect elution from the column using different characteristics of molecules. GPC instruments mostly have concentration sensitive detectors, which can measure one of the bulky properties (e.g., refractive index, density), solute properties (e.g., UV, IR absorption), evaporation, light scattering, or the viscosity of eluents. In addition to these, some GPC instruments have molar mass sensitive detectors (Striegel 2005, Susan 2009). Current instruments are available with a multidetector system to target the versatility of GPC analysis.

2.3.1.4 Solvent Selection for Various Polymers Including Textiles

Mobile phase in GPC is a solvent that transports the analyte across the column. The same solvent should be repeatedly used for a single column, otherwise column life can be reduced as well as performance. Tetrahydrofuran is the most commonly used organic solvent at room temperature. Table 2.1 shows the various solvents required to perform the GPC of different commercial polymers and fibers (Nikolay 2009, Sam 2011, Serhatli 2013, Tuan 2009).

2.3.1.5 Data Acquisition and Interpretation

The signals from various detectors are sent to software for analysis and interpretation. Three transformations have to be performed with chromatographic raw data (Trathnigg 2006):

1. Elution time to elution volume transformation is performed using an internal standard and is highly dependent on a constant and reproducible flow rate.

TABLE 2.1

Solvents for the GPC Analysis of Various Polymers and Fibers

Solvent	Temperature (°C)	Polymers/Fibers
Tertrahydrofuran	RT	Cellulose acetate, cellulose triacetate, polyvinyl chloride, polystyrene, polycarbonates, epoxy resins, polymethylmethacrylate, phenolic resins, polyurethanes (some), polysulfone
Toluene	75	Elastomers (most), polybutadiene, polyisoprene, styrene butadiene rubbers, polychloroprene, silicon oils
Dimethyl formamide (DMF)/ dimethyl acetamide with 0.05% LiBr	85	Poly(acrylonitrile/styrene butadiene) rubbers, polyacrylonitrile, polyurethanes (most)
Hexafluoroisopropanol with 0.05% sodium salt of trifluoroacetic acid	RT	Nylons (most of the polyamides), polyesters such as poly(ethyteneterephthalate), and poly(butyleneterephthalate), melamine formaldehyde
1,2,4-trichlorobenzene	135–145	Polyolefins, polyethylene, polypropylene
DMF with 6 M LiCI	85	Cellulose
Dimethylsulfoxide	50–100	Starch
N-Methyl-2-pyrrolidone with 0.05 M LiBr	100	Polyvinylidene fluoride, polyamide-imide, polyimide

RT, room temperature.

2. Elution volume to molar mass transformation is carried out using either a molar mass sensitive detector (in addition to concentration detectors) or calibration, to be discussed in a later section.
3. The third transformation is the detector response (signals) to concentration (amount of polymer in a fraction) and is of importance for copolymers, polymer blends, and oligomers. For the analysis of copolymers, a multidetector system is vital. Usually a combination of UV and refractive index detection is used (Susan 2009).

It is important to note that the standards and software demands on the instrumentation are very customized and stringent due to special calibration procedures.

2.3.1.6 Calibration of GPC Instrument

GPC instruments without molar mass sensitive detectors require calibration curves in order to estimate the average molecular masses from the chromatogram. The relationship between the molar masses and the retention volume is characterized as a calibration curve, which describes how different-sized molecules elute from the column. In order to calibrate the relationship between the retention time and the molar mass, a set of polymer standards have to be run first. Polystyrene standards are the most common. The retention volume of the polymer depends on the experimental conditions, hence also the average molecular masses calculated from chromatograms (Sadao 2011). It is important to report the standard, the solvent, as well as the temperature at which the column is heated, as all these parameters affect the hydrodynamic radius of the polymer being investigated. This technique is not able to give an absolute molar mass, but it does provide a very good approximation of molecular mass with a molecular mass distribution profile. There are three standard calibrations: (1) a relatively narrow standard calibration that provides molecular mass relative to the calibrant, (2) a broad standard calibration in which the instrument is calibrated with the same polymer being run as an unknown, and (3) universal calibration. In universal calibration, the log of the product of the intrinsic viscosity and molecular mass is plotted against retention, instead of the log molecular mass of a series of narrow standards against retention. Universal calibration delivers accurate and detailed information of molecular mass, intrinsic viscosity, and branching (Serhatli 2013).

2.3.1.7 Applications of GPC

GPC is used to determine the average molecular mass and molecular mass distribution of synthetic polymers (polyamides, polyesters), high performance polymers (polycarbonates, Nomex™, Kevlar™, aromatic polyesters), and oligomers (Nikolay 2009, Sam 2011, Tuan 2009). SEC is also used in studying the kinetics of polymerization by examining reaction mixture chromatograms after different time intervals. GPC can be applied to the observation of polymer degradation, hydrolysis, aggregation, and refolding (in the case of proteins) (Trathnigg 2006). Thermoplastic polymers such as polyesters or polyurethanes thermally degrade during the melt spinning process. The changes in molar mass and molar mass distribution of as-spun filaments assist researchers to estimate the appropriate spinning time, temperature, and shear stresses. In reaction spinning of spandex fibers, there is a molar mass buildup in spun filaments. The increase in molar mass can be estimated through GPC to determine the properties of the filaments. GPC is used to characterize copolymers by selecting a proper solvent where both components are soluble. Nonionic surfactants—used as surfactants, dispersing agents, emulsifiers, detergents, and phase transfer agents—have wide applications in many industries including textiles. They are amphiphilic compounds composed of water-soluble poly(ethylene oxide) blocks and hydrophobic fatty alcohols, fatty acids, alkylated phenol derivatives, or different synthetic polymers segments. They can be characterized by their composition, molecular mass, molecular mass distribution, and micellization behavior in selective solvents through GPC, by selecting the proper solvent where both blocks are soluble (Ivan 2009). Commercial polymeric products with complex formulation like nail varnish can be analyzed by GPC for its constituents (Agilent Technologies 2014). GPC may also be used to determine the intrinsic viscosity of polymers. For a homologous series of polymers, macromolecular architecture and conformation can be obtained from the molecular mass dependence of the intrinsic viscosity. Also, the frictional properties of a polymer in a dilute solution (i.e., associated with its molecular dimension) is the reflection of the intrinsic viscosity. Thus, SEC provides a unique opportunity to measure these phenomena and determine the intrinsic viscosity of polymers (Yefim 2009). The use of GPC for measuring the physiological properties of polymers, especially biopolymers, has emerged as an important area of research. GPC is substantially used to separate, quantify, and purify large biomolecules (molecular mass > 10,000 g/mol) like lipids, proteins, cellulose derivatives, and coal-derived substances. GPC is also used for the analysis and isolation of sugars and lipid polymers.

2.4 Thermal Analysis

Thermal analysis comprises a group of techniques in which samples are subjected to a predefined heating and cooling profile to characterize the thermal behavior of a material and to measure the physical properties of the polymer as a function of time and temperature (Lever et al. 2014). Thermal behavior includes the study of thermal transitions in polymer structure due to physical and chemical changes like glass transition, crystalline melting, recrystallization, thermoplasticity, and degradation. Thermal investigation also provides indirect information of the polymer structure and the subsequent process behavior of the polymer during melt processing. The usual techniques involved to characterize polymers thermally are DTA, DSC, TGA, and DMA as shown in Table 2.2. The International Confederation for Thermal Analysis and Calorimetry (ICTAC) continuously updates the nomenclature used for thermal analysis and provides the recommendations for collecting experimental thermal analysis data.

In order to design melt spinning process and to observe any possible chemical and physical changes, it is essential to analyze the polymers for thermal transition temperatures like glass transition temperatures (T_g), melting points (T_m), and recrystallization temperatures (T_c) and the corresponding enthalpies (ΔH). In addition, it is necessary to study the textile grade polymers for isothermal and nonisothermal thermoplasticity and degradation to evaluate their suitability for melt processing and the possible limitations that will arise directly during the processing phase. These thermal analyses also provide a picture of possible types of chemical and morphological changes that are induced in the material due to the heating and cooling cycles.

TABLE 2.2

Brief Description of Different Thermal Analysis Techniques

Thermoanalytical Technique	Property Measured	Observed Thermal Properties
TGA	Mass	Decomposition, oxidation, loss of volatiles, evaporation
DTA	Temperature	Glass transition temperature, melting point, recrystallization temperature, phase change
DSC	Heat flow or power compensation	Glass transition temperature, melting point, recrystallization temperature, enthalpies of melting and crystallization, polymerization, kinetics, polymorphism
Thermomechanical analysis	Deformation in steady mode	Softening, sintering, glass transition, thermal expansion coefficients
DMA	Deformation in dynamic mode	Storage modulus, loss modulus, glass transition

The study of thermal degradation, thermal hydrolysis, and alternation in thermal transition behavior assists the researcher to determine the process parameters and properties of products. TGA and DSC analyses provide evaluation of the stated parameters and phenomena for textile polymers. So, these two techniques will be discussed in detail.

2.4.1 Thermogravimetric Analysis

TGA measures the change in the mass of the specimen on being heated at a constant rate or isothermally as a function of temperature, time, and atmosphere (Lever et al. 2014). TGA is used to characterize a wide variety of materials, especially polymers. It provides supplementary characterization information for other most commonly used thermal techniques, like DSC. It is mainly used to investigate isothermal and nonisothermal degradation of materials, as well as to monitor the presence of volatile substances such as solvents, moisture, and oligomers.

2.4.1.1 Theory and Principle of Measurement and Instrumentation

In principle, TGA is a technique in which a change in the weight of a substance is recorded as a function of temperature or time. A typical TGA apparatus with horizontal furnace and ultramicrobalance is shown in Figure 2.3. The sample is heated under nitrogen or synthetic air with a constant heat rate in a sealed furnace with a thermocouple element, while the difference of the mass during this process is measured by an ultra-sensitive microbalance.

A mass loss indicates that a degradation of the measured substance has taken place. TGA measures the amount and rate (velocity) of change in the mass of a sample as a function of temperature or time in a controlled

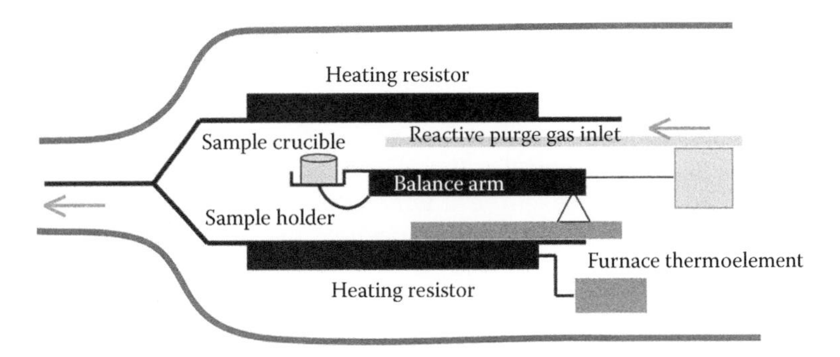

FIGURE 2.3
Schematic of TGA instrument.

atmosphere. The measurements are used primarily to determine the thermal and/or oxidative stabilities of materials as well as their compositional properties. The technique can analyze materials that exhibit either mass gain or loss due to oxidation, decomposition, or loss of volatiles (moisture, monomers, or solvents). A plot of mass loss against temperature provides information about the decomposition dynamics, while graphing the mass loss against time describes the decomposition kinetics of the specimen.

TGA curves are normally plotted with the mass change (Δm) expressed as a percentage on the vertical axis and temperature (T) or time (t) on the horizontal axis. The TGA thermogram shown in Figure 2.4 typically represents

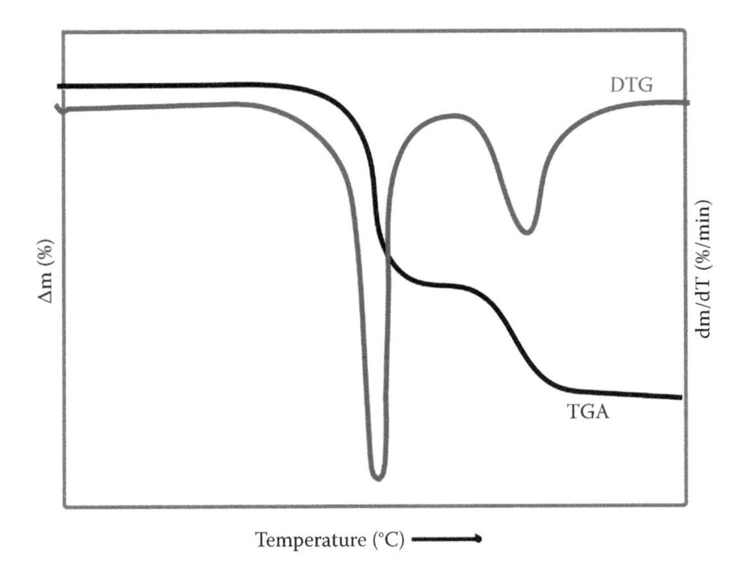

FIGURE 2.4
Typical TGA curve.

the temperature at which the polymer degrades in a nitrogen atmosphere or combusts in air. A differential thermogravimetry (DTG) curve is plotted by taking the first derivative of the TGA curve, as represented in the following equation. The peaks of the DTG curve provide information about the decomposition kinetics, and the decomposition onset and end temperature. The peak point of the DTG curve at which the slope of the TGA curve is at a maximum is taken as the mean decomposition temperature.

$$\text{Mass change}(\%) = \frac{\Delta m}{m_0}(\%)$$

$$\text{DTG} = \frac{dm/dt}{m_0}$$

2.4.1.2 Interpretation of TGA Thermograms for Various Transitions

Standards covering TGA include ISO 11358, ASTM E1131, and DIN 51006. These standard test methods define the general principles for conducting experiments, for interpreting and evaluating data, and for compiling test reports for TGA.

Figure 2.5 reports different transitions observed during TGA investigations (Hatakeyama and Quinn 1999). Identification of polymeric materials

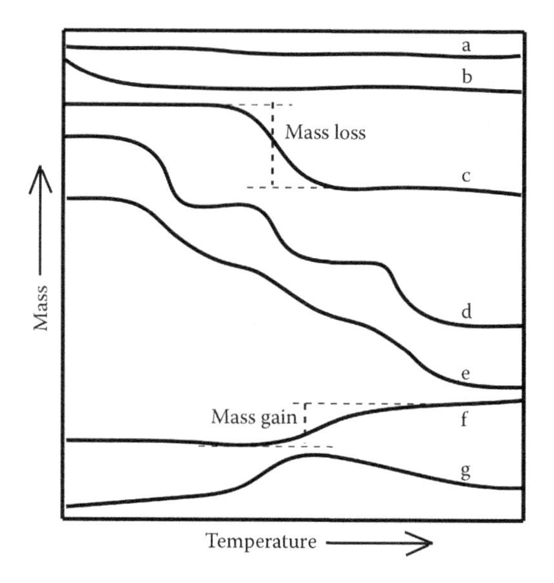

FIGURE 2.5
Possible transitions in TGA.

becomes a little complicated if it contains a large number of additives that affect their decomposition temperatures. So, by combining TGA with other analytical tools, such as mass spectrometry, gas chromatography, and FTIR, it is possible to identify the decomposition products (Giron 2002) as follows:

a. *No mass change.* If the specimen does not contain species that decompose in an inert atmosphere, such as inorganic substances, there will be no mass change. However, polymorphism (solid phase transformation) and melting could be underlying phenomena that cannot be observed by TGA.
b. *Desorption or drying.* The rapid initial mass loss in a sample could be due to desorption or the drying of water or a solvent residue. So, TGA can be used to evaluate moisture regain in various polymers. Natural fibers tend to show this transition more distinctly than those of synthetic fibers. This is a decisive transition for evaluating the predrying time and the temperature for polymer pellets like polyesters and thermoplastic polyurethanes before their melt processing.
c. *Single-step mass change.* Polymers usually decompose themselves into volatile species at certain temperatures termed decomposition temperatures. This transition is observed when a sample acquires an identical species in composition.
d. *Multiple-step mass change with stable intermediates.* This transition is usually observed in composite materials that are composed of two or more unique components in their structure. Every constituent in a composite has its own decomposition starting and end temperature. The mass loss in each step provides information about the mass percentage composition of the sample.
e. *Multiple-step mass change with no stable intermediates.* This transition is mostly observed within copolymers or the composites whose constituents have overlapping thermal transition behavior. This conversion can also be observed at high heating rates with composites with unique constituents that usually show transition type (d). Even at a very high heating rate, one can lose some information by obtaining a single-step decomposition.
f. *Gain in mass.* This transition is observed when a specimen reacts with the atmosphere, for example, the oxidation of metals to form oxides.
g. *Gain in mass followed by decomposition.* This transition is experienced when oxidized species in the sample start decomposing at a higher temperature after oxidation.

2.4.1.3 Factors Affecting the TGA Results

The steps involved in conducting a TGA experiment are

- Sample preparation
- Sample insertion into pan and weighing
- Loading the specimen on the ultrabalance
- Purging gas selection and flow setting
- Selecting the appropriate heating protocol

The factors affecting the measurement include sample preparation, sample weight, type of pan and purge gas, purge gas flow rate, and heating rate. Sample preparation has significant importance when targeting accurate and reproducible results. Polymeric and fiber samples should be chopped very fine to maximize the surface area. The sample weight should not be less than 10 mg as per the ISO 11358 recommendation (ISO11358-1 2014). A higher sample weight of 50–100 mg is considered adequate for a specimen with a high percentage of volatiles. A sample weighing more than 100 mg is not recommended because of possible volume expansion. For TGA, ceramic (alumina) or platinum pans are recommended, as they are inert toward the sample. Pans are reused again and again, so they should be cleaned thoroughly between measurements. Both inert and oxidative purge gases can be used, but the choice is based on the type of decomposition dynamics and kinetics to be studied. The water content of the purge gas must be $<0.001\%$ (wt.). Heating rates between 0.5°C/min and 20°C/min are suggested for dynamic measurements. At higher heating rates, one can lose the stepwise decomposition of the unique constituents present in composites or polymer blends. Heating may be at a constant rate or isothermal (temperature holding phase), depending on the type of kinetics to be studied. Combined heating and holding phases enhance the resolution of overlapping effects.

2.4.1.4 Calibration of TGA Instruments

A TGA instrument is a sensitive tool with an ultramicrobalance, which requires frequent calibration under the same conditions as reported in actual measurement.

1. *Blanks test (buoyancy correction).* Thermobalance has to be corrected for buoyancy and convection by performing a blank measurement.
2. *Mass calibration.* Thermobalance must be calibrated between 10 and 100 mg with no purge gas flow to avoid turbulence and buoyancy effects.

3. *Temperature calibration.* Temperature calibration has to be performed under the same conditions as the actual measurements and using the Curie temperature of ferromagnetic substances. When the ferromagnetic material is placed on the thermobalance, it causes the latter to suffer an extra force. Upon heating, the ferromagnetic substance loses its apparent weight at the Curie temperature by losing its ferromagnetic property. The TGA instrument temperature is read against the Curie temperature to calibrate it accurately. ISO 11358 defines different reference ferromagnetic materials, such as nickel (354°C), Permanorm 3 (266°C), Numetal (386°C), Permanorm 5 (459°C), and Trafoperm (754°C), for calibrating TGA instruments.

2.4.1.5 TGA Curve of Commodity and High Performance Fibers

TGA is one of the most important tools for the identification and characterization of commodity and high performance fibers (Gray et al. 2011, Liu and Yu 2006). TGA curves of several natural, synthetic, and high performance fibers are reported in Figure 2.6. TGA experiments for natural and synthetic fibers are carried out with 1 mm chopped fibers in a platinum crucible weighing 3 mg, in a temperature range between room temperature and 600°C, with a heating rate between 2.5°C/min and 15°C/min. TGA for high performance

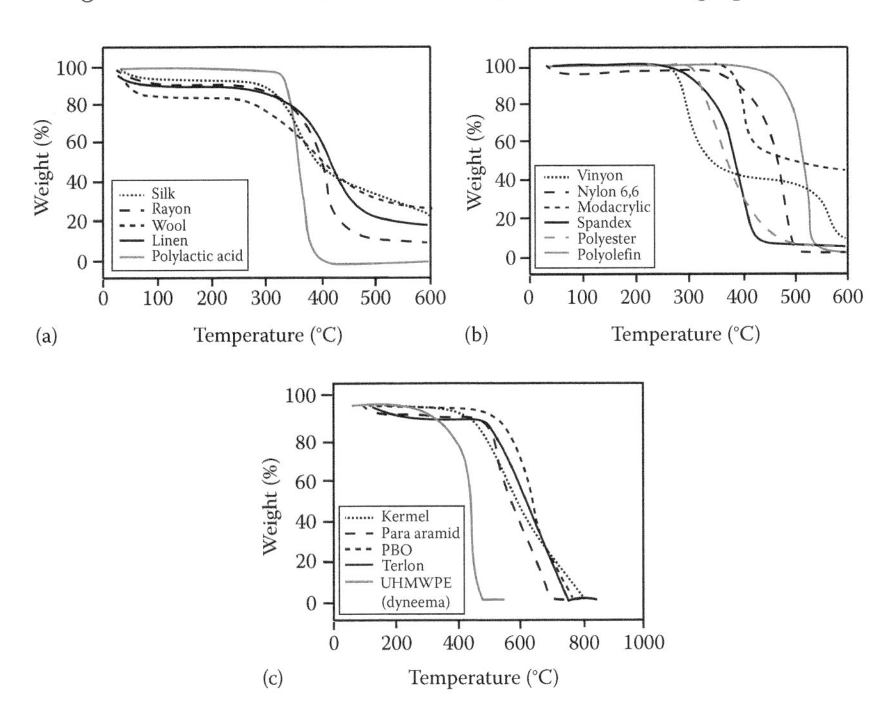

FIGURE 2.6
TGA curves of some (a) natural (b) synthetic, and (c) high performance fibers.

fibers are performed in the temperature range of 50°C–800°C at a heating rate of 20°C/min in nitrogen with a gas flow of 20 mL/min. From the results of TGA, the thermal stability of the high performance fibers can be characterized (Liu and Yu 2006).

2.4.2 Differential Scanning Calorimetry

ICTAC defines DSC as a technique where heat flow changes are recorded in comparison to a reference material subjected to the same heating and cooling protocol and presented as a function of temperature and time (Lever et al. 2014). DSC is a fundamental thermal analytical tool that allows the exploration of the thermal behavior of polymers, namely, the physical changes that take place in a polymer upon heating and cooling, that is the melting of a crystalline polymer, recrystallization, or glass transition. DSC enables us to differentiate among polymorphic structures (different forms or crystal structures of the same material) and to observe polymorphic transformations by applying different heating rates. In addition, chemical changes can be recorded with the best knowledge of chemical composition and chemical changes therein (Gregorova 2013).

Two types of DSC instrument technologies are available: heat-flux DSC and power-compensation DSC. In heat-flux DSC, the difference between the heat-flow rates into a sample and a reference material is measured. Heat-flow DSC is essentially a quantitative form of DTA where the temperature difference between a sample and a reference material is measured. In power-compensation DSC, the difference between the electrical powers into a sample and a reference material is measured. Both heat-flux DSC and power-compensation DSC provide comparable results (Hatakeyama and Quinn 1999, Lever et al. 2014). Modulated temperature DSC is a variant in which temperature modulation is applied to a temperature program (i.e., periodic heating) and is used to evaluate the glass transition of an amorphous material. The main advantage of temperature modulation is the separation of overlapping events in the DSC scans. The method is generally applied to polymeric materials (Giron 2002).

2.4.2.1 Theory and Principle of Measurement and Instrumentation

DSC is based on monitoring how a material's heat capacity (C_p) is changed by temperature in a specified atmosphere. A sample of known mass is heated or cooled and the changes in its heat capacity are tracked as changes in the heat flow. Heat capacity is the amount of energy a unit of matter can hold at a constant pressure

$$C_p = \frac{dq/dt}{dT/dt} = \frac{\text{Heat flow rate}(\Phi)}{\text{Heating rate}(\beta)}$$

where heat (q) supplied per unit time (t) is the heat flow rate and the variation of temperature (ΔT) per unit time (t) is the heating rate. In principle, DSC is the measurement of difference in heat flow or power compensation to keep the temperature of a sample pan and reference pan consistent.

A typical DSC set-up consists of two essential parts: a measurement chamber and a computer that allows us to monitor the temperature and to regulate the heat flow. In the measurement chamber, two pans are purged under either an inert atmosphere (N_2 or Ar gas flow) or an oxidative environment (air or oxygen gas flow). One is referred to as a sample pan where the sample under investigation is located; the other is the reference pan that is typically empty. Each pan (typically an aluminum pan with a lid) is positioned on the top of a heater and PT sensor (platinum resistance thermometer) as shown in Figure 2.7. Via a computer interface, it is possible to select the rate of heating for the two pans. The adsorption of heat will be different in the two pans due to the different compositions in each pan. In order to keep the temperature of the two pans constant during the experiment, the system needs to provide more or less heat to one of the two pans. The output of the DSC experiment is the additional quantity of heat that is given to the pan in order to keep the temperature of the two pans equal.

In other words, the output of the DSC is a plot of the difference in heat flow output of the two heaters against temperature. If an exothermic change occurs in the sample, more heat has to be supplied to the reference pan, which is equivalent to withdrawing energy from the sample. During an endothermic process, an additional amount of energy has to be supplied to the sample heater. This difference in the heat supplied to the sample and the reference pan is recorded as a function of temperature. This signal is

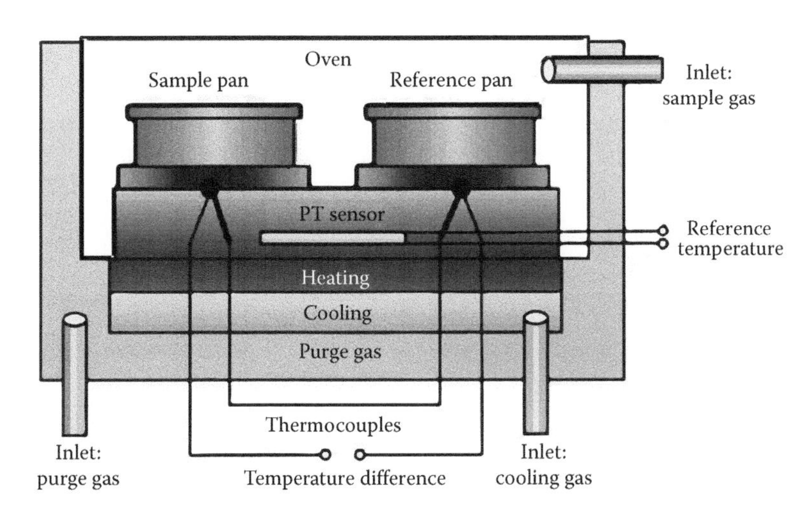

FIGURE 2.7
Schematic illustration of a DSC instrument.

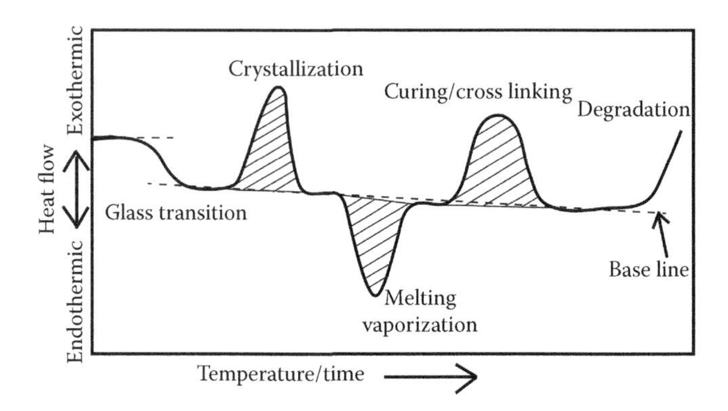

FIGURE 2.8
Typical DSC thermogram.

proportional to the sample's specific heat, which determines the amount of heat that is necessary to change the sample temperature by a given amount. As DSC is performed at constant atmospheric pressure, so specific heat or heat capacity can be calculated as

$$C_p = \left(\frac{\partial H}{\partial T}\right)_{P,N} = \left(\frac{dq}{dT}\right)_P$$

This information is then dealt with by computer software that presents it as a graph of the enthalpy changes versus the temperature, as shown in Figure 2.8.

There is a "stepwise" increase in the DSC heat flow versus temperature curve, which is labeled as a glass transition. The melting point appears as an endothermic peak on a baseline since the melting of a solid is a heat absorption process, while crystallization is an order-seeking event that necessarily releases heat and appears as an exothermic peak, as shown in Figure 2.8. The endothermic and exothermic events are evaluated for the peak temperature for a particular happening to define a melting point and a recrystallization temperature. Endothermic and exothermic peaks are integrated to the base line to discover the heat capacity, onset temperature, end temperature, and enthalpies of melting and recrystallization. These enthalpy changes may be described by the following integral:

$$dH = C_p\,dT$$

$$\int_{T_i}^{T_f} dH = \int_{T_i}^{T_f} C_p\,dT$$

$$\Delta H = \int_{T_i}^{T_f} C_p \, dT = \int_{T_i}^{T_f} \frac{K\Delta T}{q} dT$$

where the limits of integration T_i and T_f are the onset and end temperatures over which the graph is integrated, and K is the thermal conductivity of the material. This equation is used to determine the enthalpy change during a phase transition like melting, recrystallization, and decomposition. DSC could be coupled with other techniques such as X-ray diffraction or TGA to provide an understanding of the reaction kinetics and phase transitions.

2.4.2.2 Interpretation of DSC Thermogram for Various Transitions

DSC is used to observe different thermal events that happen due to an increase or decrease in enthalpy of the specimen, as indicated in Figure 2.9. Different ISO and ASTM standard test methods exist to interpret, evaluate, and report data. Standards covering DSC include ISO 11357 (ISO11357-1 2016), ASTM D3418, and DIN 53765. ISO 11357-3, 4 are test methods to determine the enthalpy temperature of melting and crystallization and the specific heat capacity respectively. ASTM D3418-03 defines the interpretation of transition temperatures of polymers by DSC. ISO 11357-1 and ASTM D 3417-99 recommend reporting endothermic peaks toward the negative y-axis, but DIN 53765 recommends reporting toward the positive y-axis.

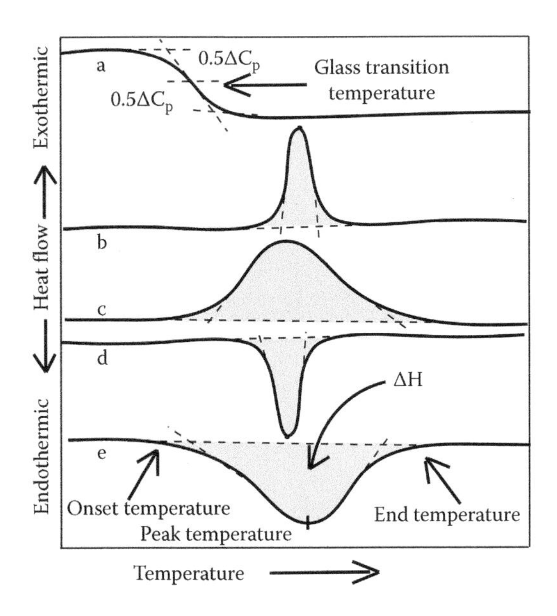

FIGURE 2.9
Possible transitions in a DSC.

Therefore, it is important to describe the test method for interpreting particular data.

Here, all the interpretation of the data is based on general assumptions, though this is not mentioned in some standard test method references.

a. Glass transition appears as a stepwise shift in the baseline. This transition in the base line is evaluated by drawing the tangents and point of tangency for the maximum slope line that is attributed to the glass transition temperature (T_g). The latter represents the temperature region at which the amorphous phase of a polymer is transformed from a brittle, glassy material into a tough rubber-like liquid.
b. Crystallization appears as an exothermic peak that is interpreted as an order-seeking event in which polymers transform from the liquid amorphous phase to a solid crystalline state. The crystallization temperature (T_c) is determined from the peak temperature at which recrystallization is at a maximum. The heat/enthalpy of recrystallization (ΔH_c) is determined by integrating the area (shaded gray) under the curve between the onset and end temperature.
c. Cross-linking, decomposition, and oxidation also appear as exothermic peaks on the DSC thermogram, though these events are broad and dynamic. From these transitions, one can find the initial reaction temperature (T_{ir}) and the final reaction temperature (T_{fr}). The heat of the reaction (ΔH_r) is determined similarly by integrating the area under the curve. The cross-linking temperature is of significant importance in curing fiber-reinforced composites.
d. Melting appears as an endothermic event on the DSC curve where the material changes its state from solid crystalline to an amorphous liquid. The melting point is the peak temperature at which state transition takes place. Semicrystalline polymers (fibers) melt over a broad range. In fibers and polymers, the shape of the melting curve heavily relies on the thermal and mechanical history of the material and the test conditions. The heat of fusion (ΔH_m) is again determined by integrating the area under the curve.
e. Evaporation and sublimation also appear as endothermic events on the thermogram. Related physical quantities, such as the heat of evaporation, temperature, and the enthalpy of sublimation, are evaluated in similar fashion.

2.4.2.3 Factors Affecting DSC Results—A Case Study for Textile Fibers

DSC experiment is the sequence of the following steps that govern the results:

- Sample preparation
- Sample insertion into the pan and weighing

- Sealing the pan with a lid having small holes
- Placement of specimen and reference in measuring chamber
- Setting the purge gas flow
- Selection of heating and cooling temperature protocols

Sample preparation is one of the critical factors in the thermal investigation of polymeric fibers. Before introducing the polymer granules and fibers into the crucible for the DSC test, the sample is chopped into small pieces so it can fit inside the crucible. It is recommended that the sample should be as thin as possible to cover the bottom of the pan, though not so small that it stresses the sample. For example, polymeric material including fibers should be chopped rather than crushed to obtain a thin sample. The physical dimension of the sample can provide different results, as depicted in Figure 2.10, so one cannot overlook the mechanical aspect of sample preparation. It is clear that best results are achieved by chopping the sample into short fibers (stripes). Two other kinds of preparation (knots and bundles) result in little altered results. The volume of the crucible also causes variation in results owing to dissimilar heat contact. Obviously, a 20 µL crucible will provide a more even thermogram than a 40 µL crucible, as illustrated in Figure 2.10 because of the better heat contact. The lid of the pan has holes in it to effect contact with the purge gas, but a lid covering is essential to prevent evaporation of volatiles (oligomers).

Interestingly, the weight of the sample inside the crucible also affects the shape of a normalized thermogram. For a small sample weight, peak sharpness increases and the overlapping effect can be clearly observed in contrast to larger sample weights, as shown in Figure 2.11 for polypropylene fibers. This could be explained on the basis that a heavier sample requires a greater amount of heat energy to make the thermal transition occur than a lighter sample (Steinmann et al. 2013).

In addition, experimental parameters like the heating rate and cooling rate also significantly influence DSC results, as shown in the thermograms

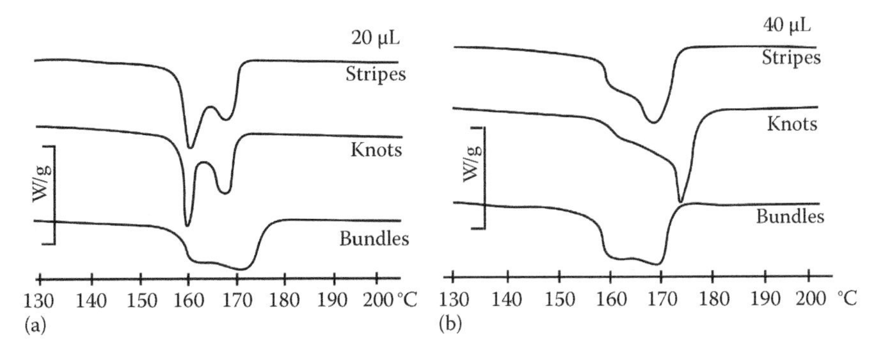

FIGURE 2.10
Effect of sample preparations and crucible size on DSC curves performed at 5°C/min.

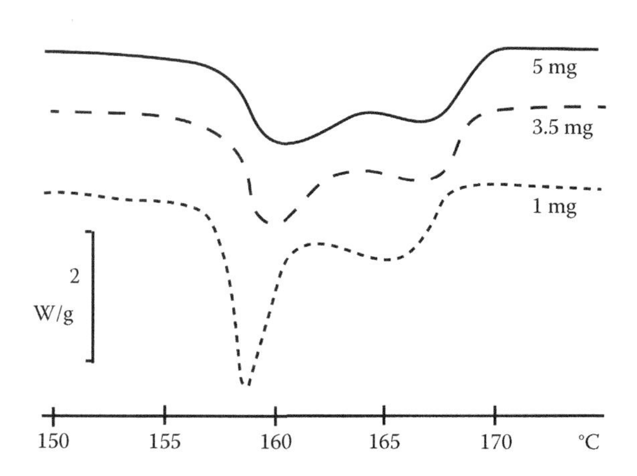

FIGURE 2.11
Normalized thermograms of polypropylene fibers at a heating rate of 10°C/min for different sample weights.

reported in subsequent section for the fibers from different commodity polymers. One can observe that peak height and width increases with an increasing heating rate. There is a decrease in height and width of the peak with positive shift for a low cooling rate, as illustrated in the thermograms in Figures 2.13 through 2.15. In the following, the DSC thermograms of commercial polypropylene (PP), polyamide 6 (PA-6), and polyester (PET) fibers will be discussed, which are produced at different heating and cooling rates. All three fibers show a similar effect on performing DSC at different heating and cooling rates, as discussed above, in addition to their peculiar behavior at different heating rates.

DSC instruments should be calibrated as per the manufacturer's instructions. Usually, such instruments are calibrated for temperature, enthalpy, specific heat (heat capacity), and heat flow.

2.4.2.4 DSC Thermograms of Natural and Synthetic Fibers

DSC is one of the important tools for the identification and characterization of textile fibers (Gray et al. 2011). DSC thermograms of several natural and synthetic fibers are shown in Figure 2.12. DSC experiments are carried out with 1 mm chopped fibers in an aluminum pan weighing 3 mg within a temperature range of 60°C–450°C with a heating rate between 2.5°C/min and 15°C/min.

DSC of PP performed at different heating rates shows various peak heights and widths, and a double peak effect as shown in Figure 2.13. A double peak is observed in PP fibers at a low heating rate due to the transition from the α to the γ phase (Andreassen et al. 1995, Cho et al. 1999).

DSC of PA-6 fibers carried out at different heating rate shows an effect similar to PP. At 25°C/min, there is an appearance of glass transition around

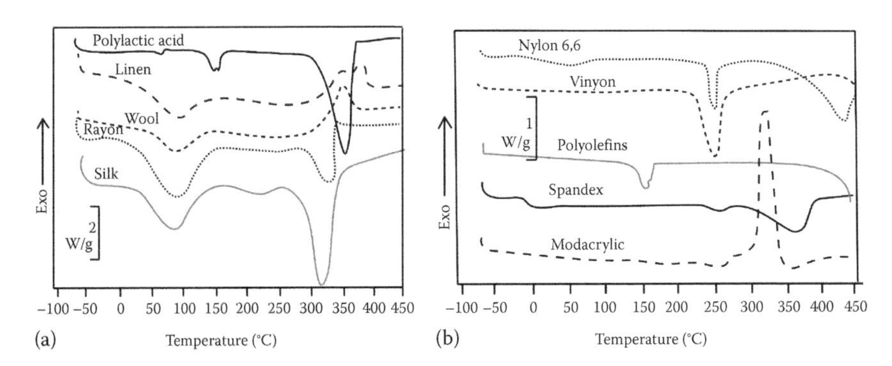

FIGURE 2.12
DSC thermograms of some selected (a) natural and (b) synthetic fibers.

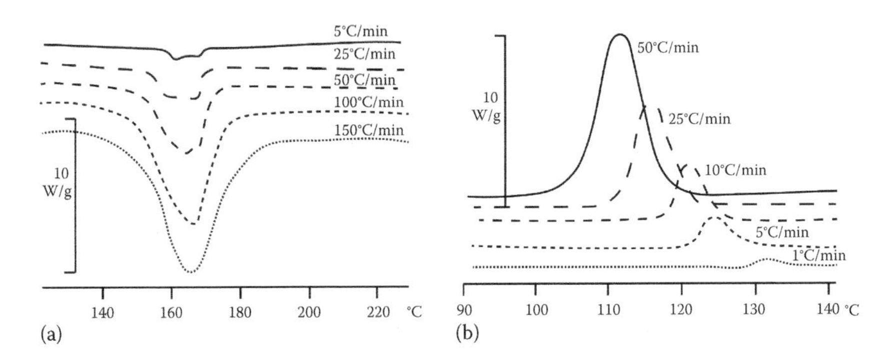

FIGURE 2.13
Effect of heating (a) and cooling (b) rates on DSC thermograms of PP fibers.

50°C, which shifts to a phase transition at a higher heating rate. One can easily observe from Figure 2.14 that there is a transition from the γ to the α phase between 120°C and 160°C for PA-6 at heating rates above 25°C/min (Liu et al. 2007).

DSC of PET fibers again depicts the peak width and height increases with an increasing heat rate. At a heating rate of 25°C/min, the glass transition appears more strongly, which becomes more pronounced at higher heating rates. Nevertheless, in PET, there is a relaxation shortly after the glass transition and afterward an exothermic recrystallization peak, as indicated in Figure 2.15. The relaxation after the glass transition occurs because of the elongated states of amorphous polymers that freeze in the cooling process of melt spinning. Recrystallization is attributed to the growth in the size of crystallites due to the increased mobility of the amorphous and mesomorphous polymer chain after T_g (Demirel et al. 2011). This phenomenon reveals the quenching of the filament during spinning.

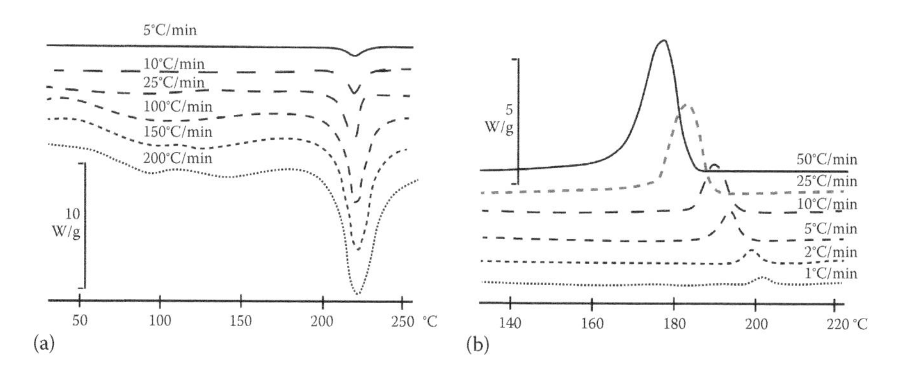

FIGURE 2.14
Effect of heating (a) and cooling (b) rates on DSC thermograms of PA-6 fibers.

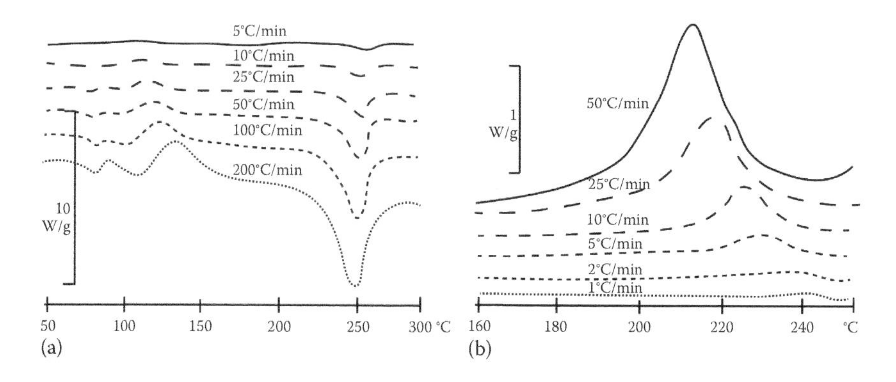

FIGURE 2.15
Effect of heating (a) and cooling (b) rates on the DSC thermograms of polyester fibers.

2.5 Rheological Analysis

Rheology is the study of the deformation and shear flow of materials and provides essential information about polymer processing. Rheology is concerned with the relationship between stress (defined as force per unit area), strain (defined as change in dimension per unit dimension), and time, both in the solid and molten state. Rheology is one of the key characterization techniques for developing materials with the desired physical properties and for controlling the manufacturing processes in order to ensure the product's quality. Product performance issues usually are associated with solid sample properties, while processability issues can be correlated with polymer melt properties. Shear flow plays an important role in both the fiber spinning and composite manufacturing industries. In order to understand polymer behavior during melt processing, it is essential to understand and characterize the rheological behavior occurring in complex flows of viscoelastic polymers as

TABLE 2.3

Test Mode Requirements for Different Polymer Processes and Phenomena

Phenomena/Process	Shear Rate (s^{-1})	Instrument/Test Mode	Properties Measured
Sagging, leveling, pouring, thermoforming	10^{-3}–10^{-1}	Rotational rheometers	Shear viscosity, yield stresses, viscoelasticity, relaxation
Dispensing, mixing, blade coating, brushing	10^{0}–10^{2}	Rotational rheometers	Shear viscosity, yield stresses, viscoelasticity, relaxation
Extrusion (films and fibers), injection molding, blow molding	10^{1}–10^{4}	Rotational rheometers, high pressure capillary rheometers	Shear viscosity, yield stresses, viscoelasticity, elongational viscosity, relaxation, wall slip
Roll coating, spraying	10^{3}–10^{6}	High pressure capillary rheometers	Shear viscosity, elongational viscosity, wall slip

depicted in Table 2.3. Figure 2.16 illustrates the different melt flow behavior of polymer melts under shear flow. Viscoelastic polymers exhibit non-Newtonian and nonisothermal melt flow, which is vital in the selection or design modification of screws and dies. Knowledge of rheological phenomena of polymers, especially with complex elastomeric behavior, is requisite for setting the right process window, such as temperature and flow rate. In addition, rheological data is needed for process simulation, which is increasingly adapted as an important part of setting up a new process (Walczak 2002).

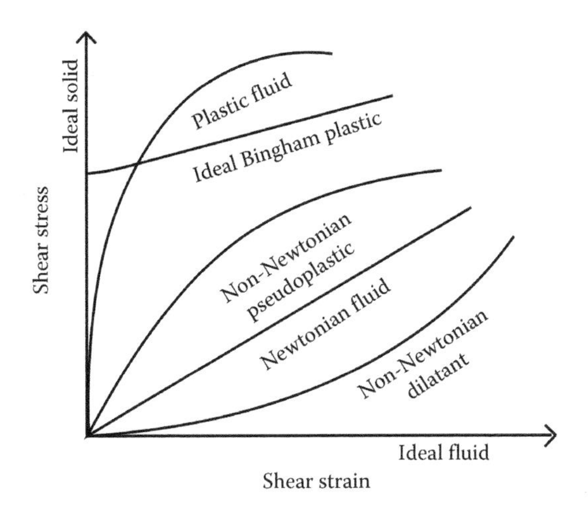

FIGURE 2.16
Melt flow behavior of liquids under shear flow.

Rheological testing provides information in three tiers, which helps researchers to establish structural characteristics (configuration, molecular mass, molecular mass distribution), to predict process parameters (shear rate, extrusion, process temperature), and to understand end user performance (strength, use temperature, dimension stability). The model depicted in Figure 2.17 shows that material structure directly dictates the rheology of the material that in turn is determinant of material process behavior and material performances and properties. Making measurements of the rheological properties of materials is termed "rheometry." For rheological characterization, knowledge of capillary rheometers and steady/dynamic shear rotational rheometers is requisite. Rotational rheometry works very well at a low shear rate, so it is helpful for understanding polymer structure, dynamic properties, and possible phase transitions occurring due to shearing and thermal treatment. Capillary rheometry is performed to provide the best simulation of high shear rate extrusion processes over a specific temperature range, as reported in Figure 2.18. In this way, rheological characterization helps to determine both dynamic properties and process parameters that regulate product end properties. In the case of thermoset samples, such as epoxy resins and adhesives, it is important not only to analyze the material properties prior to curing, but also to analyze the change in mechanical properties during and after the curing process.

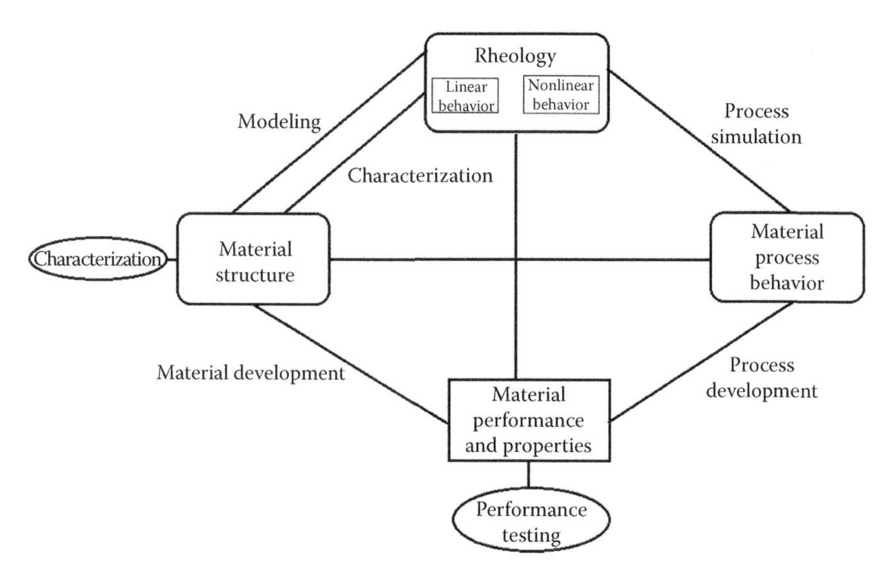

FIGURE 2.17
Model illustrating the importance of rheology.

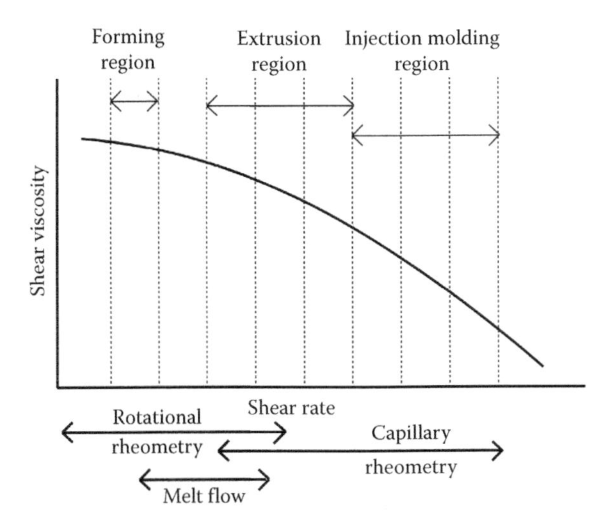

FIGURE 2.18
Range of different rheometric analyses possible at different shear rates.

2.5.1 High Pressure Capillary Rheometry

For high shear rates, the capillary rheometer is the most effective instrument for characterizing a fluid's shear flow properties by imitating the melt extrusion process, as shown in Figure 2.19. A shear rate approaching 10^3–10^4 s^{-1} is relevant to the melt spinning process. Capillary rheometry best simulates the processing conditions.

2.5.1.1 Theory and Principle of Measurement

Capillary rheometery is a technique in which a sample undergoes extrusion through a die of defined dimensions; the shear pressure drop (ΔP) across the die is recorded at set volumetric flow rates (Q). In this technique, the measurements are carried out using round-hole dies with different length and diameter (L/D) ratios. The capillary rheometer is in actual fact a "rheometer-die" with certain dimensions, say a radius (R) and length (L), for measuring the pressure difference (ΔP) that is proportional to the shear stress (τ_R) and correlated by

$$\tau_R = \frac{R}{2}\left(\frac{\Delta P}{L}\right)$$

The flow rate (Q) is proportional to the shear rate (γ_R) and is given by the following equation for Newtonian fluids:

$$\gamma_R = \frac{du}{dr} = \frac{\Delta P}{2\mu L}R = \frac{4Q}{\pi R^3} = \frac{4v_{app}}{R} = \gamma_{apparent}$$

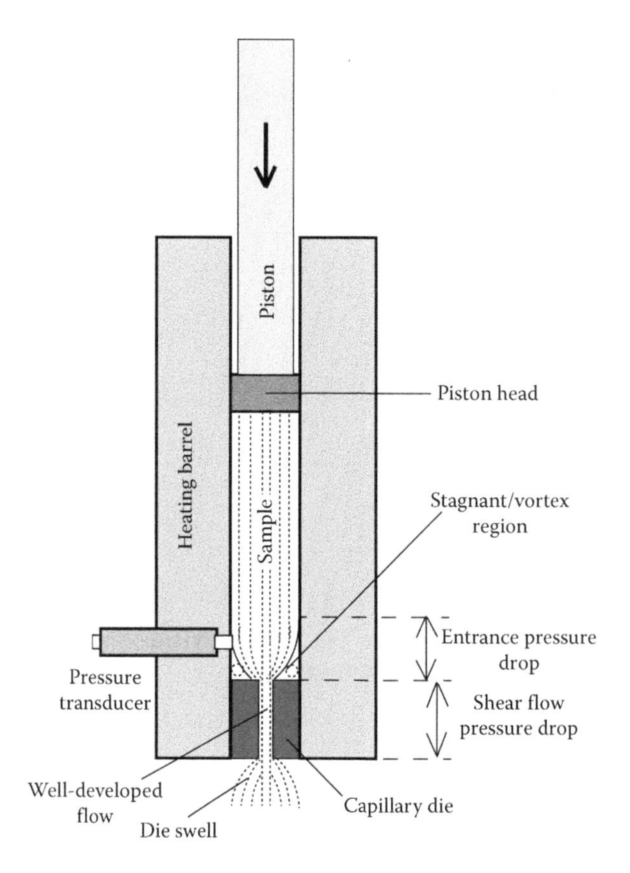

FIGURE 2.19
Schematic representation of high pressure capillary rheometer.

Thus the apparent viscosity (η_{app}) for Newtonian fluids can be calculated by

$$\eta_{app} = \frac{\tau_R}{\gamma_R\left(\gamma_{app}\right)}$$

However, it is necessary to make some assumptions to estimate the correct rheological behavior of viscoelastic polymer melts, namely

1. The flow is fully developed, steady, and isothermal.
2. There is no slip at the capillary wall—the fluid velocity at the wall is zero.
3. The fluid is incompressible and its viscosity independent of pressure.

These assumptions hold mostly for Newtonian fluids following the Hagen–Poiseuille equation given as

$$Q = \frac{\pi R^4 \Delta P}{8 \mu L}$$

Non-Newtonian fluids require corrections to estimate the actual behavior of polymer melts, that is the shear viscosity and the shear stress versus shear rate, and to eliminate induced errors of the capillary rheometer.

2.5.1.2 Correcting Capillary Flow for Non-Newtonian Polymer Melts

The following major corrections are required to the data on capillary flows (Morrison 2001).

2.5.1.2.1 *Money Analysis for Correction in Flow Rate due to Slip at Walls*

If there is a slip at the wall, it will reduce the shear rate near the wall, as shown in Figure 2.20. This will make the melt flow rate higher due to the high velocity at walls, as can be seen in

$$\gamma_{R,\text{corrected}} = \frac{4}{R}\left(v_{app} - v_{slip}\right) = \gamma_R - 4v_{slip}\left(\frac{1}{R}\right) = \text{intercept} + \text{slope}\left(\frac{1}{R}\right)$$

This correction is applied by performing rheometry with the capillaries of various radii, and plotting the data for various 1/R against apparent shear rates. The line with the positive slope has a slip velocity (v_{slip}) that can be calculated by (slope)/4.

2.5.1.2.2 *Bagley Correction for Pressure Drop due to Entrance and Exit Effects*

A correction to the measured pressure is often necessary due to losses that occur during the reduction in radius between the barrel and the capillary.

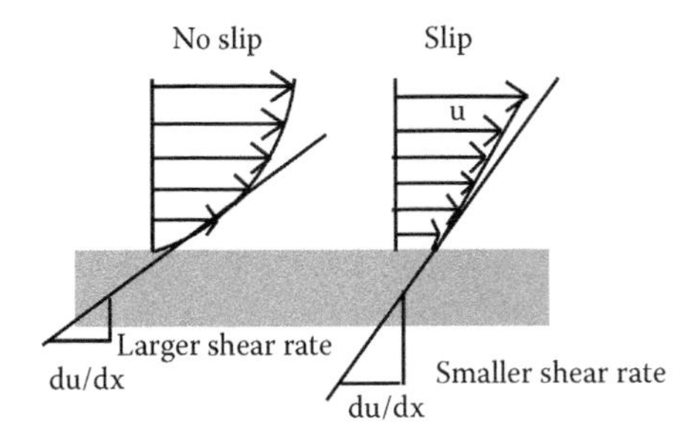

FIGURE 2.20
Effect of wall slips on shear rate.

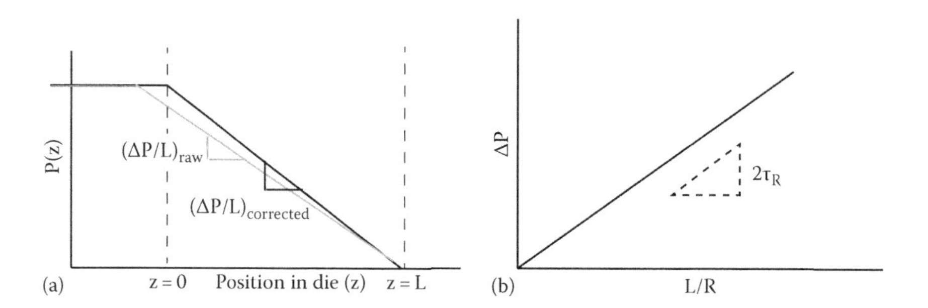

FIGURE 2.21
Effect of pressure drop at entrance and exit (a) and Bagley correction (b).

The measured pressure is the pressure inside the barrel. The raw pressure drop is calculated by dividing the measured pressure with capillary length (L). There is a loss of pressure at the entrance to the capillary, and the true pressure drop across the capillary is smaller than the raw pressure drop, as illustrated in Figure 2.21. The entrance pressure losses must be subtracted from the raw pressure in order to calculate the correct pressure drop across the capillary. In short, the entrance/exit loss correction is a correction to the pressure measurement to make the 'e' factor negligible.

$$\tau_R = \left(\frac{\Delta P}{2\left(\frac{L}{R} + e \right)} \right)$$

The Bagley correction is achieved by measuring ΔP in capillaries of different L/R (usually different lengths), and plotting the results to infer the corrected shear stress from the slope.

2.5.1.2.3 Weissenberg-Rabinowitsch Correction for Shear Rate due to Nonparabolic Velocity Profile/Non-Newtonian Effects

For an unknown, non-Newtonian fluid, it is necessary to take special steps to determine the wall shear rate, which is generally greater for a non-Newtonian than for a Newtonian fluid, as illustrated in Figure 2.22. Once the apparent shear rate is corrected to the true shear rate, the viscosity may be calculated as the ratio of the shear stress at the wall and the true shear rate at the wall of the capillary:

$$\gamma_R = \gamma_{app} \left(\frac{3}{4} + \frac{1}{4} \frac{d \ln M}{d \ln \gamma_R} \right)$$

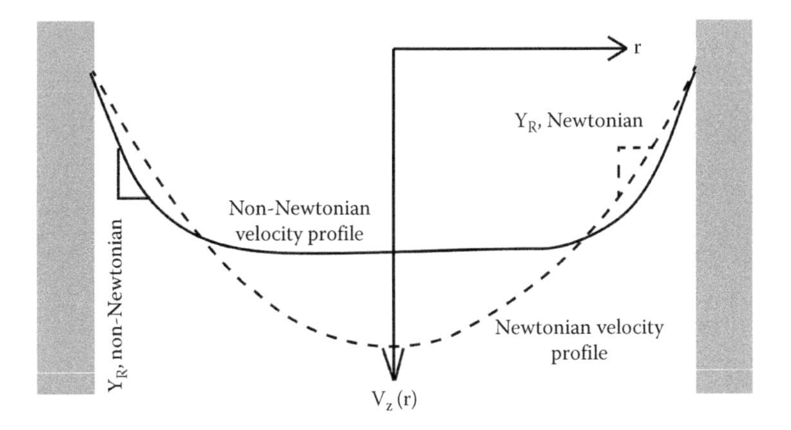

FIGURE 2.22
Non-Newtonian effects.

2.5.1.3 Protocol to Perform Capillary Rheometry

ASTM D3595 provides the norms to carry out capillary rheometer experiments. First, the polymer is introduced into a preheated barrel under a certain pressure. It is allowed to heat for a predefined time to ensure the melting of the loaded mass. The extrusion through the capillary die is carried out at different shear rates. The data of the pressure drop versus the flow rate is taken for capillaries of various lengths to correct the experimental data by performing a Bagley correction. If slip is an issue, data for capillaries of different radii is taken to perform a Mooney correction. Then, the Weissenberg-Rabinowitsch correction is carried out. After performing corrections, a plot of true viscosity versus true shear rate is made to understand the melt flow behavior of the polymer.

2.5.2 Melt Flow Index

Melt flow index (MFI) or melt mass-flow rate is the measure of the extrusion rate of polymers melted through a capillary of a specified length and diameter under a certain value of pressure and temperature. MFI is not the indicator of viscosity, even though the two values are proportional at very low shear rates. It is of industrial importance to inspect the different batches of the same material to assure quality. MFI is reported in grams of polymer melt extruded within 10 min, as given by

$$\text{MFI} = \frac{\text{m} \cdot 600}{\text{t}}$$

where
 m is the mean mass flow out in grams
 t is the time interval for extrusion

The melt volume flow rate (MVR) can also be evaluated, stated as $cm^3/10$ min

$$\mathrm{MVR} = \frac{l \cdot 427}{t}$$

where
l is the distance traveled by the piston
t is the time for extrusion

The capillary used for defining MFI has an L/D ratio of 10:1. The pressure at the piston is reached by adding weights. Standards ISO 1133 and ASTM D1238 define the testing parameters for a given material to determine the MFI values. The testing conditions are chosen to yield low shear rates to represent the effect of the molecular mass of polymers, for example, polypropylene requires 230°C and a 2.16 kg load. MFI is used to designate the grades to commercial materials in terms of their molecular weight. The value of MFI and the molecular weight of a certain polymer are inversely related. Injection molding and fiber melt spinning polymer grades usually have a high MFI, while extrusion, thermoforming, and blow molding polymer grades require a low MFI to perform (Grellmann and Seidler 2013).

2.5.3 Rotational Steady and Dynamic Mode Rheometry

Rotational rheometers essentially consist of two parts with a testing polymer in-between. The shearing of polymer melt is produced by rotating one part relative to other. This technique involves the control or measurement of torque, angular displacement, angular velocity, normal force, and temperature. Rotational rheometers are more versatile than capillary one as they can deal with non-Newtonian fluids with no difficulties. However, working with rotational rheometers requires some correction of data to obtain accurate measurements because of some limitations that arise from the gap and edge. Usually, rotational rheometers are specified in terms of their torque range, temperature range, frequency range, angular velocity range, normal or axial force range, and angular resolution. Rotational rheometers can work in both steady state and oscillatory modes. Dynamic mechanical analysis can be accomplished on these types of rheometers with little adjustment and will be discussed at the end of the section (Donald 2004). ASTM D4440 defines the norms to carry out rotational rheometry.

2.5.4 Concentric Cylinder or Couettes Rheometers

A coaxial/concentric cylinder is the earliest form of rotational viscometer, which consists of two cylinders (i.e., cup and bob) separated by a testing fluid, as shown in Figure 2.23. These types of rheometers are preferred for very low viscosity and highly volatile liquids. Shear stress applied

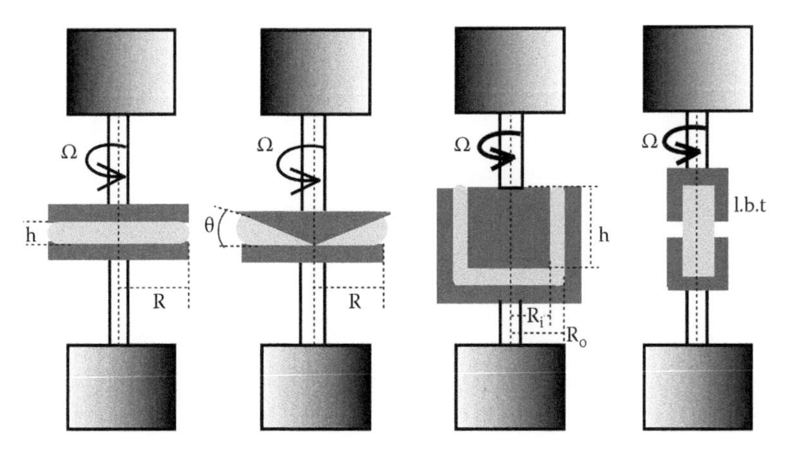

FIGURE 2.23
Possible configurations of parallel plate rheometers.

on a fluid can be calculated from the torque (M) applied to the rotating cylinder from

$$\tau_R = \frac{M}{2R_a^2 h}$$

where

h is the height of the bob

R_a is the average radius of the inner R_i and the outer R_o radius of cylinders

The shear rate on the fluid can be determined from the following formula, where Ω is angular velocity:

$$\gamma_R = \frac{2R_o^2 R_i^2 \Omega}{R_a^2 \left(R_o^2 - R_i^2\right)}$$

The shear viscosity, which is the ratio of shear stress to shear rate, of Newtonian fluids is given by

$$\eta(\gamma_R) = \frac{M}{\gamma_R \cdot 2R_a^2 h}$$

So, the shear viscosity of fluids can be estimated by determining the torque and angular velocity. For non-Newtonian liquids, the viscosity requires a correction factor, which can be negligible if the ratio between the outer and inner radius approaches unity. Practically, it is not possible to work with such a low gap, so it requires a correction factor within acceptable limits. In addition to non-Newtonian correction, the end effect also needs to be addressed, as the cylinders are of finite height (h). This correction is made by introducing an additional height (h_o) in height (h). Term h_o is established by extrapolating the curve between M/Ω and h for $M/\Omega = 0$.

2.5.5 Cone and Plate Rheometers

In the cone and plate setup, a low angle cone (≤3°) rotates against a flat plate with fluid between them, as shown in Figure 2.23. Rotational rheometers with flat profiles have a varied shear rate from the axis of rotation. But in cone plate rheometers the shear rate is constant along the radius of plates because the conical profiles both of linear velocity and the gap increases proportionately. This is the reason why such rheometers do not require corrections for non-Newtonian fluids. The shear stress (τ_R), shear rate (γ_R), and shear viscosity (η) in this geometry are given by

$$\tau_R = \frac{3M}{2\pi R^3}$$

$$\gamma_R = \frac{\Omega}{\theta}$$

$$\eta(\gamma_R) = \frac{3M}{\gamma_R \cdot 2\pi R^3}$$

where

M is the torque
Ω is the relative angular velocity
R is the radius of the cone plate
θ is the cone angle

These rheometers are very sensitive to temperature change.

2.5.6 Parallel Plate Rheometers

A parallel plate (plate–plate) rheometer belongs to the category of rotational rheometer suitable for the rheological testing of highly viscoelastic fluids at low shear rates. The gap is usually widened and can be adjusted freely in these devices. The gap adjustment is not as critical here as in cone plate rheometers and the wide gap means there is less temperature sensitivity of the system. Parallel plate systems, when used to determine the shear rate dependency of the viscosity of the melts showing a distinct non-Newtonian flow behavior, require the Weissenberg correction to take care of the variation of the shear rate as a function of the plate radius.

The shear rate in a parallel plate system is determined by angular velocity (Ω), the thickness of the sample layer (i.e., the gap between the plates (h)), and the distance from the center of the plate (R). Because of the rotation kinematics, parallel plate geometry produces an uneven velocity field: the shear rate is at its highest on the rim, and zero at the center of the plate. The shear rate (γR) at the rim is

$$\gamma_R = \frac{R\Omega}{h}$$

The shear stress (τ_R) can be calculated from the measured torque (M). When the geometry is used in rotational mode for measuring nonlinear properties, a correction procedure must be applied in order to overcome the error caused by the nonconstant shear rate profile. Then the shear stress has the form

$$\tau_R = \frac{2M}{\pi R^3}\left(\frac{3}{4} + \frac{1}{4}\frac{d\ln M}{d\ln \gamma_R}\right)$$

where the term in brackets is the correction factor analogous to the Rabinowitsch correction, which is applied in capillary rheometry for the true wall shear rate. The corrected viscosity (η) is thus

$$\eta(\gamma_R) = \frac{2M}{\gamma_R \pi R^3}\left(\frac{3}{4} + \frac{1}{4}\frac{d\ln M}{d\ln \gamma_R}\right)$$

For Newtonian fluids, the term in brackets yields 1 and the shear viscosity simply becomes

$$\eta(\gamma_R) = \frac{2M}{\gamma_R \pi R^3}$$

The results can be presented as viscosity versus shear rate (as shown in Figure 2.24), time sweep, frequency sweep, and the temperature dependence of viscosity and modulus.

2.5.7 Torsional Dynamic Mechanical Analyzers

Dynamic mechanical analyses are a wide range of techniques that depend on the type of mechanical loading used to reveal the mechanical behavior of polymeric materials over a range of temperature, frequency, and amplitude sweeps. ISO 6721 and ASTM D5279 standardize the various methods of DMA. Dynamic mechanical tests provide information about thermal properties such as the softening point, glass transition temperature, service temperature, viscoelastic behavior, and the effect of temperature on the modulus of a material. The most common type of DMA is the torsion test or the twisting of a rectangular sample or rod. In the torsional dynamic analyzer, the sample is rigidly clamped between two ends, one is fixed and the other is twisted

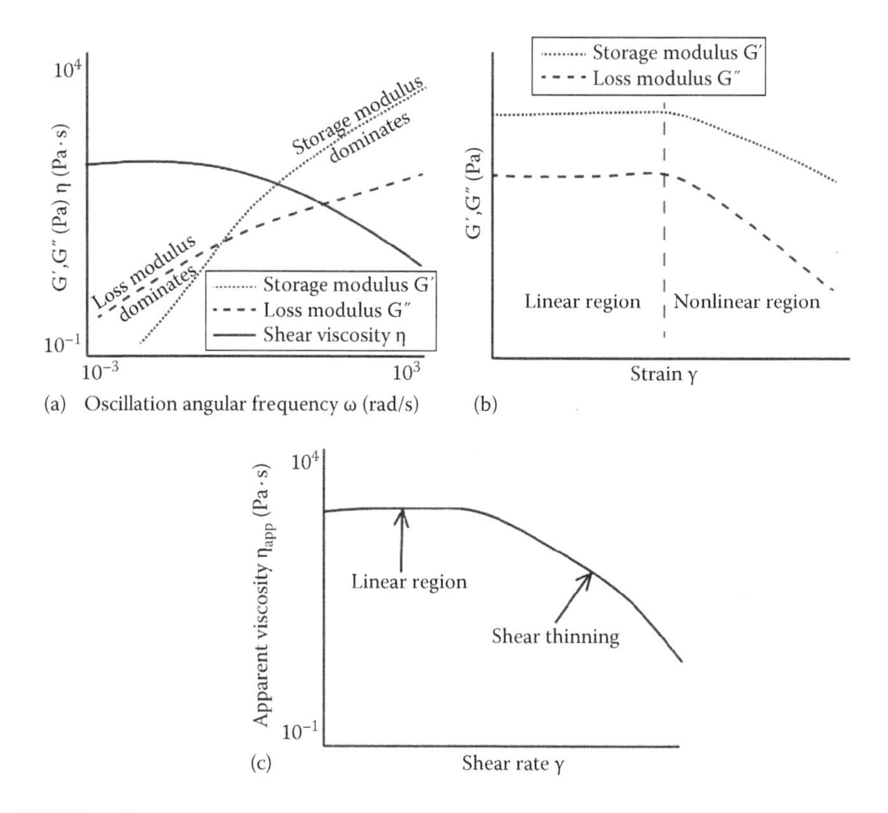

FIGURE 2.24
Graphs from parallel plate rheometry. Effects of (a) angular frequency on viscosity, storage and loss modulus, (b) strain on storage and loss modulus and (c) shear rate on apparent viscosity.

back and forth in an oscillatory motion as shown in Figure 2.23. The degree of the twist and force required to induce the twist are related to strain and stress respectively. A phase shift or time lag is observed between the signal to twist the sample and the actual response of the sample that characterizes its viscoelastic behavior, which can be elucidated graphically in sine wave form as shown in Figure 2.25. The dynamic shear modulus (G*) of a viscoelastic material may be resolved into two components, the elastic (G′) and the viscous (G″):

$$G^* = \sqrt{(G')^2 + (G'')^2}$$

The storage or elastic modulus (G′) is the in-phase response, while the loss or viscous modulus (G″) is the out-of-phase complex response of the sample. The ratio of G″/G′, referred to as tan δ, shows a peak when the sample experiences a thermal transition, such as a glass transition or melting. From a physical viewpoint, the shear modulus is a pure elastic property like Young's

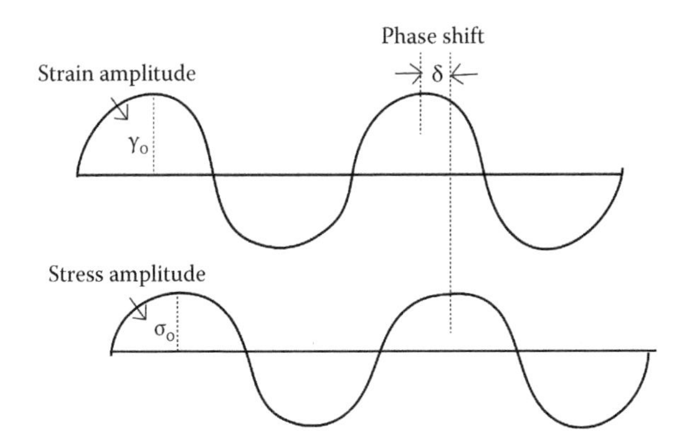

FIGURE 2.25
Phase shift due to the viscoelastic behavior of polymers.

modulus but with shearing strains instead of elongation ones. The shear modulus (G) of a pure elastic material is given by

$$G = \frac{\sigma}{\gamma}$$

where
- σ is the shear stress
- γ is the shear strain

In dynamic shear testing mode, the strain (γ_o) is proportional to the angle of twist (θ) in a rectangular bar, while the stress (σ_o) is proportional to torque (M_o) over a cross-sectional area of the sample, and can be evaluated by

$$\gamma_o = \frac{t \cdot \theta}{2l}$$

$$\sigma_o = \frac{8M_o}{bt^2}$$

where
- l is the sample length
- t is the sample thickness
- b is the sample width

The elastic or storage modulus (G′) and the viscous or loss modulus (G″) can be computed from the following equations. The ratio of the two is termed tan δ, where δ is the phase shift of the two components

$$G' = \left(\frac{\sigma_o}{\gamma_o}\right)\cos\delta$$

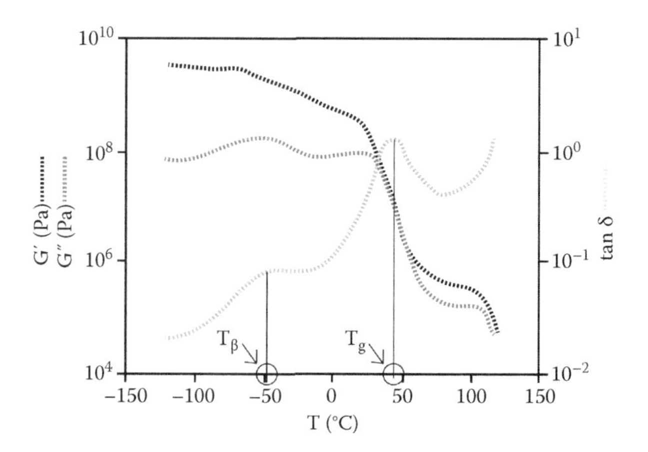

FIGURE 2.26
Typical DMA plot for thermoplastic polymers.

$$G'' = \left(\frac{\sigma_o}{\gamma_o}\right)\sin\delta$$

$$\tan\delta = \frac{G''}{G'}$$

It can be deduced that, if there is no phase shift ($\delta = 0°$), the loss modulus (G'') becomes zero and the cos δ component of G' becomes unity. This demonstrates that material behaves purely elastically without a viscous component when there is no phase shift. When this shift is completely out of phase (i.e., $\delta = 90°$), cos δ becomes zero, thus the elastic component disappears and the material behaves as a pure viscous liquid. When the elastic component is greater than the viscous component, the material will behave more as a solid, and if the viscous component is greater, the material behaves more as a liquid. Peaks in the tan δ curve represent thermal transitions such as relaxation, glass transition, and melting, as indicated in Figure 2.26. Below T_g, the polymer is rigid and brittle. Above T_g, the polymer is less rigid and ductile. A material has better impact properties above rather than below T_g (Campbell et al. 2000).

References

Agilent Technologies. 2014. An introduction to gel permeation chromatography and size exclusion chromatography, Agilent Technologies, Inc., Santa Clara, CA.

Agilent Technologies. 2015. Quick guide to selecting columns and standards for gel permeation chromatography and size exclusion chromatography, Agilent Technologies, Inc., Santa Clara, CA.

Andreassen, E., K. Grøstad, O.J. Myhre, M.D. Braathen, E.L. Hinrichsen, A.M.V. Syre, and T.B. Løvgren. 1995. Melting behavior of polypropylene fibers studied by differential scanning calorimetry. *Journal of Applied Polymer Science* 57(9): 1075–1084. doi: 10.1002/app.1995.070570906.

Campbell, D., R.A. Pethrick, and J.R. White (eds.). 2000. Thermal analysis. In *Polymer Characterization: Physical Techniques*, 2nd edn., Stanley Thornes (Publishers) Ltd, Gloucester, U.K.

Cho, K., F. Li, and J. Choi. 1999. Crystallization and melting behavior of polypropylene and maleated polypropylene blends. *Polymer* 40: 1719–1729.

Demirel, B., A. Yaraş, and H. Elçiçek. 2011. Crystallization behavior of PET materials. *Balikesir University Institute of Science and Technology Magazine* 13(1): 26–35.

Donald, C.H. (ed.) 2004. Viscous flow properties. In *Understanding Plastics Testing*, 1st edn., Carl Hanser Verlag GmbH & Co. KG, Munchen, Germany, pp. 45–60.

Giron, D. 2002. Applications of thermal analysis and coupled techniques in pharmaceutical industry. *Journal of Thermal Analysis and Calorimetry* 68: 335–357.

Gray, F.M., M.J. Smith, and M.B. Silva. 2011. Identification and characterization of textile fibers by thermal analysis. *Journal of Chemical Education* 88(4): 476–479.

Gregorova, A. 2013. Application of differential scanning calorimetry to the characterization of biopolymers. In *Applications of Calorimetry in a Wide Context—Differential Scanning Calorimetry, Isothermal Titration Calorimetry and Microcalorimetry*, edited by Amal A.E., InTech, Rijeka, Croatia. DOI: 10.5772/53822.

Grellmann, W. and S. Seidler. (eds.) 2013. Determining process-related properties. In *Polymer Testing*, 2nd edn., Hanser, Munchen, Germany, pp. 39–71.

Gu, X.X., C. Lai, F. Liu, W.L. Yang, Y.L. Hou, and S.Q. Zhang. 2015. A conductive interwoven bamboo carbon fiber membrane for Li-S batteries. *Journal of Materials Chemistry A* 3(18): 9502–9509. doi: 10.1039/c5ta00681c.

Hatakeyama, T. and F.X. Quinn. 1999. *Thermal Analysis: Fundamentals and Applications to Polymer Science*, 2nd edn., John Wiley & Sons, Inc, West Sussex, England.

Holding, S. 2011. Characterisation of polymers using gel permeation chromatography and associated techniques. *Chromatography Today*, Vol. Nov/Dec 2011, pp. 4–8, U.K.

ISO11357-1. 2016. Plastics—Differential scanning calorimetry (DSC)—Part 1: General principles by technical committee ISO/TC 61/SC 5 Physical-chemical properties. Geneva, Switzerland.

ISO11358-1. 2014. Plastics—Thermogravimetry (TG) of polymers—Part 1: General principles by technical committee ISO/TC 61/SC 5 Physical-chemical properties. Geneva, Switzerland.

Ivan, G. 2009. Nonionic surfactants: GPC/SEC analysis. In *Encyclopedia of Chromatography*, edited by Jack C., 3rd edn., Taylor & Francis, Boca Raton, FL, pp. 1598–1600.

Lafita, V.S. 2011. GPC/SEC: Introduction and principles. In *Encyclopedia of Chromatography*, 2nd edn., Taylor & Francis, pp. 695–698.

Letot, L., J. Lesec, and C. Quivoron. 1980. The errors in the treatment of gel permeation chromatography data: Origins and corrections. *Journal of Liquid Chromatography* 3(11): 1637–1655. doi: 10.1080/01483918008064757.

Lever, T., P. Haines, J. Rouquerol, E.L. Charsley, P. Van Eckeren, and D.J. Burlett. 2014. ICTAC nomenclature of thermal analysis (IUPAC Recommendations 2014). *Pure and Applied Chemistry* 86(4): 545–553. doi: 10.1515/pac-2012-0609.

Liu, X. and W. Yu. 2006. Evaluating the thermal stability of high performance fibers by TGA. *Journal of Applied Polymer Science* 99(3): 937–944. doi: 10.1002/app.22305.

Liu, Y., L. Cui, F. Guan, Y. Gao, N.E. Hedin, L. Zhu, and H. Fong. 2007. Crystalline morphology and polymorphic phase transitions in electrospun Nylon-6 nanofibers. *Macromolecules* 40(17): 6283–6290. doi: 10.1021/ma070039p.

Moore, J.C. 1964. Gel permeation chromatography. I. A new method for molecular weight distribution of high polymers. *Journal of Polymer Science Part A: General Papers* 2(2): 835–843. doi: 10.1002/pol.1964.100020220.

Morrison, F.A. (ed.) 2001. Rheometry. In *Understanding Rheology*, Oxford University Press, Inc, New York.

Nikolay, V. 2009. Polycarbonates: GPC/SEC analysis. In *Encyclopedia of Chromatography*, edited by Jack. C., 3rd edn., Taylor & Francis, Boca Raton, FL, pp. 1850–1852.

Peter, K. 2011. High-speed SEC methods. In *Encyclopedia of Chromatography*, edited by Jack. C., 2nd edn., Taylor & Francis, Boca Raton, FL, pp. 752–756.

Sadao, M. 2011. GPC/SEC: Effect of experimental conditions. In *Encyclopedia of Chromatography*, edited by Jack. C., 2nd edn., Taylor & Francis, Boca Raton, FL, pp. 689–690.

Sam, J.F. 2011. Polyesters: Analysis by GPC/SEC. In *Encyclopedia of Chromatography*, edited by Jack. C., 2nd edn., Taylor & Francis, Boca Raton, FL, pp. 1312–1314.

Serhatli, İ.E. 2013. Chapter 10: Gel permeation chroramatography (GPC). In *Synthesis and Characterization of Macromolecules*, edited by Department of Polymer Science and Technology, Istanbul Technical University, Istanbul, Turkey, pp. 61–77.

Steinmann, W., S. Walter, M. Beckers, G. Seide, and T. Gries. 2013. Thermal analysis of phase transitions and crystallization in polymeric fibers. In *Applications of Calorimetry in a Wide Context—Differential Scanning Calorimetry, Isothermal Titration Calorimetry and Microcalorimetry*, edited by A.A. Elkordy, InTech, pp. 277–302.

Striegel, A.M. 2005. Multiple detection in size-exclusion chromatography of macromolecules. *Analytical Chemistry* 77(5): 104 A–113 A.

Susan, V.G. 2009. SEC with on-line triple detection: Light scattering, viscometry, and refractive index. In *Encyclopedia of Chromatography*, 3rd edn., Taylor & Francis, pp. 2120–2123.

The Fiber Year Consulting. 2015. *The Fiber Year 2015—World Survey on Textiles & Nonwovens*, The Fiber Year Consulting, Geneva, Switzerland.

Trathnigg, B. 2006. Size-exclusion chromatography of polymers. In *Encyclopedia of Analytical Chemistry*, edited by Meyers, R.A., John Wiley & Sons, Ltd, New York.

Tuan, Q.N. 2009. Polyamides: GPC/SEC analysis. In *Encyclopedia of Chromatography*, 3rd edn., Taylor & Francis, pp. 1846–1849.

Walczak, Z.K. 2002. *Processes of Fiber Formation*, 1st edn., Elsevier, Oxford, U.K.

Witten, E. 2014. Composites market report 2014—European GRP and global CRP markets. In *Market Developments, Trends, Challenges and Opportunities: AVK Federation of Reinforced Plastics and Carbon Composites*. by, the European Composites Industry Association (EuCIA), Brussels, Belgium.

Yefim, B. 2009. Polymers: GPC determination of intrinsic viscosity. In *Encyclopedia of Chromatography*, 3rd edn., Taylor & Francis, pp. 1882–1884.

3

Advanced Characterization Techniques: Conventional and Technical Textiles

Usman Zubair, Madeha Jabbar, Abher Rasheed, and Sheraz Ahmad

CONTENTS

3.1 Introduction

This chapter introduces some of the advanced material characterization techniques such as scanning electron microscopy (SEM), energy dispersive X-ray spectroscopy (EDX), and X-ray diffraction (XRD). In later sections, advanced characterization techniques particular to the textile industry, like moisture management and IR thermography, will be discussed.

3.2 Scanning Electron Microscopy

SEM employs an extremely finely focused electron beam in order to yield high-resolution imaging of samples and their surfaces. Optical microscopes are limited in their resolution to 0.2 μm by the intrinsic properties of light. Owing to electrons' smaller wavelengths, scientists began working with electron microscopes that provide a two to three times higher resolution than light microscopes in the 1930s. The wavelength of a subatomic particle moving with velocity (v), mass (m), and momentum (p) is given by the equation derived by de Broglie

$$\lambda = \frac{h}{p} = \frac{h}{mv}$$

where h is Planck's constant.

Knoll (1935) and von Ardenne (1938) are considered the pioneers in the electron microscopy field. Currently, there are more than 50,000 SEM equipped laboratories and businesses working across the globe (Breton, 2009).

3.2.1 Theory and Principle of SEM

In SEM, an electron beam is directed onto the sample under investigation. When electrons are accelerated under a potential difference (V), the potential energy becomes equal to the kinetic energy of the accelerated electrons

$$eV = \frac{m_o v^2}{2}$$

$$v = \sqrt{\frac{2eV}{m_o}}$$

where

e is the charge of the accelerated electrons

v is the mass

m_o is the rest mass of the accelerated electrons

Substituting v in the de Broglie equation, wavelength becomes

$$\lambda = \frac{h}{\sqrt{2m_o eV}}$$

At a very large accelerating potential, the velocity of an electron will approach the velocity of light, and relativistic effects become quite significant. By incorporating the relativistic effects and replacing the constant terms, the above expression transforms into

$$\lambda = \frac{h}{\sqrt{2m_0eV\left(1+\frac{eV}{2m_0c^2}\right)}} = \frac{1.5}{\sqrt{V+10^{-6}V^2}}\,\text{nm}$$

Even with a potential difference of 10 kV, the wavelength of electrons becomes 14.9 picometer (pm), which is smaller than the size of any atom. For generating such accelerated electrons, an electron gun is situated at the upper side of the microscope. The gun projects a ray of highly concentrated electrons, and a series of magnetic lenses direct the beam onto the sample to attain maximum efficiency. The more electrons focused onto the specimen, the higher will be the magnification. The whole assembly is placed inside a vacuum chamber to avoid obstruction of the electron beam. When the test sample is impacted on by the ray of incident electrons, it will emit characteristic X-rays and three types of electrons: primary backscattered electrons, secondary electrons, and Auger electrons. SEM employs primary backscattered electrons and secondary electrons to image the topography and morphology of the sample. An electron receptor gathers the rebounding electrons and reports their trajectories. This data is then converted and displayed on a monitor as clear three-dimensional images (Reimer, 1998).

3.2.2 Electron Beam and Specimen Interaction (Electron Scattering)

When a specimen is bombarded by a ray of electrons, most of the latter interact with the atoms of the specimen rather than simply penetrating them. The volume of the specimen that interacts with the electron beam is usually termed the "interaction or excitation volume," which could be semi-spherical or semi-dumb-bell shape, having dimensions of usually 1–5 μm depth. The interaction between beam and specimen is both elastic and inelastic. These interactions create the distinct nature of the signals. Elastic scattering causes changes in the electrons' trajectories without them losing their kinetic energy. Inelastic scattering shows the transfer of energy from the incident ray to the atoms present in the test sample. Therefore, the electrons suffer energy loss and a small trajectory deviation. In addition to the scattering events, elastic and inelastic interactions also give rise to some secondary effects that are a unique source of signals that can be used to analyze the sample in SEM, as shown in Figure 3.1.

The effects involved in the operation of SEM are secondary electrons, backscattered electrons, cathodoluminescence (CL), and electron beam–induced current. Other effects are Auger electrons, continuum X-ray radiation

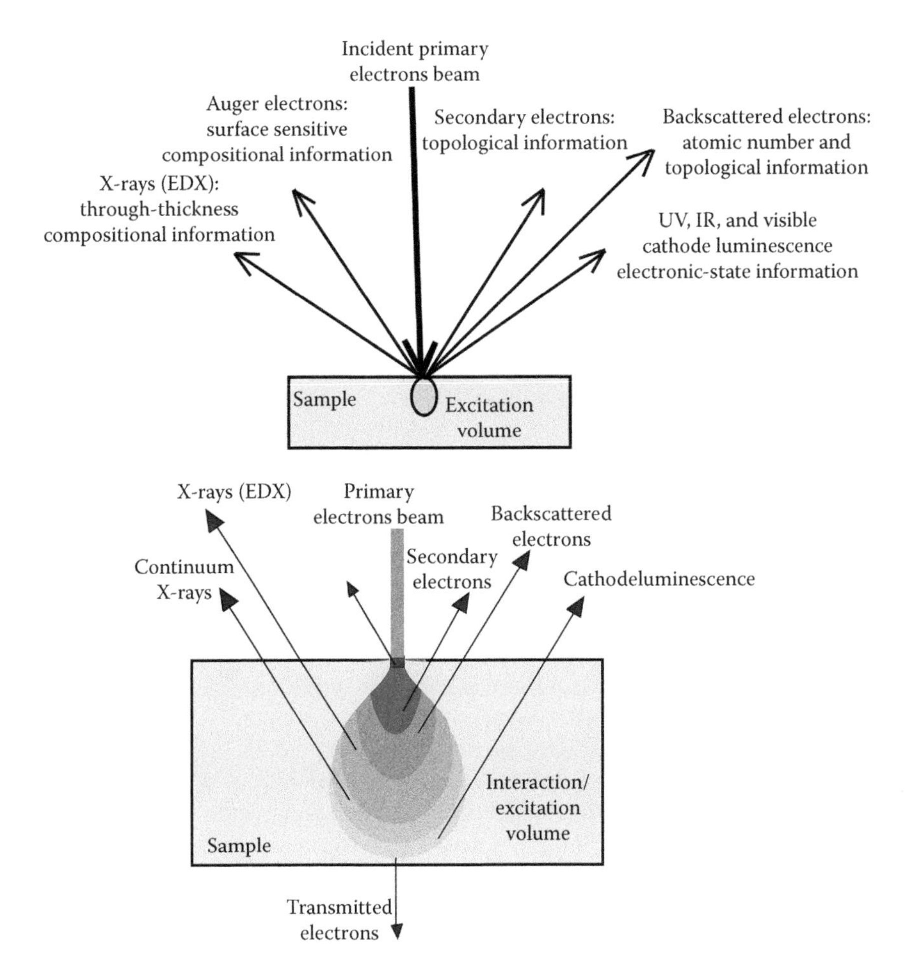

FIGURE 3.1
Interaction of an electron beam with a sample's surface and the resultant signals.

(Bremsstrahlung), and characteristics of X-rays, which will be discussed later (Oleshko et al., 2000).

3.2.2.1 Emission of Backscattered and Secondary Electrons

Secondary electrons are produced in abundance and form the basis of SEM. The secondary electrons are dislodged or discharge from the atoms in the test sample and have energies lower than 50 eV. Such electrons can therefore include the primary electrons that lose their energies through successive scattering and reach the surface of the specimen. Backscattered electrons are the primary electrons that do retain substantial energy before escaping the specimen's surface. Backscattering is related to the atomic number wherein samples with a larger atomic number give brighter signals.

3.2.2.2 Electron Beam–Induced Current and Cathodoluminescence (CL)

An electron can knock off a valence electron from the colliding atom, creating an electron–hole pair. An electron falls back into the hole, releasing the excess energy as light. CL provides a method of distinguishing phases and elemental concentrations and of localizing specific fluorescent stains (Goldstein et al., 2003).

3.2.3 Basic Components of SEM Instrumentation

3.2.3.1 Electron Guns

The purpose of the electron gun is to yield electrons. These electrons are accelerated into a narrow beam. Electron guns (emission sources) can be categorized into two classes: thermionic electron guns and field-emission guns. The basic distinction among the scanning electron microscope and the field emission scanning electron microscope (FESEM) is the source of the electron emission.

A thermionic electron gun comprises a cathode filament, a Wehnelt cup, and an anode. Tungsten (W) and lanthanum hexaboride (LaB_6) are the two most common cathode filament materials used in a thermionic gun. A tungsten thermionic cathode usually appears as a wire bent into a hairpin, and lanthanum hexaboride rods are assembled in a tip form by squeezing them between carbon electrodes, as represented in Figure 3.2.

Tungsten filaments are low cost and easy to maintain, while lanthanum hexaboride crystals produce a slightly brighter and cohesive beam with a longer lifetime. Cathode filaments are heated up by applying electrical current. When there is sufficient amount of heat, the electrons can get away from the Fermi level of the cathode material. Thermionic cathodes have the problems of relatively low brightness, evaporation of cathode material, a limited lifetime of the filament, and thermal drift during operation. These limitations are resolved by using field emission phenomena to generate electrons.

A field emission gun (FEG) or cold field emitter will not require heating of the filament. A large potential gradient (electrical) of the filament is required for emission. FEG guns comprise a cathode filament and two anodes, as indicated in Figure 3.3.

The FEG cathode is usually a wire of tungsten crafted into a sharp point, as shown in Figure 3.2. The small tip's radius of around 100 nm allows the electric field to be concentrated to an extreme level, becoming so big that the work function of the material is lowered and electrons can leave the cathode through *tunneling* (Goldstein et al., 2003). A FESEM uses these FEGs to produce a cleaner image with higher brightness, less electrostatic distortion, and a spatial resolution <2 nm. FESEM produces an image three to six times better than SEM with a thermionic gun. In Table 3.1, a comparison of the operating parameters and performance of three different emitters is shown at 100 kV (Michler, 2008).

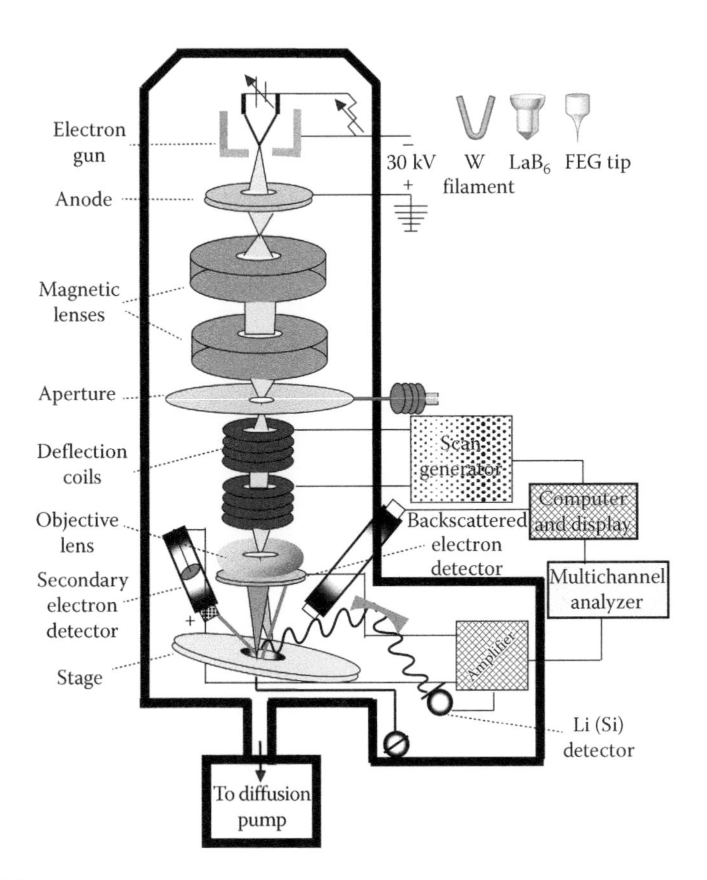

FIGURE 3.2
Schematic representation of scanning electron microscope and cathode filaments.

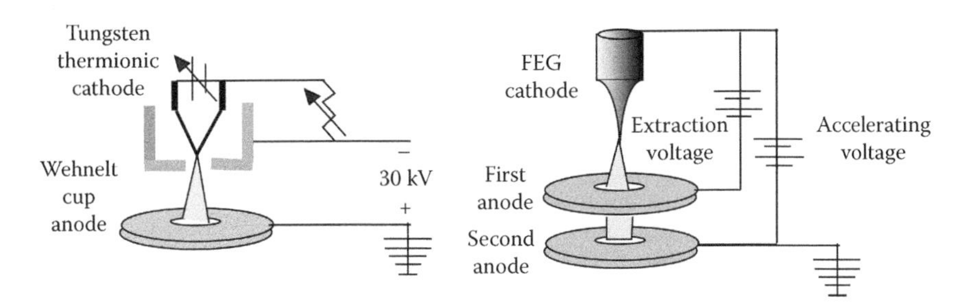

FIGURE 3.3
Schematic representation of thermionic and field emission guns.

TABLE 3.1

Comparison of Three Types of Emission Sources Operating at 100 kV

Type of Gun	W	LaB_6	Field Emission
Work function (eV)	4.5	2.4	4.5
Operating temperature (K)	2700	1700	300
Probe diameter (Å)	1×10^6	2×10^5	$<1 \times 10^2$
Current density (A/m^2)	5×10^4	10^6	10^{10}
Crossover size (μm)	50	10	<0.01
Energy spread (eV)	3	1.5	0.3
Emission current stability (%/h)	<1	<1	5
Brightness (A/m^2 sr)		10^5	10^{13}
Vacuum (Pa)	10^{-2}–10^{-3}	10^{-4}	10^{-8}
Lifetime (h)	100	500	>1000

Source: Michler, G.H., *Electron Microscopy of Polymers*, Springer-Verlag, Berlin, Germany, 2008.

3.2.3.2 Anodes

Thermionic scanning electron microscopes have a Wehnelt cup and an anode. The Wehnelt cup acts as a control grid and a convergent electrostatic lens. The filament is positioned directly above the Wehnelt aperture and the anode is located below, as represented in Figure 3.3. The Wehnelt is biased to a negative voltage of −200 to −300 V relative to the emitter. This forms a repulsive electrostatic field that concentrates the cloud of primary electrons produced by the filament. The V-shaped filament tip that is located above the aperture emits the electrons from a very small area in the form of a circular beam. The anode is biased to a high, positive, accelerating voltage of 1–50 kV relative to the cathode filament to accelerate electrons from the emitter towards the anode, thus creating an electron beam that passes through the Wehnelt aperture. The accelerating voltage combined with the beam diameter limits the resolution. As voltage increases, better point-to-point resolution can be achieved.

The FESEM has two anodes for electrostatic focusing: one regulates the potential difference at the tip to emit electrons, while the second accelerates the electrons as shown in Figure 3.3. A voltage of 0–5 kV, termed the extraction voltage, is applied across the field discharge tip and the primary anode to control the current emission at 1–20 μA. An accelerating voltage (1–50 kV) across the cathode and the secondary anode will increase the electron beam's energy and control the movement of electrons into a column with a certain velocity (Reimer, 1998).

3.2.3.3 Vacuum and Pumping System

A scanning electron microscope requires a high vacuum to ensure the prevention of discharging inside the gun zone, the long lifetime of cathode filaments, and electron movement along the column without scattering. Modern scanning electron microscopes have a differential vacuum along the column with the highest vacuum in the area around the filament. This is because of a very high scattering of electrons by the molecules present in the air. Vacuums better than 10^{-3} Pa and 10^{-4} Pa are required to prevent formation of volatile oxidation products for tungsten and lanthanum hexaboride filaments, respectively. FEG filaments require an ultra-high vacuum in the range of 10^{-7}–10^{-8} Pa, because otherwise the tip radius is destroyed by ion bombardment from the residual gas. Diffusion and turbo-molecular pumps can create the high pressure of vacuum required in normal scanning electron microscopes. For FEGs, ion pumps are often used to reach such a high level of vacuum. An environmental scanning electron microscope (ESEM) is an exception that operates at reduced vacuum levels and uses only backscattered electrons for imaging, hence the poor resolution. Rotary pumps are used in ESEMs to attain a relatively low vacuum.

3.2.3.4 Electromagnetic Lenses

An electron beam diverges as it passes across the anode plates; it should be condensed and refocused to resolve the features on the specimen's surface. In order to do the latter, the diameter of the beam should be smaller than the features to be examined. Both the electric and magnetic fields can be employed to bend moving electrons. To achieve demagnification of the beam, electromagnetic lenses and condensers are used. The Lorentz force law gives the deflection experienced by a fast moving electron with velocity (v) in a magnetic field (B)

$$\vec{F} = -e\left(\vec{v} \times \vec{B}\right)$$

A magnetic lens comprises two circularly symmetric iron pole pieces having copper windings with a hole in the center through which the electron beam passes. The two pole pieces are separated by an "air gap" where focusing actually takes place as depicted in Figure 3.4.

Electrons travel nearly parallel to a magnetic field that has radial and axial components. When electrons enter the magnetic field, they encounter the radial component of B that pushes them into a circular path. Meanwhile the axial component comes into play and forces the electrons down into a spiral path. Thus, electrons focus themselves into a narrow beam. Magnetic lenses are not good at focusing as they also confront aberration defects. The crossover diameter in the field emission source is much less in comparison to a thermionic emission source, as is clear from Table 3.1. Therefore, a smaller level of beam condensation is required for a useful probe that can be employed for the processing of an image. This makes the FESEM the highest resolution instrument (Goldstein et al., 2003).

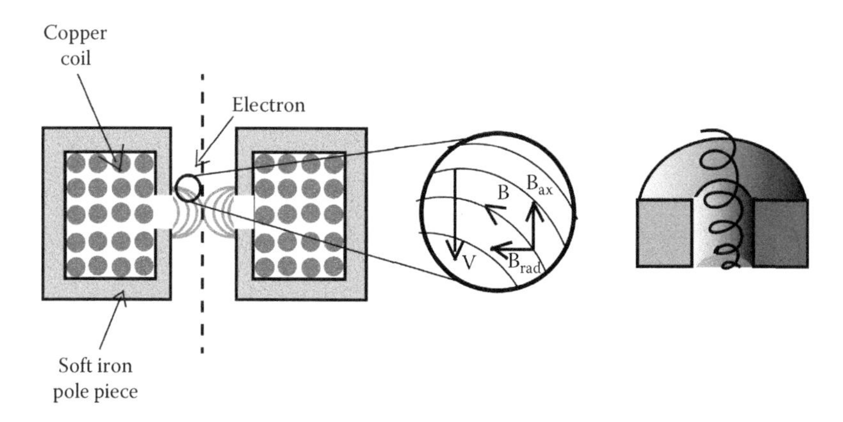

FIGURE 3.4
Working principle of magnetic lenses to focus electron beam.

3.2.3.5 Aperture and Deflection (Scanning) Coil

The purpose of the aperture is to abandon off-axis and off-energy electrons going down the column with the electron beam. The aperture plate acquires holes of various sizes to process the beam. The beam with the better resolution is produced with the help of small hole sizes in the objective aperture. It also produces a minimum amount of charge as well as good depth of field. Therefore, the correct aperture size is imperative. The deflecting or scanning coils perform raster of electron beam across the sample surface for textural imaging. Coils located within the objective lens carry out deflection of the electron beam. The coils consist of four magnets oriented radially that produce a field perpendicular to the optical axis of the electron beam (Wittke, 2016).

3.2.3.6 Electron Detectors

Principally, it is the signals produced at the surface of the specimen that are detected. A detector is biased with positive potential to attract the low energy (<50 eV) secondary electrons, which are emitted from the specimen and detected using a scintillator-photomultiplier (Everhart-Thornley) detector. The detector used for backscattered electrons is not biased. Such electrons are detected using interface between positively and negatively doped semiconductor called as PN junctions positioned at the base of the objective lens at a high take-off angle. CL is also detected using a photomultiplier tube (Michler, 2008).

3.2.3.7 Signal Processing and Image Contrast

The focused beam of electrons is scanned across the surface in a raster mode. The numbers of secondary electrons produced by the specimen at each

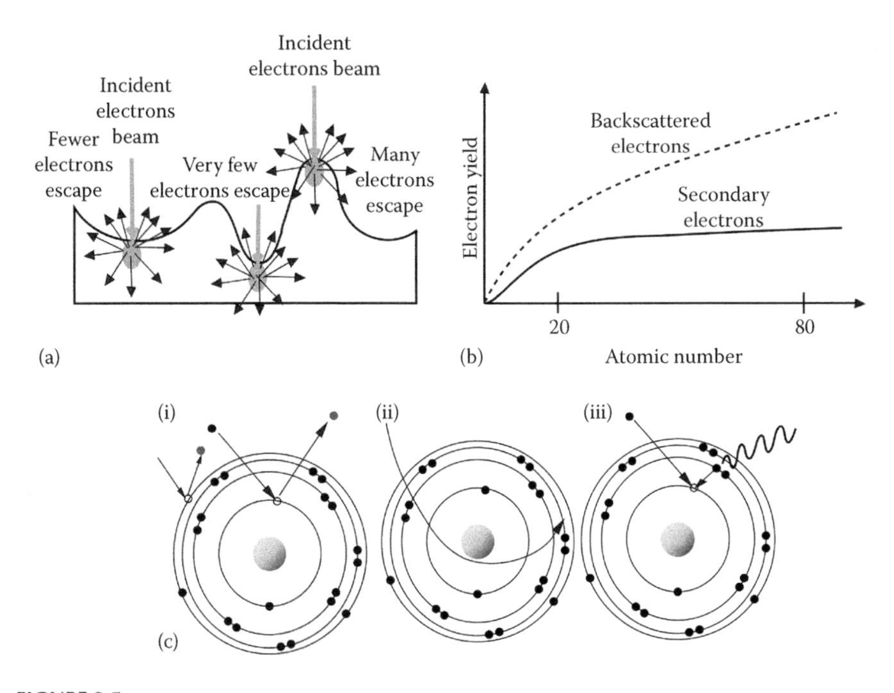

FIGURE 3.5
(a) Signals for topographical contrast in SEM, (b) effect of atomic number on yield of secondary and backscattered electrons, and (c) electron interaction with atoms to generate various signals.

scanned point are plotted to give a two-dimensional image. Topographic contrast can be constructed either by secondary electrons or by backscattered electrons at low take-off angles. In both cases, the contrast depends on the selected angular range of the electrons collected. The secondary electrons escaping from a point depend on the work function; its vicinity in the sample is called the secondary electron yield, as represented in Figure 3.5a. Spots with larger electrons escaping than for the rest of the surface appear brighter in display. Secondary electrons can also distinguish different materials by virtue of the differences in their yield. The amplified electrical signals coming from photomultiplier detectors are displayed as a two-dimensional intensity distribution for the rastered area. Material contrast in backscattered electron (BSE) images corresponds to an increase in intensity with increasing mean atomic number owing to the increase of the backscattering coefficient, as indicated in Figure 3.5b.

Backscattered electrons detector positioned on top optimally records this contrast at high take-off angles. As backscattered electrons come from a significant depth within the sample, as labeled in Figure 3.1, they do not provide much information about the specimen's topology.

3.2.3.8 Magnification, Resolution, and Depth of Field

Magnification in a scanning electron microscope can be controlled over a range of about 6 orders of magnitude, from about 10 to 500,000 times. The resolution of light microscopes is determined using

$$d_{min} = \frac{1.22\lambda}{2n\sin\alpha}$$

In SEM, the refractive index (n) can be assumed to be unity ($n \approx 1$) because of the high vacuum. As electrons are deflected by very small angles, so $\sin\alpha \approx \alpha$. The equation for the resolution reduces to

$$d_{min} = \frac{0.61\lambda}{\alpha}$$

Assuming $\alpha = 5°$ (0.1 rad), the theoretical resolution of a scanning electron microscope operating at an accelerating voltage of 100 kV turns out to be 2.26 pm. In practice, however, resolutions better than 0.2 nm are rarely achieved, largely due to lens aberration.

3.2.4 Sample Preparation for SEM

Sample preparation is an important and first step in any microscopy technique. The sample is usually mounted firmly on a sample holder termed as stub. For SEM, the sample should be electrically conductive, at least at the surface, and it should be earthed to avoid accumulation of a static charge at the surface. To achieve depletion of the charge from the surface, the sample is fixed to the stub using two-faced conductive carbon tape. As metals are electrically conductive, this requires no preparation except cleaning and mounting. However, nonconductive specimens tend to accumulate charge at the surface when exposed to an electron beam, which results in scanning faults. Therefore, nonconductive samples have to be coated with an ultra-thin conductive layer using a sputter coater, usually gold. The metallic coated layer is acquired using an electric field and argon gas. Very soft samples are cryogenically frozen into solids and then cut apart with a sharp knife to observe the morphology under SEM (Echlin, 2009).

3.2.5 Applications of SEM

SEM reveals very comprehensive three-dimensional contrast imaging of textile and polymer substrates at very high magnifications (up to 500,000) at a scale of 1–5 nm. SEM can be applied to evaluate the microscopic features and microstructures of nanofibers, nanoparticles, and polymer nanocomposites. Figure 3.6 shows vertically grown carbon nanotube (CNT) fibers at different length scales, revealing the arrangement of CNT in bulk.

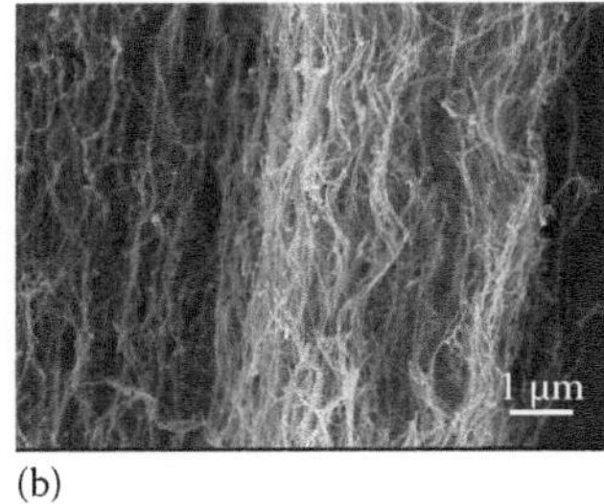

FIGURE 3.6
Scanning electron micrographs of vertically grown CNTs at (a) 300 μm and (b) 1 μm.

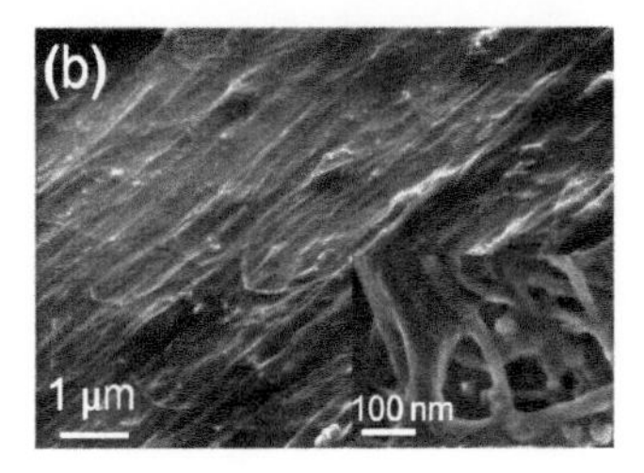

FIGURE 3.7
(a) SEM image of electrospun nylon 6 nanofibers coated with antibacterial silver nanoparticles and (b) SEM image of PLA-CNT nanocomposites.

Electrospun nanofibers are extensively used as scaffolds in biomedical applications, as filtration media, and in many other technical textile applications. For example, Figure 3.7a reveals electrospun nylon 6 nanofibers coated with antibacterial silver nanoparticles to make an effective sterilizing air filter (Hong Dong, Dong Wang, Gang Sun et al., 2008).

Polymer nanocomposites have a broad range of microstructures and applications that are extensively studied by SEM, for example, Figure 3.7b illustrates the coating of CNT fibers by polylactic acid (PLA) as evidence of their affinity in PLA-CNT fiber composites (Yue et al., 2015). Very thin coatings on textile substrates at the nanoscale can be evaluated by observation under electron microscopes. Wang et al. (2011) reported functional cotton fabrics coated with SiO_2 coated ZnO nanorods involve hydrothermal growth of ZnO nanorods followed by SiO_2 shell deposition using a layer-by-layer technique. Figure 3.8 shows this uniform and dense multifunctional coating under SEM.

SEM analysis can provide forensic evaluation of textile materials through surface contamination investigations. Composite failure analysis and fracture characterization has also emerged as an application field of electron microscopy (Ehrenstein and Raue, 2005). Figure 3.9a shows the SEM micrograph of an ultra-high molecular weight polyethylene (UHMWPE) (Dyneema) fiber reinforced polymer matrix composite for ballistic protection application (Fejdyś et al., 2016). Figure 3.9b shows the SEM of the tensile failure of a 4 mm basalt fiber polyester matrix composite (Amuthakkannan et al., 2013).

FIGURE 3.8
Scanning electron micrographs of (a, c, d) SiO_2-coated ZnO nanorods on a cotton substrate at different length scales, (b) on uncoated cotton fiber.

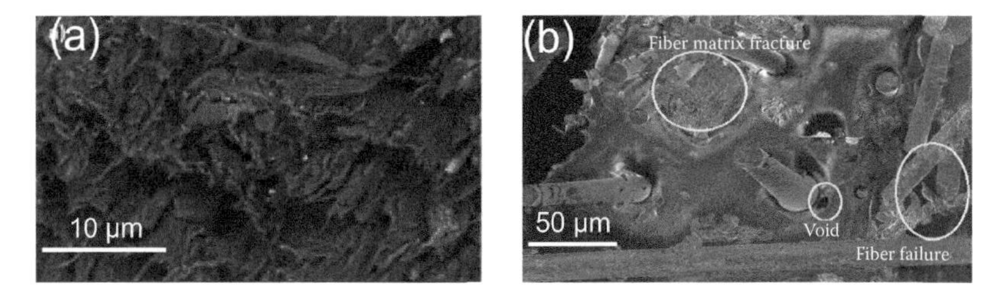

FIGURE 3.9
SEM images of (a) a UHMWPE (Dyneema) fiber reinforced polymer matrix composite and (b) the tensile failure of a 4 mm basalt fiber polyester matrix composite.

3.3 Energy Dispersive X-Ray Spectroscopy

EDX appears as a complementary elemental analysis or chemical characterization technique in modern electron microscopes that are equipped with an X-ray spectrometer.

3.3.1 Theory and Principle of EDX Measurement and Instrumentation

EDX measurements rely on the generation and detection of characteristic X-rays when high-energy accelerating electrons interact with the constituent elements of a specimen. When a ray of high-energy electrons strikes a surface, it knocks out a small amount of electrons from the inner orbitals of atoms, characterized as "inner-shell ionization." The critical amount of energy required to bring about the inner shell transition is termed the "critical excitation/ionization potential." Within a short lifespan of ~10^{-14} s, outer shell electrons occupy inner-shell vacancies by radiating characteristic X-rays. These transitions of inner shell electrons obey the quantum selection rule

$$\Delta m \geq 1, \Delta l = 1, \Delta j = 0 \text{ or } 1$$

where
m is the magnetic quantum number
l is the angular momentum quantum number
s is the spin quantum number
j is the total angular momentum quantum number (i.e., $l \pm s$)

The innermost atomic orbitals demand a higher value of excitation potential to eject electrons. Therefore, incident electrons with a lower potential to excite Kα radiation have sufficient energy to excite Lα or Mα radiation; 0.1% of beam electrons are involved in K-shell ionization. In 1923, M. Siegbahn introduced the nomenclature to identify the X-ray spectrum that is still in use today; for example, Fe-Kα_1, where Fe is the element emitting the characteristic X-rays, K is the shell bearing vacancies, α is the next shell L from where the electron came to fill the vacancy, and the subscript symbolizes the orbital in the returning shell. K could be L or M if the vacancy belongs to these shells, and α could be either β or γ if the occupying electrons come from the second next and third next shell, respectively, as indicated in Figure 3.10.

Each atom can be repeatedly ionized to generate detectable signals for elemental characterization.

3.3.1.1 Energy Dispersive X-Ray Spectrometer

An EDX instrument is of the same setup as SEM for bombarding electrons onto the sample. In addition, EDX instruments have an X-ray spectrometer for the ultra-detection of emitted characteristic X-rays, as illustrated in Figure 3.2. There are various X-ray detecting technologies, such as gas scintillator detectors and solid state (semiconductor) detectors. Semiconductor detecting technology such as Si(Li) is a major breakthrough in energy dispersive X-ray spectrometry. Semiconductor detectors used in EDX are proportional type detectors. They are comprised of Si or Ge single crystals. Pure and perfect Si crystals, termed "intrinsic semiconductors," are ideal candidates for detectors. However, semiconducting materials usually have imperfect crystal structures because of lattice defects or inevitable impurities like B or Al that give rise to electron efficient and electron deficient (hole) regions. B has less valence electrons than Si that cause a "hole" in the valence band of the lattice termed "p-type semiconductors." In contrast,

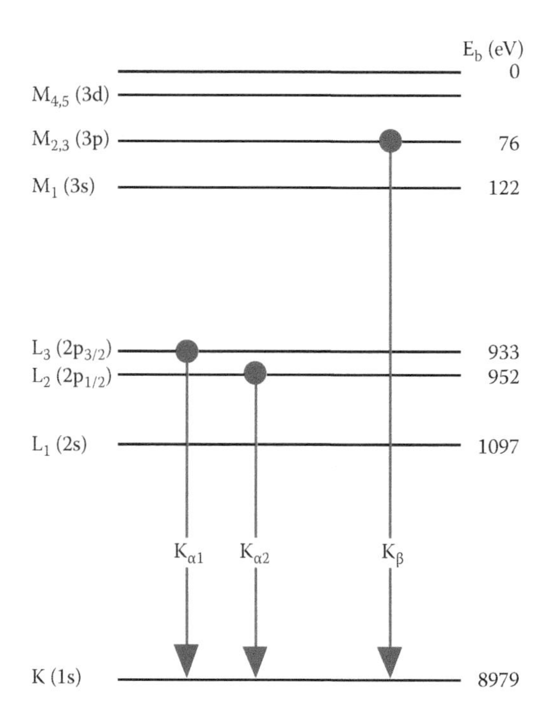

FIGURE 3.10
Schematic representation of different categories of characteristic X-rays generated by inner shell transitions.

elements like Li or P add electrons to the conduction band of the lattice, generating free electrons labeled "n-type semiconductors." The process of adding traces of impurities in a crystal lattice of a pure semiconductor is termed "doping." Doped semiconducting materials are abundant in charge carriers (i.e., holes and free electrons), known as "extrinsic semiconductors." It is obvious that doped semiconductors allow more current to pass through on applying an electric field than that of pure semiconductors, owing to more charge transferring sites.

3.3.1.1.1 Detector Crystals

Detector crystals used in an EDX spectrometer are fabricated by drifting Li in 2–5 mm thick Si crystal labeled Si(Li). Li drifting induces differentially doped regions that are free of Li regions (n-type); Li drifted intrinsic and p-type regions form a p-i-n junction. When an Si(Li) crystal with gold contacts is voltage biased, a current is caused to flow. The detector crystal is kept at lower temperatures to inhibit Li diffusion between doped and intrinsic regions.

3.3.1.1.2 Detection Process

When a detector crystal is subjected to X-rays, it absorbs and yields high-energy photoelectrons, which interact with valence band electrons to push them into the conduction band and generate electron–hole pairs. Because of this shift, photoelectrons lose a certain amount of energy (i.e., 3.8–3.9 eV per electron–hole pair for Si crystal) and keep on interacting until all their energy is dissipated. The voltage biased detector crystal sets off the electron–hole pairs to travel and increase the sample conductivity. The higher the energy of the X-rays, the more electron–hole pairs will form and the greater will be the increase in the conductivity (or decrease in resistivity). Thus, the current conducted by the detector crystal is directly proportional to the energy of characteristic X-rays. Signal noise can be high due to leakage of current because of thermal excitation of the electrons in the detector crystal. The detector crystal is housed in a liquid N_2 cryostat to avoid noise in the detection signal. Present-day detectors require a reduced amount of liquid N_2 and can be configured for light elemental analysis down to beryllium (Garratt-Reed and Bell, 2003).

3.3.2 Data Processing and Analysis Sensitivity

The pulses produced by the Si(Li) crystal require amplification before their construction, analysis, and interpretation. The preamplifier with a field-effect transistor is positioned near the detector crystal, and the whole assembly is cooled with liquid N_2 that helps to minimize electronic noise. Then, the resultant signal is sent to a main amplifier, which delivers linear, low noise amplification of the preamplifier signal. The main amplifier should be able to process the pulses quickly to avoid overlay of the signals. The amplitude of the pulse is proportional to the energy of the incident X-ray photon. So, large pulses are converted into larger numbers. These pulses are counted and accumulated to display the result of the analysis.

3.3.3 Elemental Analysis and Imaging with X-Rays

The resulting energy spectra can be interpreted qualitatively or quantitatively. The elements having concentration levels of about 0.5% by weight create distinctive detectable EDX peaks. The height of the peaks can be employed for making a rough approximation of the element's proportion in a sample. Precise quantification of the spectrum involves a technique called "peak fitting," in which reference peaks are used in variable proportions to reconstruct the spectrum (Goldstein et al., 2003).

3.3.4 Applications of EDX Analysis

EDX analysis assists in the rapid identification of material surfaces, phases, and their distribution across the substrate. It can be also used to analyze small component materials, foreign material, and coating compositions. Figure 3.11a exhibits the SEM image and EDX images of multiwalled

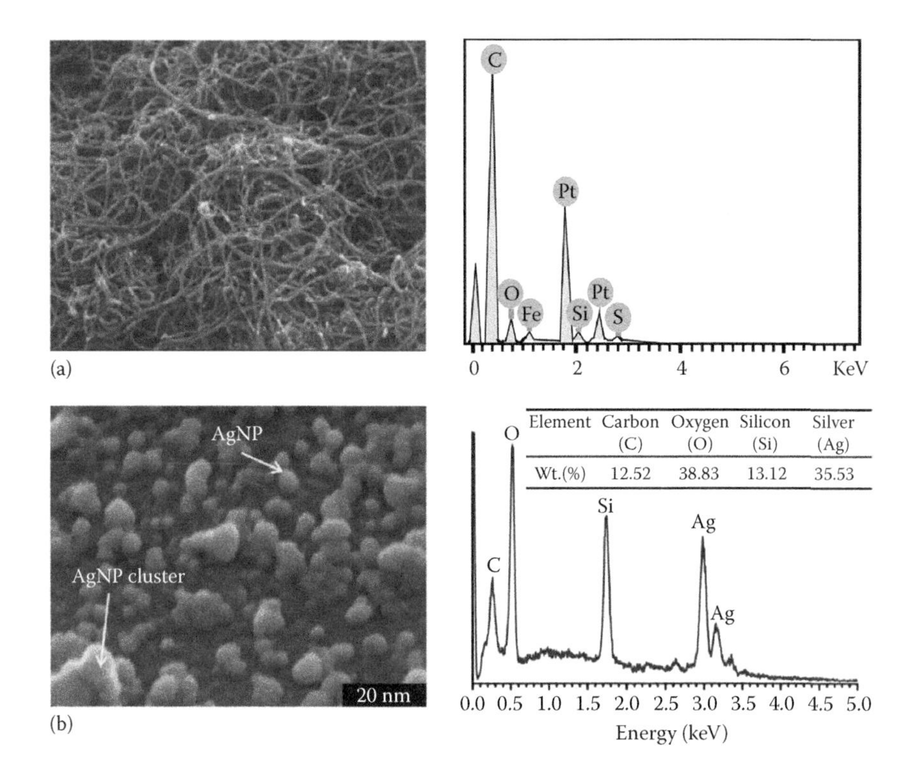

FIGURE 3.11
(a) SEM image and EDX images of pristine multiwall nanotube MWCNTs produced by chemical vapour deposition (CVD) and (b) FESEM image and EDX analysis of silver nanoparticles immobilized on the amine-modified glass surface.

CNT (MWCNT) (Lee et al., 2011). Figure 3.11b shows the FESEM image and the EDX analysis of silver nanoparticles (Agnihotri et al., 2013).

3.4 X-Ray Diffraction Analysis

Since 1912, when Max von Laue discovered the diffraction of X-rays by crystals, the X-ray study of textile materials and polymers emerged as the ultimate choice for revealing the amount of crystallinity, crystallite size, lamellar folding dimension, orientation and texture in fibers or fabricated items, and eventually unique morphologies and morphological changes. Figure 3.12 depicts the hierarchical arrangement of a fiber that has been probed by X-ray analysis (Murthy, 2004).

Depending on the desired structural information, different X-ray scattering methods—wide angle X-ray diffraction (WAXD) or small-angle X-ray scattering (SAXS)—can be used.

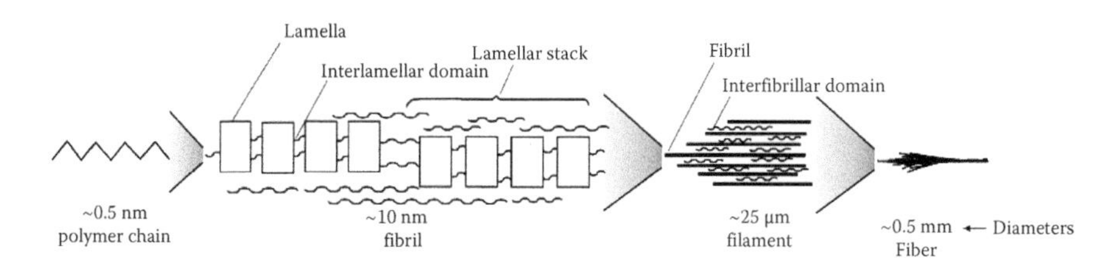

FIGURE 3.12
Schematic representation of the crystalline-based hierarchical structure of a fiber.

3.4.1 Generation of Characteristic X-Rays and Their Interaction with Matter

X-rays are high energy electromagnetic radiation that lies between ultraviolet and gamma radiation, with wavelengths ranging from 0.01 Å (hard X-rays) to 100 Å (soft X-rays), which corresponds to energies ranging from ~120 eV to ~120 keV. X-rays are generated by decelerating fast-moving electrons and transforming their kinetic energy into high-energy photons. The wavelength of the resultant X-ray radiation depends upon the kinetic energy of the electrons. Generally, X-rays are generated by striking thermally emitted high-energy electrons from a cathode filament against a heavy metal anode target at a very high potential in a vacuum, known as an X-ray tube. Incoming electrons remove those from the inner orbits of an atom, for example, the K orbit. Then, electrons from higher energy levels (e.g., L, M shells) fall to a lower energy level to fill the vacancy. The energy difference of this inner shell transition is radiated as high-energy X-ray photons, which are labeled K_α, K_β, and so on, if the radiated photons are due to the descent of electrons from the L and M shells to the K shell, respectively, as shown in Figure 3.10. As electron shells in an atom of each element have their own quantized level, so the wavelength, frequency, and energy of emitted X-ray photons are unique to each metal and are called "characteristic X-rays." The energy variation among the various orbits increases with the increasing atomic number, hence metal anodes with a higher atomic number emit radiation with a shorter wavelength and higher energy than those with lower atomic number targets. In addition to these high intensity characteristic X-rays, there is a continuous X-ray radiation spectrum, called Bremsstrahlung or white radiation, that arises due to the deceleration of highly accelerated electrons at the nucleus with a minimum wavelength (λ_{min}) given as

$$\lambda_{min} = \frac{12.4}{V_{acc}} \text{Å}$$

where V_{acc} is the high accelerating voltage (Pecharsky and Zavalij, 2003).

When X-rays come across a material, they interact with the matter in various ways by dissipating energy to electrons or by scattering, as illustrated in Figure 3.13.

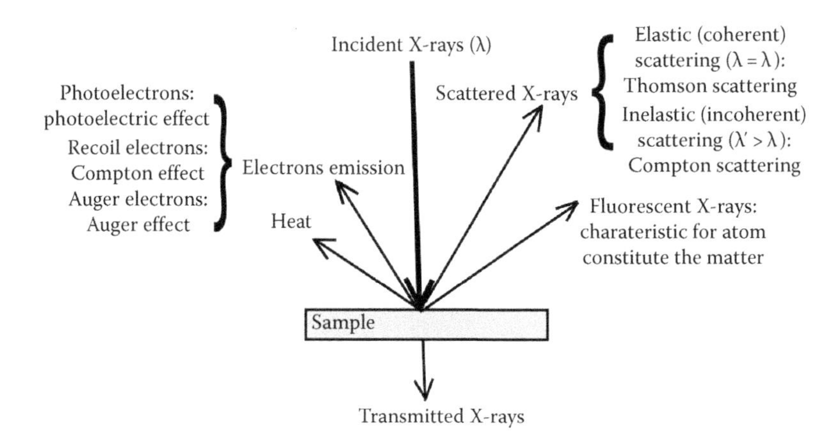

FIGURE 3.13
Interaction of X-rays with matter, and the resultant signals.

The absorption of X-rays in a material is given by

$$I = I_0 e^{-\mu\tau}$$

where

μ is the linear absorption coefficient
τ is the path length through the solid
I_0 and I are the intensity of X-rays before and after absorption

True absorption is accompanied by the emission of electrons due to the absorption of X-ray energy, such as Auger, photo, and Compton electron emission. In the photoelectric effect, the photoelectron release is accompanied by the emission of characteristic X-rays called fluorescent X-rays. But in Auger and Compton electron discharge, there is no fluorescent X-ray emission (Seeck and Vainio, 2014).

3.4.2 Theory and Principle of Measurement

X-rays interact with materials to generate different phenomena. But the signals coming from a Thomson coherent scattering of X-ray photons by the electrons and atoms of the material are of interest in XRD analysis. X-ray beams interact with the material's crystalline domain and get reflected/scattered when projected at certain angles to provide an interference pattern away from the material. X-ray interference or diffraction pattern is the result of crystallite atomic layer spacing and the wavelength of the incident X-ray beam. Crystalline materials comprise a regular three-dimensional arrangement of atoms in certain crystallographic planes. The English crystallographer William H. Miller designated Miller (hkl) indices to these different

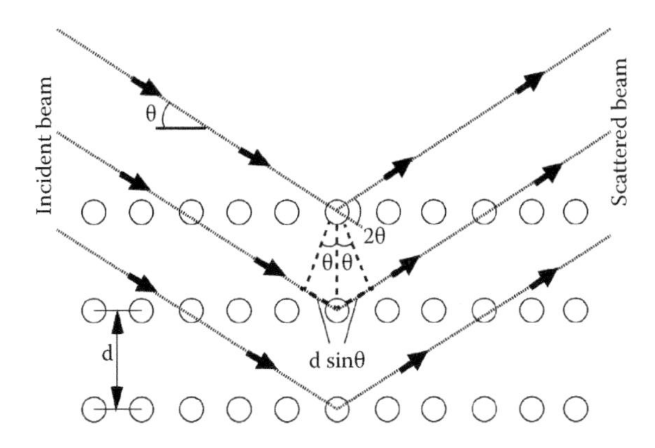

FIGURE 3.14
X-ray scattering from various layers of crystals, and Bragg's law.

orientations of planes in a lattice. In a crystal lattice, there exists an assembly of parallel equidistant planes with a spacing (d) to the planes defined by Miller (hkl) indices. When an incident X-ray beam passes through a specimen, it penetrates several layers of the crystalline lattice, as shown in Figure 3.14.

Bragg's law provides the conditions required for the constructive interference of X-rays from successive crystallographic planes (hkl) of an ordered sample using

$$n\lambda = 2d\sin\theta$$

where
- λ is the wavelength of the X-rays
- d is the spacing between the diffracting planes of the crystal
- n is the so-called "order of diffraction"
- θ is the angle between the scattering planes and the incident X-ray beam

In principle, both X-ray scattering methods (WAXD and SAXS) are the same with only a small difference in the experimental setup to target the different scattering angles. WAXD is a technique used for the determination of the crystal structure and the degree of crystallinity of materials with a well-defined periodicity in their structure. Information concerning distances in the range 1–10 Å can be obtained using WAXD. SAXS covers angles between $0.5° < 2\theta < 6°$ (Kasai and Kakudo, 2005).

3.4.3 Experimental Setup and Instrumentation

3.4.3.1 Sample Preparation for XRD Analysis

XRD analysis could be carried out in either powder form or as a bulk. Each of the sample forms requires some preparation to achieve reliable results.

Powder samples should be ground to particle sizes between 0.1 and 40 mm. Particle sizes less than 0.1 mm cause peak broadening, while particles larger than 40 mm exhibit less diffraction. The specimen is placed on a glass slide or metallic (aluminum) mount using double-sided tape, which is then positioned on the sample holder inside the instrument. XRD of powder does not require any specific orientation of the specimen, though its surface should be smooth and level with the top of the cavity in the sample holder. XRD analysis of bulks requires a smooth surface through polishing, though such a polished specimen requires thermal annealing to eliminate any induced surface deformations. Uniaxially oriented fibrous samples require a special holder to align the bundle of fibers in a certain direction; this experimental setup will be discussed later.

3.4.3.2 Instrumental Setups

XRD instruments have five key components: an X-ray tube, sample holder, collimator, monochromator, and detector. The X-ray tube and the detector are attached to one of the two arms of the goniometer, as depicted in Figure 3.15a.

The X-ray source is a vacuum-sealed tube. As only less than 1% of accelerated electrons convert their kinetic energy into X-rays and most of the energy goes in the form of heat, so an X-ray tube requires a continuous cooling mechanism. The collimator slit devices are placed on either side of the sample for both incident and diffracted X-rays. Collimator devices align and focus the X-ray beam onto the sample and detector. For X-ray diffraction analysis, a monochromatic X-ray source is required as much as possible. The monochromator, attached to the same armrest as the detector, is an optical mechanism to produce X-rays at a single wavelength for a distinctive diffraction pattern. Both X-ray tube and detector move together and follow the measuring diffractometer circle to uphold a $\theta/2\theta$ relation at all times. The sample mount is fixed in a position on the sample holder for powder diffraction analysis.

The principle of measurement for analyzing the diffraction pattern is the same, but the experimental setup varies for fiber and powder/bulk substrate to carry out X-ray diffraction measurement. For polymer powder or granulate measurement, the polymer is exposed to an X-ray at a certain angle and the diffracted X-rays are reflected due to the crystalline structure. A series of slits are used to direct the X-ray beam onto the sample, as discussed above. Then the diffracted X-ray beam is directed toward the detector through the monochromator, as illustrated in Figure 3.15a (Kasai and Kakudo, 2005).

In another setup, the X-ray beam is directed toward the sample, such as a bundle of fibers, at a certain angle and position. The specimen in such a setup rotates along its axis, while the X-ray source and image plate are fixed in position. The X-rays diffracted from the crystal are allowed to expose an image plate to provide an observation of the crystal structure and degree of crystallinity, as shown in Figure 3.15b.

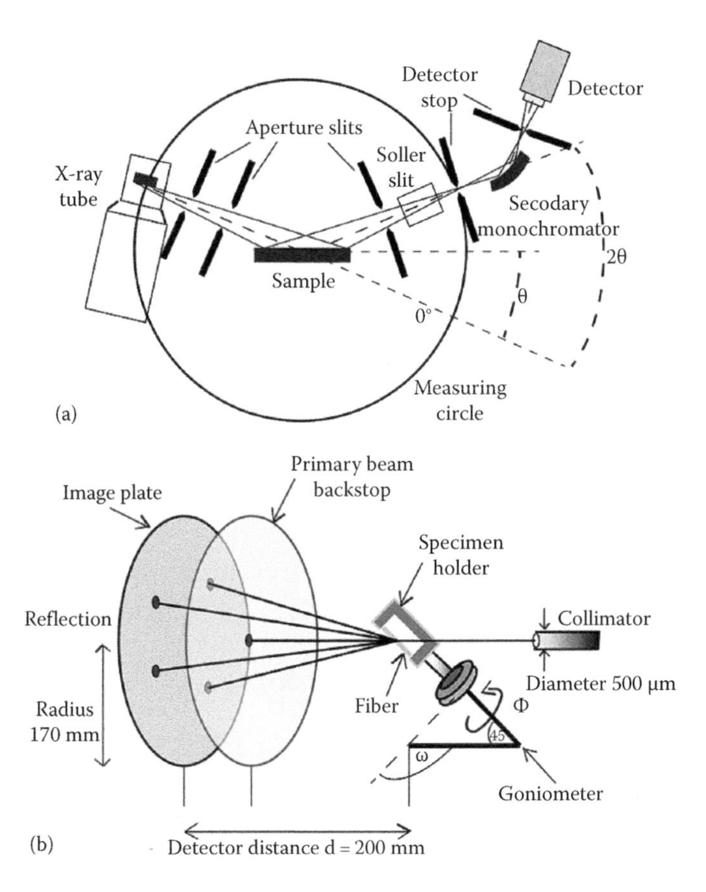

FIGURE 3.15
Schematic representation of XRD instrumentation used for (a) powder diffraction and (b) fiber diffraction analysis.

3.4.4 XRD Spectra of Crystalline Assemblies in Powder Form

Powder X-ray diffraction data lack three dimensionality: the recorded diffraction pattern is a one-dimensional representation of a three-dimensional reciprocal lattice of a crystal. The outcome of an X-ray experiment is typically plotted as scatter intensity versus the diffraction Bragg angle (θ) or the wave vector (q), to determine the diffracted Bragg peaks, which indicate the crystalline phases present in the material. The resultant plot is typically reported as a histogram in terms of intensities versus 2θ, labeled as a "diffractogram," which offers three kinds of information as depicted in Figure 3.16.

3.4.4.1 Peak Position

Peak positions in a diffractogram are the characteristics of crystal lattice parameters. In-phase diffracted X-rays from crystal planes will contribute to

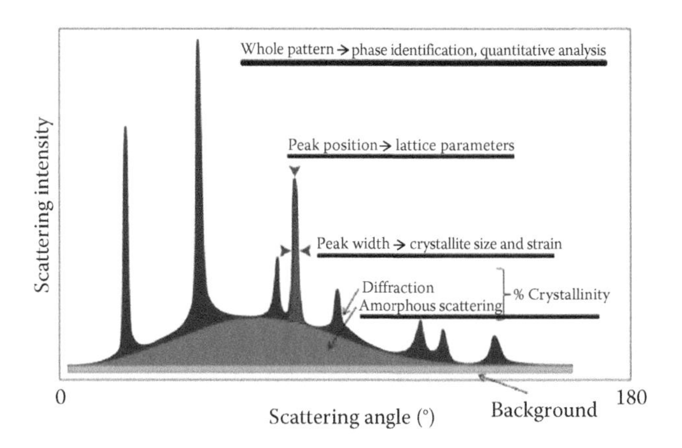

FIGURE 3.16
XRD diffractogram and the acquired information.

constructive interference at certain angles and to destructive interference at other angles. A detector records the diffracted X-ray at each angle (2θ). Peak position can be used to calculate the unit cell dimension and crystal lattice parameters (Kasai and Kakudo, 2005).

3.4.4.2 Peak Intensity

X-ray intensity is generally recorded as "count per second." Intensity modulation arises from different phase shifts and different amplitudes owing to various factors. The information regarding unit cell contents is coded by the relative intensities of diffraction peaks.

3.4.4.3 Peak Profile

Specific information about sample imperfections, like crystallite size, microstrain, or stacking faults, can be acquired from the profile of the peak. A peak profile bears two important aspects, that is the peak shape and peak width. The width of the peak profile is usually described as full width, half maximum (FWHM) (i.e., the width of the peak at half of its height).

3.4.5 WAXD Analysis of Polymeric (Fibrous) Structures

Polymers are semi-crystalline in their morphology. Both crystalline and amorphous regions diffract X-rays, the difference between these two types of X-ray scattering is the ordering of the phases. In the case of polymers, the crystals have very small sizes, hence a definitive structural analysis is difficult using only unoriented diffraction patterns. The crystalline parts give sharp narrow diffraction peaks and the amorphous component gives a very broad peak (halo) as shown in Figure 3.17.

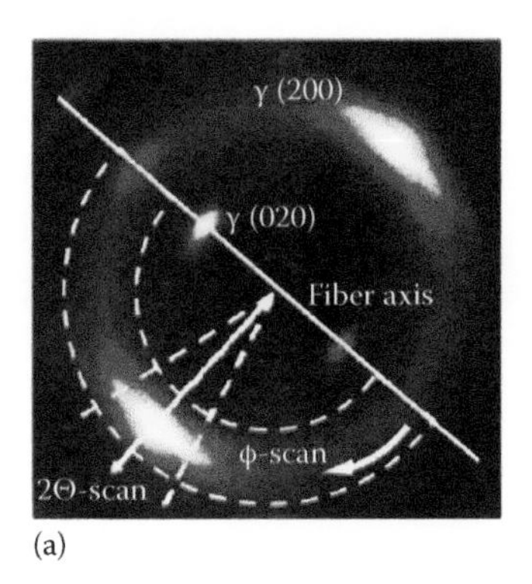

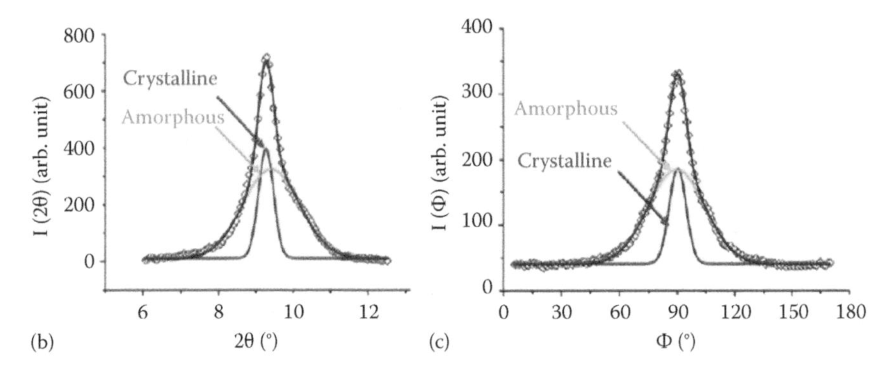

FIGURE 3.17
Principle of WAXD analysis (a) two-dimensional WAXD intensity distribution and radial 2θ scans and azimuthal φ scans around the Bragg reflection, (b) 2θ scan, and (c) φ scan with intensity contributions from the crystalline and the amorphous part of the polymer.

As a result, the WAXD spectrum of a polymer contains a mixture of Bragg peaks (very distinctive, high intensity, sharp peaks attributed to crystalline organization) which contain information about its "atomic arrangement" and diffuse scattering peaks from the amorphous part of the sample called the "amorphous halo" (Vad et al., 2013). Figure 3.18 shows the diffraction pattern of thermoplastic polyurethane extruded filaments at different drawing ratios and stretches.

It can be observed that, on drawing and stretching, polyurethane filament yarn induces molecular chain alignment, hence crystallinity. The concept of the crystallinity in semi-crystalline polymers is based on the two-phase approximation of the polymeric structures. The percentage of crystallinity using XRD measurement, taken as the ratio of intensity of crystalline peaks to the summation of intensities of both crystalline and amorphous regions, is given by

$$\%\text{Crystallinity} = \frac{I_{\text{crystalline}}}{I_{\text{crystalline}} + I_{\text{amorphous}}}$$

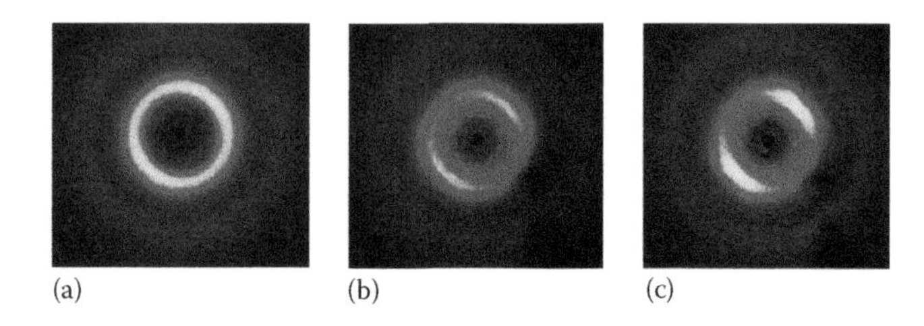

FIGURE 3.18
WAXD intensity distributions for thermoplastic polyurethane filaments (a) strand from spinneret, (b) spun filaments with drawing ratio 1, and (c) filaments with drawing ratio 1 stretch at 250%.

The size (thickness) of crystallites can be calculated by using the Debye-Scherrer formula in the direction perpendicular to a plane (hkl).

$$D_{hkl} = \frac{K\lambda}{\beta \cos\theta}$$

where
K is the Scherrer constant
β is FWHM at the specific angle θ
λ is the wavelength of the X-rays

XRD analysis of untreated and bleached (hydrolyzed) cotton fibers exhibits no change in crystallinity of the cellulose as depicted in Figure 3.19 (Zhao et al., 2007; Parikh et al., 2007).

El-Nahhal reported XRD patterns of ZnO-coated cotton fibers in a study to acquire antibacterial properties, as shown in Figure 3.20a (El-Nahhal et al., 2013). Figure 3.20b reports the diffractogram of silver nanoparticles—another highly reported antibacterial agent.

Diéval et al. (2004) reported the XRD pattern of three different types of polyester fibers, as shown in Figure 3.21, and studied the effect of heat setting on crystallinity.

3.4.6 SAXS Analysis of Polymeric Structures

SAXS analysis in a polymeric structure is based on heterogeneous electron density distribution over distances that are larger than the wavelength of X-rays. SAXS allows the evaluation of various structures such as high order polymeric structures, liquid crystals, rubbers, and micelles in the range of 5–50 nm. The study of SAXS is based on the same principles as that of WAXD discussed above. SAXS analysis works on the dimension related to the supramolecular structure and phase morphology of polymers, rather than

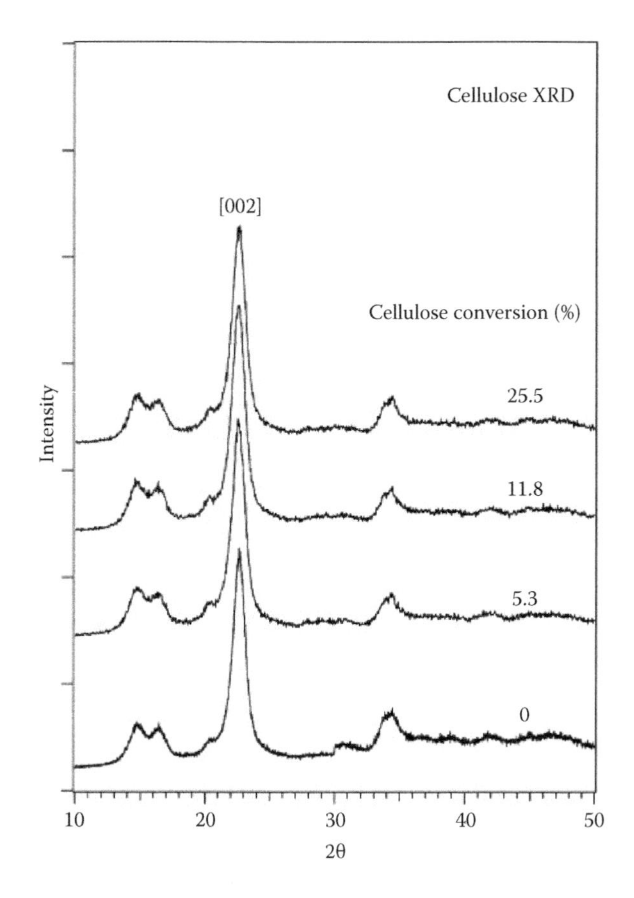

FIGURE 3.19
WAXD analysis of untreated and hydrolyzed cotton fibers to various degrees that exhibited no change in crystallinity of the cellulose.

the molecular structure, and therefore requires different approaches to deal with the data. SAXS instrumentation also differs significantly; hence it has a special status in analytical laboratories. SAXS analysis requires synchrotron radiation that is produced by accelerating electrons to several GeV in an electron storage ring.

In Figure 3.22, (a) represents spherically symmetric assemblies of crystallites and (b) symbolizes the unidirectional stacked lamellar crystal. (c) and (d) exemplify the scattering pattern for the stacked lamellar crystals that are inclined in the fiber's direction, and the scattering pattern for the stacked lamellar crystals inclined and with mirror image symmetry, respectively. (e) denotes the scattering pattern for double oriented structures (Li et al., 2000; Seeck and Vainio, 2014).

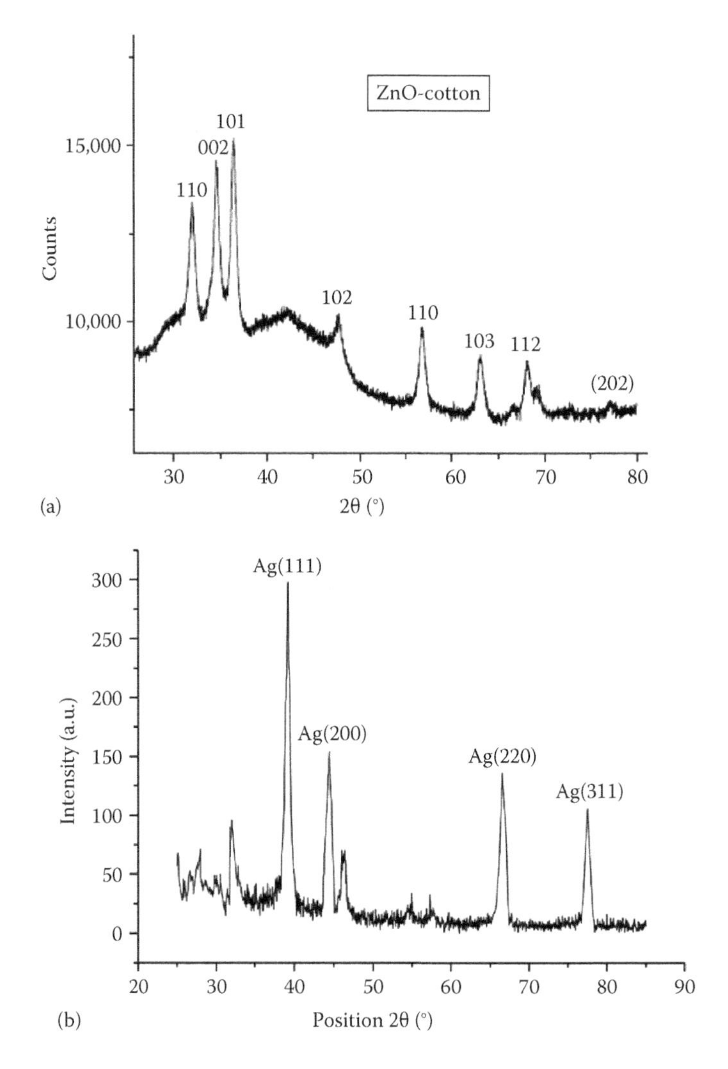

FIGURE 3.20
X-ray diffractogram of (a) ZnO-coated cotton fabrics, (b) silver nanoparticles.

3.5 Fourier Transform Infrared Spectroscopy

Fourier transform infrared spectroscopy (FTIR) is an analytical technique used to identify organic and, in some cases, inorganic materials. FTIR involves the measurement of absorption of infrared (IR) radiation by a material at certain wavelengths.

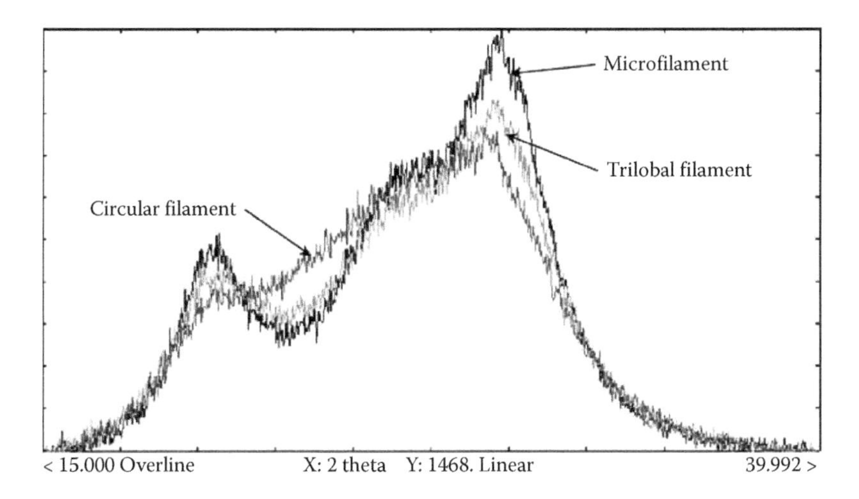

FIGURE 3.21
XRD pattern of three differently extruded polyester fibers.

3.5.1 Theory and Principle of Measurement

IR microscopy greatly involves the use of mid-infrared (MIR) radiation, which ranges from 4000 to 400 cm^{-1}, to investigate chemical structures and processes. IR spectroscopy measures the absorption of IR radiation by chemical bonds in a material. When IR irradiates a material, the absorbed radiations excite the molecules to higher vibrational states at a particular wavelength, as indicated in Table 3.2. IR radiation triggers the twisting, rotating, bending, vibrating, rocking, scissoring, stretching, and wagging of the chemical bonds. The absorbed wavelength is the function of the energy difference between the rest and excited vibrational states that is in turn the characteristic of molecular structure. Molecular components (i.e., functional groups) tend to absorb IR radiation in the same frequency range, regardless of the structure of the rest of the molecule. Thus, an IR absorption profile can identify functional groups, chemical structures, and process changes.

3.5.2 Transmission–Absorption IR Spectroscopy

Conventionally FTIR works in the transmission–absorption mode in which radiation propagates through a sample and becomes absorbed at certain frequencies where resonant energy transfer occurs. Transmittance (T) and absorbance (A) of IR radiation are computed for every wavelength by taking the ratio between the incident intensity (I_o) and the transmitted intensity (I_1) to acquire an FTIR spectrum by

$$A = -\log T = -\log \frac{I}{I_o}$$

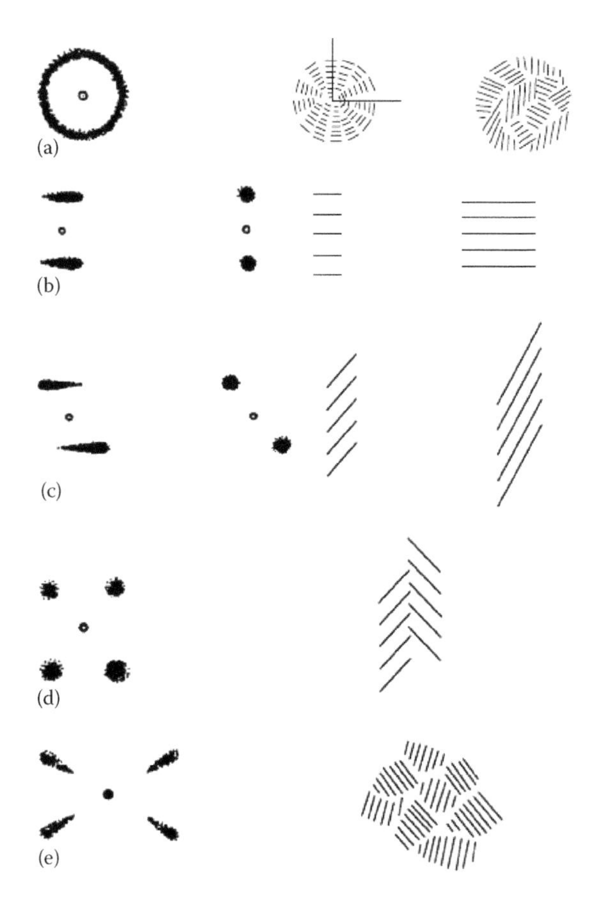

FIGURE 3.22
Set of common SAXS patterns, and their corresponding supramolecular structure and phase morphology in direct space.

In absorption spectroscopy, the profile (i.e., height and area) of the peak determines the intensity of the spectral bands in an absorbance spectrum. The Beer–Lambert–Bouguer law relates the concentration of the analyte to the intensity of the peak. The law provides the absorbance at any wave number, for an analyte at low concentration according to

$$A = \log\frac{1}{T} = a \cdot c \cdot l$$

where
- A is absorbance
- T is transmittance
- a is absorptivity at the wave number
- l is the path length
- c is the concentration of the analyte

TABLE 3.2

Some of the Functional Groups with Their Corresponding IR Absorption Frequency Range

Major Functional Groups	Absorption Frequency Regions
O–H	3650–3590
N–H	3500–3300 1650–1590 900–650
$=CH_2$	3100–3070 1420–1410 900–880
=CH	3100–3000 2000–1600
CH	2900–2700 1440–1320
–CH3	2880–2860 2970–2950 1380–1370 1470–1430
OH	2700–2500 1320–1210 950–900
C≡C	2140–2100
C=O	1750–1700
C=C	1600–1500
C–N	1340–1250
C–O–C	1200–1180
C–H	770–730

3.5.2.1 Instrumentation

A typical FTIR spectrometer has three basic components: a source, a detector, and a Michelson interferometer. There are two types of MIR radiation sources, which operate at a temperature of 1100 °C. Ceramic-based sources are made of ceramic-coated nichrome wire that produces heat radiation on the passing of a current. Silicon carbide–based sources are robust and efficient and most commonly employed in FTIR spectrometers. Two types of detection technologies are considered ideal for MIR spectroscopy: a thermal detector and a pyroelectric detector.

A Michelson interferometer is a central component of an FTIR instrument for modulating wavelengths from a broad-band IR source. An interferometer is an arrangement of mirrors to split a beam of light into two and then to make these travel different optical path lengths, as shown in Figure 3.23.

The assembly consists of a beam splitter inclined at an angle of 45° that splits up the light beam into two perpendicular beams, which are reflected back from the surfaces of a fixed mirror and a moving mirror. The translational movement of the moving mirror introduces a path difference between the reflected beams to produce constructive and destructive interference at the beam splitter. Light intensity versus optical path difference is plotted, resulting in an interferogram (Stuart, 2004; Simmons and Ng, 2000).

The interferogram is Fourier transformed to obtain a single beam IR spectrum. The FTIR spectra are usually presented as plots of intensity versus wave number (in cm^{-1}) as shown in Figure 3.24.

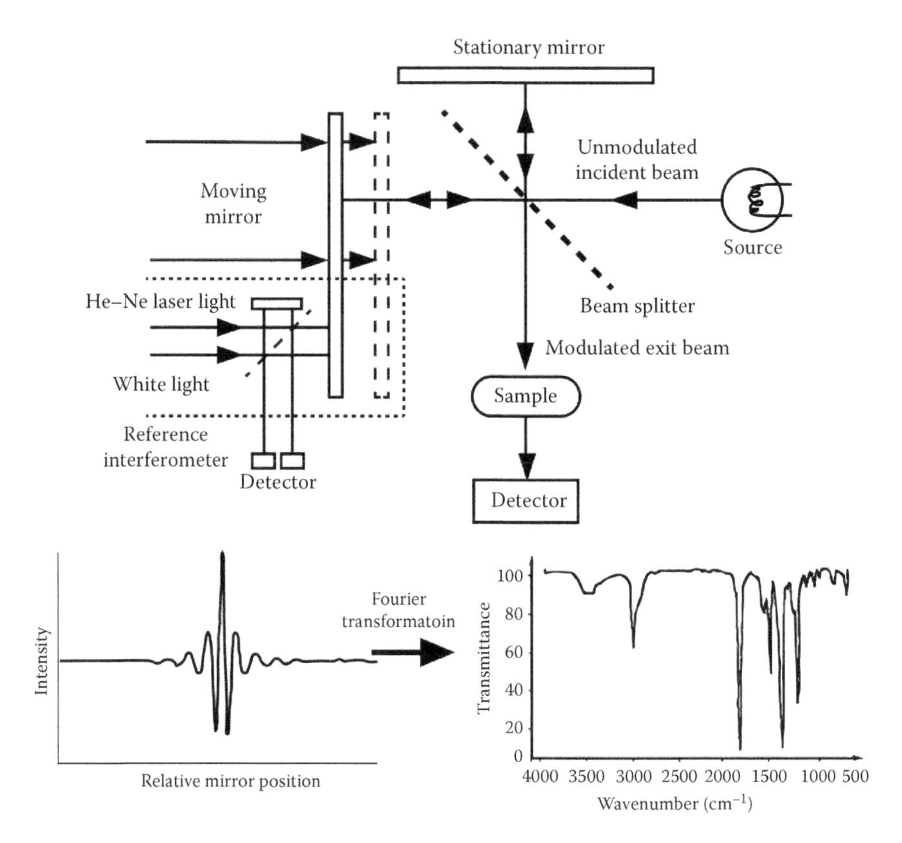

FIGURE 3.23
Schematic diagram of FTIR instrument with Michelson interferometer and output signals.

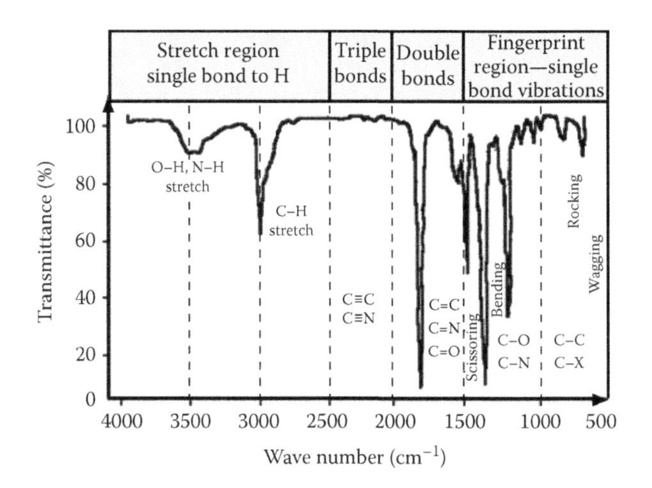

FIGURE 3.24
A typical FTIR spectrum and various bond vibration regions.

3.5.2.2 Sample Preparation

FTIR analysis can be carried out for solid, liquid, and gaseous samples. Each form of a sample requires some preparation to acquire a reliable spectrum. Solid samples can be prepared in the following forms:

- *Alkali halide (KBr) disks.* The most established practice for analyzing solids is the preparation of a KBr disk. Alkali halides are inactive and optically transparent to IR over the range typically used for IR analysis. A few milligrams of a finely grounded sample is blended with 50–100 parts of dry KBr powder. This mixture is then compressed into a transparent disc at high pressure.
- *Mulls.* In this technique, the finely grounded solid sample is dispersed into 1–2 drops of a mulling agent, which is further ground to obtain a smooth paste for analysis.
- *Films.* Solids can be analyzed in the form of thin films by casting a thin layer of solution and then evaporating off the solvent. Films can be cast onto an IR transparent window or onto a suitable support from which it can be readily peeled. Then the film is mounted on a suitable holder for analysis.

Samples in the liquid state can be analyzed by either placing them on IR transmitting plates or pouring them into fixed or variable path length cells. Viscous samples can be squeezed between the two IR transparent plates and then examined as thin films. For volatile liquids, fixed cells are used because there is no need of cleaning. For other nonvolatile, nonviscous liquids, variable path length cells are used as they are demountable and the windows can be cleaned easily. These cells can be filled with a liquid sample using a syringe. Insoluble polymer film of various thickness is used as a spacer to achieve variable path lengths.

3.5.3 Reflective IR Spectroscopy: Attenuated Total Reflection (FTIR-ATR)

By combining MIR spectroscopy with theories of reflection, it is possible to perform IR analysis easily, faster, accurately, and in a versatile way. IR reflectance can be divided into specular, diffusive, and internal. Internal reflectance or attenuated total reflectance is the basis of this advanced technique established by Fahrenfort and Harrick in the early 1960s. When light enters from a denser (high refractive index n_1) to a rarer (low refractive index n_2) medium, it bends away from the perpendicular; but for a certain angle, the critical angle (θ_c), it is totally reflected back from the interface of the two media, termed "total internal reflection":

$$\theta_c = \sin^{-1}\left(\frac{n_2}{n_1}\right)$$

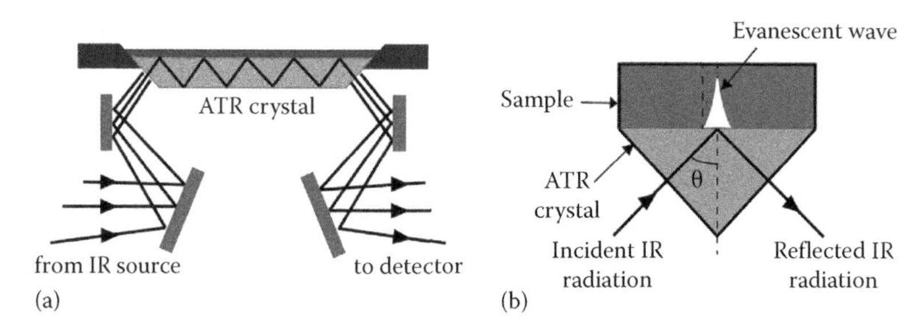

FIGURE 3.25
Schematic of evanescent wave with (a) multiple reflections and (b) single reflection ATR.

In ATR, IR passes through an IR transmitting crystal of high refractive index, allowing the radiation to reflect back into the crystal, once or more. The evanescent waves penetrate 1–4 μm into the sample at each reflection point, as shown in Figure 3.25.

The IR interacts with the sample through a series of standing waves, called evanescent waves (Kwan, 1998).

3.5.3.1 Instrumentation

The instrument used for ATR analysis is principally similar to that used for FTIR transmission spectroscopy. However, ATR instrument requires an additional attenuated total reflection accessory. The ATR accessory bears an IR transmitting ATR crystal of high refractive index. ATR crystals are made of zinc selenide (ZnSe), germanium (Ge), zinc sulfide (ZnS), silicon (Si), diamond, and germanium–arsenic–selenium (GeAsSe). Diamond-based crystals are the most versatile. There are two types of ATR crystals: multibounce and single bounce. There are some limitations related to ATR crystal in that it absorbs energy at lower energy levels between 650 and 400 cm^{-1} and that the data will be noisy. Figure 3.25 shows the path of the infrared beam from the IR source, through the crystal, and on to the detector. The detector records the attenuated beam as an interferogram signal, which generates an infrared spectrum (Simmons and Ng, 2000).

3.5.3.2 Sample Handling

ATR analysis is much quicker than transmission analysis and can be used easily to analyze a wide range of sample types. But ATR needs to be used with some caution as the technique amends by sample type, for example, rigid samples should be cut to the correct size in order to make good contact with the crystal, or a powder sample should be in sufficient amount to cover the crystal properly. The pressure applied to make the contact should be optimum: neither too low that the sample does not make proper contact, nor too high to avoid sample crushing. Also the optimum position of the beam will change over time, so the beam splitter alignment is necessary to realize good spectra.

3.5.4 FTIR Spectrum Interpretation and Manipulation

FTIR peaks appeared at certain wave numbers depending on the functional groups are plotted against the reflection (transmittance) percentage, as shown in Figure 3.24. The raw data acquired from an FTIR spectrometer requires some spectral manipulations to extrapolate more interpretations and quantitative information, so comprehending and working with IR data is quite significant. Various approaches have been developed and employed for this purpose, like base line correction, smoothing, deconvolution, and curve fitting (Stuart, 2004).

3.5.4.1 Spectral Subtraction

Spectral subtraction involves a set of point-to-point subtract operations to get rid of the absorbance values of the reference from the sample spectrum. After performing spectral subtraction, some peaks will appear that can be referred to the sample that was previously hidden.

3.5.4.2 Spectral Derivation

The second-derivative Fourier transform spectrum presents the additional information in comparison to the original spectrum. It extends in a direct way to distinguish the peak frequencies of characteristic components and thus permits much more detailed qualitative and quantitative findings.

3.5.4.3 Peaks Deconvolutions

Deconvolution is one of the most useful operations in interpreting the featured information from the raw spectra. Deconvolution involves identifying and resolving a composite peak, which most likely is composed of two or more smaller peaks. These peaks are often broad and asymmetric and may contain some shoulder, giving the hint of some hidden bands. Resolution enhancement helps in locating such composite bands. Deconvolution results can be challenged, so careful interpretation of data is imperative.

3.5.5 Applications of FTIR-ATR

FTIR analysis can be effectively employed to identify the chemical constitution of various materials in any form (gas, liquid, or solid). It can be employed to identify the varieties of textile fibers (Wu and He, 2010). It can assist in distinguishing the constituents in multilayered materials. FTIR is frequently used in polymer testing and forensic analysis. It can be used to study the effects of temperature on dynamics and the kinetics of various processes and reactions. Here, some of the examples put forth are to enable the comprehension of possible applications in textiles. Figure 3.26a reports a comparison of FTIR spectra of activated carbon fibers by KOH activation, as obtained by carbon fiber and liquefied wood-based precursor fibers (Huang et al., 2015). Ugur et al. (2010) used FTIR-ATR to verify the deposition of nanocomposites, as shown in

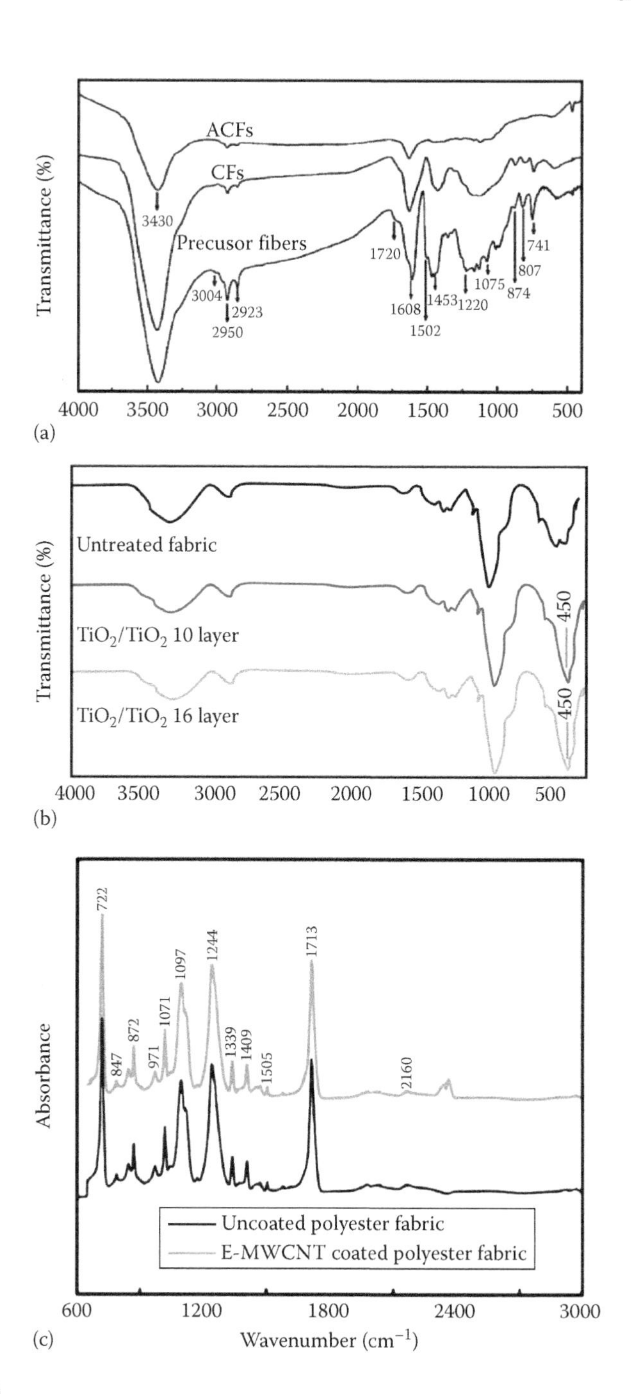

FIGURE 3.26
FTIR spectra of (a) activated carbon fibers by KOH activation, as obtained from carbon fiber and liquefied wood-based precursor fibers, (b) TiO_2 nanoparticles coated with multifunctional cotton fabric, and (c) polyester fabric, and MWCNT-coated polyester fabric (used as textile based flexible counter electrode).

Figure 3.26b. Figure 3.26c illustrates the FTIR spectrum of polyester fabric and MWCNT-coated polyester fabric produced by Arbab et al. (2015).

3.6 Atomic Absorption Spectroscopy

The analytical method that measures the chemical elements of a solution quantitatively is termed atomic absorption spectroscopy (AAS). The free atoms, while in the gaseous state, absorb light (optical radiation), giving the spectroscopy. This technique is employed in analytical chemistry and used to conclude on the concentration level of any element in a given solution or sample. It can be employed to measure the concentration level of more than 70 distinct elements present in a solution. The same method can also be employed to measure the element concentration level in solid form.

AAS helps to conclude on the elemental composition of a material with the help of its mass or electromagnetic spectrum. Optical atomic spectroscopy is the study of the mass or electromagnetic spectrum of the elements. Electrons have a distinct level of energy while they are within the atom. The concentration of different elements is measured by using this technique. It has a very good sensitivity level and can measure down to parts per billion of a gram ($\mu g\ dm^{-3}$) in a sample. The specific light absorption property of an element is the most important factor used in this technique. It employs the phenomenon of change in the energy level of an electron as it absorbs energy and jumps to a higher level with the absorption of energy. It can be employed in many aspects of chemistry as well as in the material sciences.

3.6.1 Working Principle

The wavelengths absorbed by an element is dependent on the characteristics of that element. In order to analyze a sample to check if it has a certain kind of element, the light used should have the characteristics of that element. For example, in order to analyze if a sample contains traces of lead, a light source having the characteristics of lead should be used. The light source will produce lead atoms having a higher energy level and which will then be absorbed by any lead atoms present in the test specimen. In this way, the presence of lead can be determined in any given sample. The test specimen is first converted into a gaseous or vapor state that allows the atoms to be in a free state. Then the specimen is bombarded with the excited atoms of lead. The higher energy of the bombarded atoms will be absorbed by the atoms in the vaporized form. The amount of radiation absorbed will depend upon the number of trace elements present in the test specimen. In other words, it can be said that the amount of radiation absorbed will be directly proportional to the number of atoms present in the specimen. The instrument is calibrated with the help of a calibration curve, which is obtained by analyzing the samples with a known

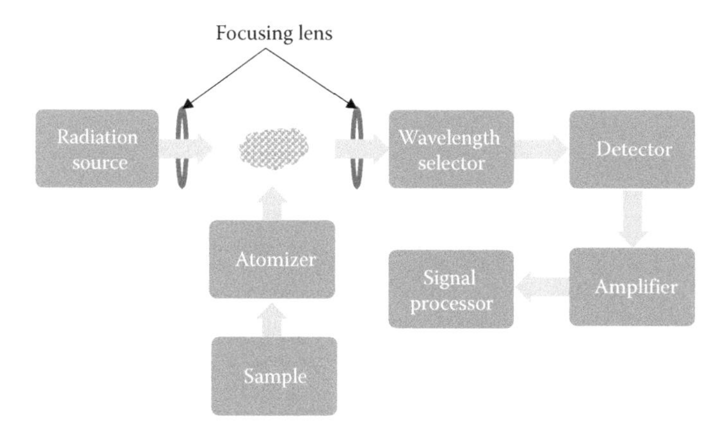

FIGURE 3.27
Block diagram of AAS.

level of lead. The sample with this known concentration is analyzed several times and compared with a sample having an unknown concentration of lead. The curves obtained from the unknown sample and the known sample are also compared. The calculation on the basis of these curves will give the amount of the element in the test specimen. So the atomic absorption spectrometer should have the following three essential parts: a light source; a sample cell to produce gaseous atoms; and a means of measuring the specific light absorbed (Welz and Sperling, 2007). The block diagram of AAS is shown in Figure 3.27.

3.6.2 Applications of AAS

AAS can be used in many areas, ranging from clinical laboratory analysis to material science analysis. It helps to find the presence of heavy metals in body fluids like urine and blood and to diagnose diseases. It can also be used to monitor the environment efficiently, including water as free from hazardous materials. The level of heavy metals or hazardous materials can be analyzed in drinking water. AAS also helps to monitor the suitability of different soft and hard drinks.

AAS can also be used to check the presence of metals in final products. In many chemical reactions, metals are used as a catalyst. AAS can be used to verify that no metal is present in the final product. The raw materials are tested to check that they are free of any hazardous element. AAS is also used to eliminate the possibility of toxic elements in any material or final product. It can also help miners to detect the presence of precious metal like gold in rocks and to locate the place where the mining should be done. For textiles, it can be used to monitor the waste water from textile units, the efficiency of the effluent treatment plan, and the presence of trace elements of metals in the final product, particularly in the case of rotary printing rollers.

3.7 Thermal and Water Vapor Resistivity

The use of textile materials as thermal insulators is very common, particularly in the case of buildings and transport (Marsh, 1930). The thermal properties (insulation) of textile materials should be carefully studied at different atmospheric conditions. Thermal resistance can be defined as the resistance of fabric with thickness (d) to thermal conduction (λ), from the region with a higher temperature to the region with a lower temperature (Saville, 1999). It is denoted by R and expressed in m^2 K/W or Clo (where, 1 Clo = 0.155 m^2 K/W). Heat can be transferred in three ways (conduction, convection, and radiation) that must be controlled to promote effective thermal insulation. To minimize convective heat transfer, material must be relatively impermeable to air (Epps, 1988). Heat transfer through fabrics is a complex phenomenon, depending on a number of parameters like fabric geometry, fabric thickness, fabric density, yarn structure, weave design, and the number of fabric layers (Abdel-Rehim et al., 2006; Ahmad et al., 2014).

The sweating guarded hotplate (SGHP) instrument is used to measure the thermal resistance (R_{ct}) and water vapor resistance (R_{et}) of fabrics. The hotplate is a porous bronze plate, providing the best simulation of human skin as compared to plates having a distinct pattern of drilled holes. Hence SGHP is also referred to as the "skin model" as it simulates the heat and mass transfer processes that occur next to human skin.

The SGHP instrument consists of a temperature controller, water supply unit, and measuring unit, as shown in Figure 3.28.

It consists of a metal chamber with heating element and a square porous metal plate (3 mm thick). The test area lies in the center of the plate and is surrounded by the guard heaters (to prevent lateral heat leakage) and beneath by a bottom heater (to prevent downward heat loss). This arrangement drives heat or moisture to transfer only along the specimen's thickness. The plate temperature is measured by the sensor (thermocouple) directly below the plate surface. The whole apparatus is housed in a chamber for controlled environmental conditions.

$$R_{ct} = \frac{A(T_s - T_a)}{H} \tag{3.1}$$

where

A is area of the test section (m^2)
T_s is the temperature at the plate surface (K)
T_a is the temperature of the air (K)
H is the electrical power (W)

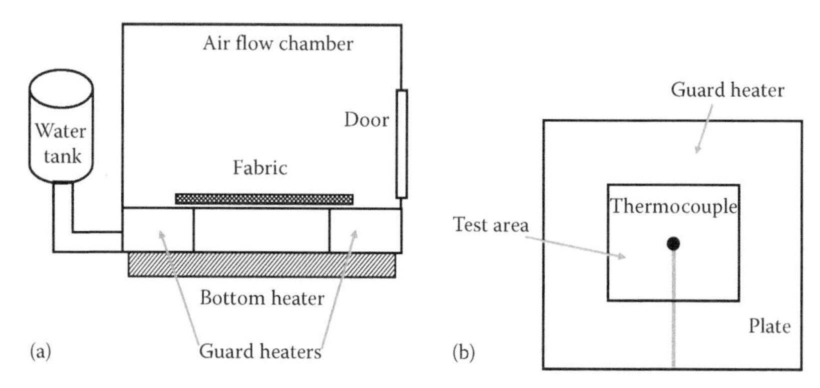

FIGURE 3.28
Schematic of SGHP instrument: (a) side view and (b) top view.

The SGHP is based on the standard test method ISO 11092:2014. According to this test, the samples are placed on the hot plate (250 × 250 mm) and enclosed in a controlled environment (air temperature $20 \pm 0.1\,°C$, relative humidity $65 \pm 3\%$, thermal plate temperature $35 \pm 0.1\,°C$, air speed 1.00 ± 0.05 m/s, and measuring unit temperature $35 \pm 0.1\,°C$). The accuracy of measurement is determined by a number of factors like air velocity, leading edge effect, bubbles and wrinkles, and membrane effect.

Water-vapor resistance (R_{et}) is defined as the water-vapor pressure difference between two faces of a material divided by the resultant evaporative heat flux per unit area. Distilled water is fed to the surface of the porous plate from the water tank. The supply is activated when the water level in the plate is about 1 mm below the plate's surface. This water is preheated by passing it through the guard heater section, and a constant rate of evaporation is supplied to the measuring unit using a level switch. A water vapor permeable and liquid water impermeable membrane is fitted over the plate to allow only water vapor to pass. The test fabric is placed above the membrane and a controlled environment is maintained (air temperature 35°C and relative humidity 40%).

$$R_{et} = \frac{A(P_s - P_a)}{H} \tag{3.2}$$

where

A is the area of the test section (m^2)
P_s is the water vapor pressure at the plate surface (kPa)
P_a is the water vapor pressure of the air (kPa)
H is the electrical power (W)

The thermal resistance (R_{ct}) and water vapor resistance (R_{et}) of the fabric is subsequently used to determine the water vapor permeability index (i_{mt}) and water vapor permeability (W_d) using the following relations:

$$\text{Water vapor permeability index, } i_{mt} = S \cdot \frac{R_{ct}}{R_{et}} \tag{3.3}$$

where S = 60 Pa/K.

W_d is the water vapor permeability, in grams per square meter hour Pascal

$$\text{Water vapor permeability, } W_d = \frac{1}{R_{et} \cdot \varphi T_m} \tag{3.4}$$

where φT_m is the latent heat of vaporization of water at temperature (T_m) of the measuring unit. It equals 0.672 W·h/g at T_m of 35 °C.

3.8 Moisture Management

Moisture management characteristics are determined according to the AATCC 195-2011 test method, which not only covers the measurement of liquid moisture management properties, but also classifies textile fabrics on the basis of these properties. Moisture management characteristics are largely dependent on fabric structure and the wicking characteristics of the constituent material. The moisture management tester (MMT) is used to test the transport and delivery behavior of a liquid solution in the fabric samples. The schematic of an MMT is shown in Figure 3.29.

The measuring principle involves information on the transportation of moisture through a fabric. It is measured by placing the fabric sample

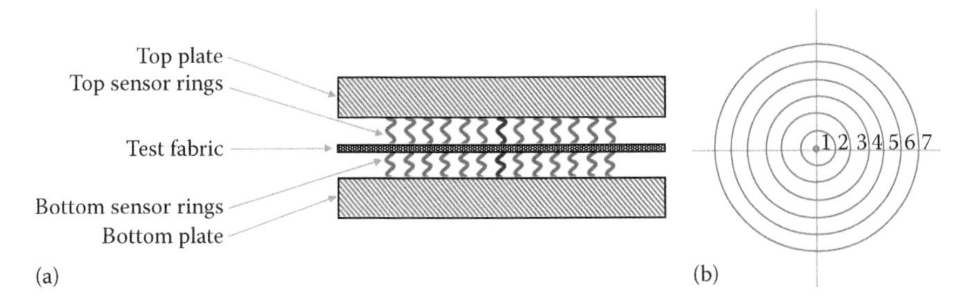

FIGURE 3.29
Schematic of MMT: (a) structure and (b) censor rings.

between horizontal (upper and lower) electrical sensors. The change in voltage will give the difference of the moisture content in the upper and lower surfaces of the fabric, which can then be calculated.

A fabric sample of 80 × 80 mm is placed in the tester, keeping the fabric face downwards. Now, the upper portion of the fabric is considered to be the surface next to human skin, while the bottom surface of the fabric is closest to the environment. The test solution is obtained by dissolving 9 g of sodium chloride in 1 L of distilled water, with its electrical conductivity adjusted to 16 ± 2 millisiemens at 25 °C. A set amount (0.15 g) of this solution is dropped onto the upper face of the fabric to replicate a drop of sweat. The signal for electrical resistance of the fabric samples is processed by the MMT's software. The measured parameters include wetting time (top and bottom surface), absorption rate (top and bottom surface), maximum wetted radius (top and bottom surface), spreading speed (top and bottom surface), one-way transport, and overall moisture management capability.

For ease of interpretation, all the parameters measured are graded and converted from value to grade based on a five-grade scale (1–5), as shown in Table 3.3.

The fabrics can be classified into seven types based on the results obtained from the MMT, for example, waterproof fabric and moisture management fabric.

TABLE 3.3

Grading System of All Parameters

		Grade				
Index		**1**	**2**	**3**	**4**	**5**
Wetting time (s)	Top	≥120 No wetting	20–119 Slow	5–19 Medium	3–5 Fast	<3 Very fast
	Bottom	≥120	20–119	5–19	3–5	<3
Absorption rate (%/s)	Top	0–9 Very slow	10–29 Slow	30–49 Medium	50–100 Fast	>100 Very fast
	Bottom	0–9	10–29	30–49	50–100	>100
Max wetted radius (mm)	Top	0–7 No wetting	8–12 Small	13–17 Medium	18–22 Large	>22 Very large
	Bottom	0–7	8–12	13–17	18–22	>22
Spreading speed (mm/s)	Top	0.0–0.9 Very slow	1.0–1.9 Slow	2.0–2.9 Medium	3.0–4.0 Fast	>4.0 Very fast
	Bottom	0.0–0.9	1.0–1.9	2.0–2.9	3.0–4.0	>4.0
One-way transport capability (R)		<–50 Poor	–50–99 Fair	100–199 Good	200–400 Very good	>400 Excellent
Overall moisture management capability		0.0–0.19 Poor	0.2–0.39 Fair	0.4–0.59 Good	0.6–0.8 Very good	>0.8 Excellent

3.9 Air Permeability

Air permeability can be measured by using distinct test protocols, for example, ASTM D 737 and ISO 9237. According to these test methods, the airflow through a given area of fabric is measured at a constant pressure drop across the fabric. The fabric is clamped over the air inlet, and the air is drawn through this fabric sample by means of a suction pump. The rate of air flow at this point is measured using a flow meter (Booth, 1968). The value of permeability may differ considerably across the entire area of the fabric due to irregularities in the yarn (Havlova, 2014).

The permeability findings are not influenced by fabric orientation because the measured area is always circular. The test conditions for the measurement of air permeability are the clamping area of the sample (cm^2) and the pressure difference (Pa). According to the ASTM standard, the recommended test area is 38.3 cm^2, while alternate areas are 5 and 100 cm^2. The air is drawn perpendicularly through the fabric and the airflow rate is adjusted to provide a pressure difference of between 100 and 2500 Pa between the two fabric surfaces (minimum pressure drop of 125 Pa). The pressure drop is 100 Pa for apparels and 200 Pa for industrial fabrics. The air permeability in mm/s can be calculated as

$$\text{Air permeability, } R = \frac{Q_V}{A} \times 167 \tag{3.5}$$

where

Q_V is the flow rate of air in "dm^3 per min" or "liters per min"
A is the fabric area in cm^2

3.10 IR Thermography

Electromagnetic radiation is a form of energy. It has specific electrical and magnetic properties. Different particles behave differently when they are exposed to electromagnetic radiation. The wavelength range of such radiation is termed the electromagnetic spectrum (Sankaran and Ehsani, 2014). It has two parts, that is visible radiation and invisible radiation, depending on whether it is visible to the human eye. Other systems exist to identify the presence of the invisible part of the spectrum, which consists of radiowaves, microwaves, the infrared, the ultraviolet, X-rays, and gamma rays. The different constituents of the electromagnetic spectrum are shown in Figure 3.30 (Stuart, 2004).

Any object with a temperature above 0 K emits IR (Ghassemi, 2012). The pattern of this radiation can be recorded by using an infrared camera. As infrared relates to the heat, this technique is known as thermography

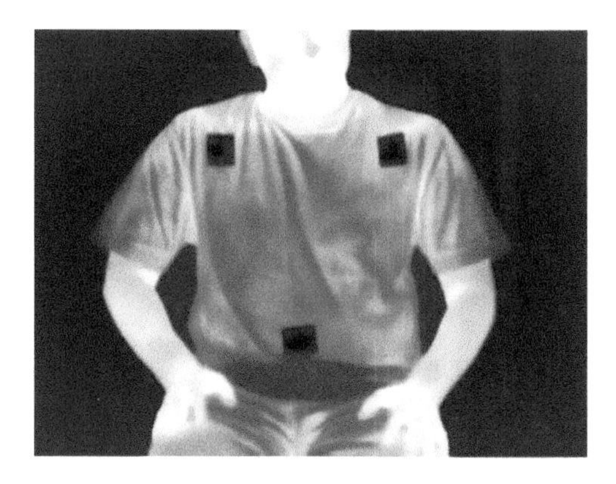

FIGURE 3.30
Thermograph of a human being.

or IR thermography. The use of thermography has increased dramatically within commercial and industrial applications during recent years (FLIR, 2009). It helps firefighters during incidents with a lot of smoke. Thermography, as its name indicates, is a technique used to map the temperature of a scene. It actually allows us to see thermal energy or heat. It is a rapid, efficient, and full-field technique for non-destructive testing and evaluation.

Special cameras are used to take thermal images that are generally called IR cameras. These cameras use special focal plane arrays (FPAs) as sensors. Previously, such cameras used a bolometer as a sensor, which required a cooling mechanism. The new methods involve the use of a cheap, uncooled micro-bolometer, such as an FPA. IR cameras have several applications, for example, night vision, research, non-destructive testing, process control, medical, surveillance, and health care. The biggest use, right now, of IR thermography is for night vision devices. In addition to that, it is also used for research and development. IR thermography can also be used for maintenance. Any kind of leakage of gas can be detected with an IR camera. On the other hand, the excessive heating of the parts of an electronic circuit can also be spotted with an IR camera.

A new application of these cameras is the study of interface pressure distribution (IPD) in a car seat. One of us has introduced the protocol to study IPD in a car seat by using IR thermography (Rasheed et al., 2011). As soon as a human body comes in contact with a car seat, heat exchange starts between them due to the temperature difference between the bodies. However, the amount of heat exchanged decreases as the temperature difference between both bodies reduces and finally comes to equilibrium. Further, the pressure exerted by the human body on the car seat is not the same at all points. There were two hypotheses in this study:

1. Heat transferred at different points in the car seat is not the same.
2. Heat transferred has a correlation with the pressure exerted.

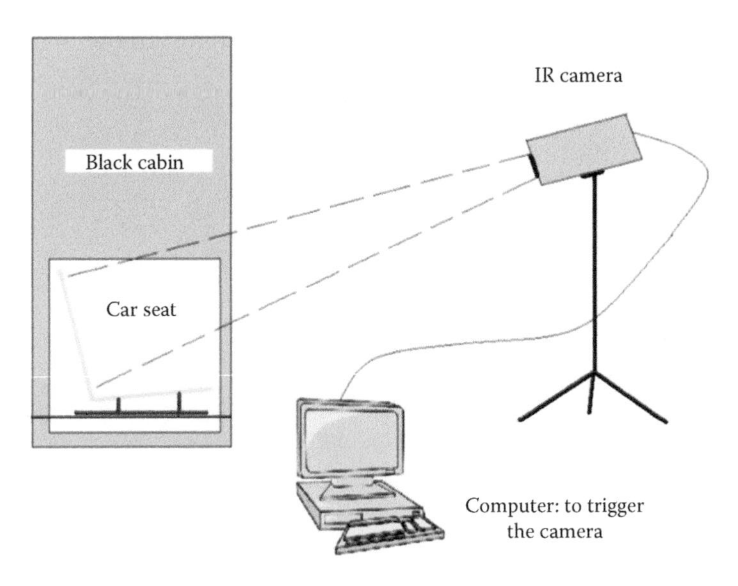

FIGURE 3.31
Experimental setup to study IPD using thermography.

A special experimental setup was arranged to conduct the study, which is illustrated in Figure 3.31.

A totally black cabin was designed, keeping in view that the results of thermography may be affected by daylight, which contains IR. Initially a protocol was developed by conducting a study in which nine subjects with different weights and heights participated. It was found that equilibrium was achieved after 60 s. Therefore, it was decided that the optimum sitting time is 60 s for further experimentation.

After finalizing the protocol, it was validated. Twenty-two subjects participated in a validation exercise. By observing five thermographs for each subject, it was observed that the results are reproducible. Further, the first hypothesis was also confirmed, as is shown in Figure 3.32.

We designed a system named the interface pressure distribution measurement system to study pressure and heat transfer simultaneously. For this purpose, six pressure sensors were embedded in the car seat at different positions to determine the exerted pressure. Data was taken at a frequency of 20 Hz. Additionally, an IR camera was used to record the thermographs. Forty-one subjects participated in this experiment with different BMIs and body shapes. Three repetitions of each subject were taken. The data obtained from the sensors were compared with the data obtained from the IR camera. The statistical analysis of the results showed that there is a direct correlation between the heat transferred and pressure exerted. Keeping in view the results, we concluded that thermography can be used to study IPD.

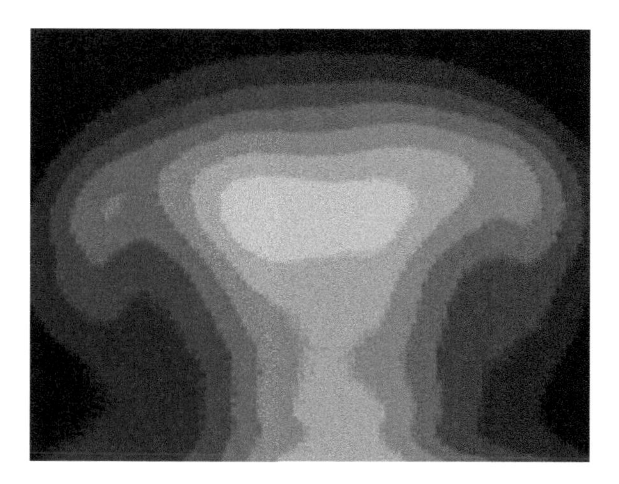

FIGURE 3.32
Thermograph of a car seat: different temperature zones.

References

AATCC 195. n.d. *Liquid Moisture Management Properties of Textile Fabrics*. American Association of Textile Chemists and Colorists. Colorado, USA.

Abdel-Rehim, Z.S., Saad, M.M., El-Shakankery, M., and Hanafy, I. 2006. Textile fabrics as thermal insulators. *Autex Research Journal* 6(3): 148–161.

Ahmad, S., Ahmad, F., Afzal, A., Rasheed, A., Mohsin, M., and Ahmad, N. 2014. Effect of weave structure on thermo-physiological properties of cotton fabrics. *Autex Research Journal* 15(1): 1–5. http://doi.org/10.2478/aut-2014-0011.

Agnihotri, S., Mukherji, S., and Mukherji, S. 2013. Immobilized silver nanoparticles enhance contact killing and show highest efficacy: Elucidation of the mechanism of bactericidal action of silver. *Nanoscale* 5(16): 7328–7340. doi: 10.1039/C3NR00024A.

Amuthakkannan, P., Manikandan, V., Winowlin Jappes, J.T., and Uthayakumar, M. 2013. Effect of fibre length and fibre content on mechanical properties of short basalt fibre reinforced polymer matrix composites. *Materials Physics and Mechanics* 16: 107–117.

Arbab, A.A., Sun, K.C., Sahito, I.A., Qadir, M.B., and Jeong, S.H. 2015. Multiwalled carbon nanotube coated polyester fabric as textile based flexible counter electrode for dye sensitized solar cell. *Physical Chemistry Chemical Physics* 17(19): 12957–12969. doi: 10.1039/C5CP00818B.

ASTM D737-04(2016), 2016. *Standard Test Method for Air Permeability of Textile Fabrics*, ASTM International, West Conshohocken, PA, doi: 10.1520/D0737-04R16.

Booth, J.E. 1968. *Principle of Textile Testing*, 4th edn. Chemical Publishing, London, U.K.

Bramson, M.A. 1968. Chapter 1: Introduction. In *Infrared Radiation*, 1st edn., edited by Plenum Press, Spring Science and Business Media, New York, pp. 1–30.

Breton, B.C. 2009. The early history and development of the scanning electron microscope. http://www2.eng.cam.ac.uk/~bcb/history.html. Accessed April 24, 2016.

Diéval, F., Mathieu, D., and Durand, B. 2004. Comparison and characterization of polyester fiber and microfiber structure by x-ray diffractometry and visco-elasticimetry. *The Journal of The Textile Institute* 95(1–6): 131–146. doi: 10.1533/joti.2002.0031.

Dong, H., Wang, D., Sun, G., and Hinestroza, J.P. 2008. Assembly of metal nanoparticles on electrospun nylon 6 nanofibers by control of interfacial hydrogen-bonding interactions. *Chemistry of Materials*, 20(21): 6627–6632. doi: 10.1021/cm801077p.

Echlin, P. 2009. *Handbook of Sample Preparation for Scanning Electron Microscopy and X-Ray Microanalysis*. Springer Science+Business Media, New York, LLC.

Ehrenstein, G.W. and Raue, F. 2005. Fractography. In *Encyclopedia of Polymer Science and Technology*, edited by H.F. Mark. John Wiley & Sons, Inc, Hoboken, NJ.

El-Nahhal, I.M., Zourab, S.M., Kodeh, F.S., Elmanama, A.A., Selmane, M., Genois, I., and Babonneau, F. 2013. Nano-structured zinc oxide–cotton fibers: Synthesis, characterization and applications. *Journal of Materials Science: Materials in Electronics* 24(10): 3970–3975. doi: 10.1007/s10854-013-1349-1.

Epps, H.H. 1988. Insulation characteristics of fabric assemblies. *Journal of Industrial Textiles* 17(3): 212–218.

Fejdyś, M., Łandwijt, M., Kucharska-Jastrząbek, A., and Struszczyk, M.H. 2016. The effect of processing conditions on the performance of UHMWPE-fibre reinforced polymer matrix composites. *Fibres & Textiles in Eastern Europe* 24(4(118)): 112–120.

FLIR. 2009. FLIR infrared cameras help detect the spreading of swine flu and other viral disease. Press release, April 28, 2009. [Online]. Available at http://www.flir.com. Accessed October 2016.

Garratt-Reed, A.J. and Bell, D.C. 2003. *Energy-Dispersive X-Ray Analysis in the Electron Microscope*, edited by M. Rainforth, Microscopy Handbooks. BIOS Scientific Publishers Ltd, Oxford, U.K.

Ghassemi, A. 2012. Solar and infrared radiation measurement. In *Infrared Measurements*, edited by F. Vignola, J. Michalsky, T. Stoffel, 1st edn., Ch. 10. Taylor & Francis, Boca Raton, FL, pp. 205–210.

Goldstein, J.I., Lyman, C.E., Newbury, D.E., Lifshin, E., Echlin, P., Sawyer, L., Joy, D.C., and Michael, J.R. 2003. *Scanning Electron Microscopy and X-Ray Microanalysis*, 3rd edn. Springer Science+Business Media, Kluwer Academic Publishers, New York, LLC.

Havlova, M. 2014. Detection of fabric structure irregularities using air permeability measurements. *Journal of Engineered Fibers & Fabrics* 9(4): 157–164.

Huang, J. 2006. Sweating guarded hot plate test method. *Polymer Testing* 25(5): 709–716. http://doi.org/10.1016/j.polymertesting.2006.03.002.

Huang, Y., Ma, E., and Zhao, G. 2015. Thermal and structure analysis on reaction mechanisms during the preparation of activated carbon fibers by KOH activation from liquefied wood-based fibers. *Industrial Crops and Products* 69: 447–455. doi: http://dx.doi.org/10.1016/j.indcrop.2015.03.002.

ISO 11092:2014. 2014. Textiles—Physiological effects—Measurement of thermal and water vapour resistance under steady-state conditions (sweating guarded-hotplate). ISO/Technical Committee 38 Textiles, 2nd edn.

Joshi, M., Bhattacharyya, A., and Ali, S.W. 2008. Characterization techniques for nanotechnology applications in textiles. *Indian Journal of Fiber and Textile Research* 33(3): 304–317.

Kasai, N. and Kakudo, M. 2005. *X-Ray Diffraction by Macromolecules*, edited by Jr. A.W. Castleman, J.P. Toennies, and W. Zinth, Vol. 80, Chemical Physics. Springer-Verlag, Berlin, Germany.

Kwan, K.S. Jr. 1998. Chapter 4—FTIR-ATR diffusion study in the role of penetrant structure on the transport and mechanical properties of a thermoset adhesive, PhD dissertation, Materials Science and Engineering, Virginia Tech, Blacksburg, VA.

Lee, E.-J., Yoon, J.-S., Kim, M.-N., and Park, E.-S. 2011. Preparation and applicability of vinyl alcohol group containing polymer/MWNT nanocomposite using a simple saponification method. In *Carbon Nanotubes—Polymer Nanocomposites*, edited by S. Yellampalli. InTech, Rijeka, Croatia. DOI: 10.5772/20444.

Li, C.Y., Wang, B., and Cheng, S.Z.D. 2000. X-ray scattering in analysis of polymers. In *Encyclopedia of Analytical Chemistry—Applications, Theory and Instrumentation*, edited by R.A. Meyers. John Wiley & Sons Ltd, Chichester, U.K. DOI: 10.1002/9780470027318.a2039.

Marsh, M.C. 1930. The thermal insulating properties of fabrics. *Proceedings of the Physical Society* 42(5): 577–588. http://doi.org/10.1177/109719638400800205.

Mazeyar, P.G., Alimohammadi, F., Song, G., and Kiumarsi, A. 2012. Characterization of nanocomposite coatings on textiles: A brief review on microscopic technology. In *Current Microscopy Contribution to Advances in Science and Technology*, edited by A. Méndez-Vilas, FORMATEX, Badajoz, Spain, pp. 1424–1437.

Michler, G.H. 2008. *Electron Microscopy of Polymers*, Springer-Verlag, Berlin, Germany.

Murthy, N.S. 2004. Recent developments in polymer characterization using x-ray diffraction. *The Rigaku Journal* 21(1): 15–24.

Simmons, R. and Ng, L.M. 2000. Infrared spectroscopy in clinical chemistry. In *Encyclopedia of Analytical Chemistry*, edited by R.A. Meyers. Wiley, Chichester, U.K. DOI: 10.1002/9780470027318.a0527.

Oleshko, V., Gijbels, R., and Amelinckx, S. 2000. Electron microscopy and scanning microanalysis. In *Encyclopedia of Analytical Chemistry—Applications, Theory and Instrumentation*, edited by R.A. Meyers. John Wiley & Sons Ltd.

Parikh, D.V., Thibodeaux, D.P., and Condon, B. 2007. X-ray crystallinity of bleached and crosslinked cottons. *Textile Research Journal* 77(8): 612–616. doi: 10.1177/0040517507081982.

Pecharsky, V.K. and Zavalij, P.Y. 2003. *Fundamentals of Powder Diffraction and Structural Characterization of Materials*. Springer Science+Business Media, Inc, New York.

Rasheed, A., Drean, J.Y., and Osselin, J.F. 2011. Study of pressure distribution on a car seat by using IR technology. In *Proceedings of AUTEX-2011*, Mulhouse, France, June 2011, pp. 1257–1262.

Reimer, L. (ed.) 1998. *Scanning Electron Microscopy—Physics of Image Formation and Microanalysis*, edited by P.W. Hawkes, Optical Sciences. Springer-Verlag, Berlin, Germany.

Sankaran, S. and Ehsani, R. 2014. Introduction to electromagnetic spectrum. In *Imaging with Electromagnetic Spectrum*, edited by A. Manickavasagan, H. Jayasuriya, 1st edn., Ch. 1. Springer, New York, pp. 1–3.

Saville, B.P. 1999. *Physical Testing of Textiles*, 1st edn. CRC Press, Cambridge, U.K.

Seeck, O.H. and Vainio, U. 2014. *X-Ray Diffraction—Modern Experimental Techniques*, edited by B.M. Murphy and O.H. Seeck. Pan Standford, Taylor & Francis Group, Boca Raton, FL.

Stuart, B. 2004. *Infrared Spectroscopy: Fundamentals and Applications, Analytical Techniques in the Sciences*. John Wiley & Sons, West Sussex, England.

Ugur, S.S., Sariişik, M., and Aktaş, A.H. 2010. The fabrication of nanocomposite thin films with TiO_2 nanoparticles by the layer-by-layer deposition method for multifunctional cotton fabrics. *Nanotechnology* 21(32): 325603.

Vad, T., Wulfhorst, J., Pan, T.-T., Steinmann, W., Dabringhaus, S., Beckers, M., Seide, G. et al. 2013. Orientation of well-dispersed multiwalled carbon nanotubes in melt-spun polymer fibers and its impact on the formation of the semicrystalline polymer structure: A combined wide-angle x-ray scattering and electron tomography study. *Macromolecules* 46(14): 5604–5613. doi: 10.1021/ma4001226.

Wang, L., Zhang, X., Li, B., Sun, P., Yang, J., Xu, H., and Liu, Y. 2011. Superhydrophobic and ultraviolet-blocking cotton textiles. *ACS Applied Materials & Interfaces* 3(4): 1277–1281. doi: 10.1021/am200083z.

Welz, B. and Sperling, M. 2007. *Atomic Absorption Spectrometry*, 3rd edn. Wiley Publisher, Hoboken, NJ.

Wittke, J.H. 2016. Electron microprobe. http://nau.edu/cefns/labs/electron-microprobe/. Accessed April 25, 2016.

Wu, G.F. and He, Y. 2010. Identification of varieties of textile fibers by using Vis/NIR infrared spectroscopy technique. *Guang Pu Xue Yu Guang Pu Fen Xi* 30(2): 331–335.

Yao, B.g., Li, Y., Hu, J.Y., Kwok, Y.L., and Yeung, K.W. 2006. An improved test method for characterizing the dynamic liquid moisture transfer in porous polymeric materials. *Polymer Testing* 25(5): 677–689. http://doi.org/10.1016/j.polymertesting.2006.03.014.

Yue, H., Monreal-Bernal, A., Fernández-Blázquez, J.P., Llorca, J., and Vilatela, J.J. 2015. Macroscopic CNT fibres inducing non-epitaxial nucleation and orientation of semicrystalline polymers. *Scientific Reports* 5: 16729. doi: 10.1038/srep16729. http://www.nature.com/articles/srep16729#supplementary-information.

Zhao, H., Kwak, J.H., Zhang, Z.C., Brown, H.M., Arey, B.W., and Holladay, J.E. 2007. Studying cellulose fiber structure by SEM, XRD, NMR and acid hydrolysis. *Carbohydrate Polymers* 68(2): 235–241. doi: http://dx.doi.org/10.1016/j.carbpol.2006.12.013.

Section I

Testing of Conventional Textiles

4

Textile Fibers

Ali Afzal and Azeem Ullah

CONTENTS

4.1 Introduction

The fabric- and cloth-making industry is one of the most essential industries. Its raw materials are fibers. So, in making a textile product the parameters of the basic raw material, fiber, are very important [1]. In this chapter, we will discuss some basic testing for fibers used in conventional textiles.

A fiber that has the ability to be processed into yarn which can further be woven or knitted by certain interlacing methods is termed a textile fiber. Textile fiber comprises of polymeric chains. Polymers consist of long molecular chains which are further divided into monomers. These monomers are joined together by a process called polymerization. The length of the polymer chain is represented as the degree of polymerization [2].

The use of textiles for clothes and furnishing hinges on an exceptional combination of properties, such as warmth, softness, and pliability. These properties depend upon the raw materials used to make these products. Thus for a fiber to be useful for textile purposes, it should have certain properties: the fiber length must be several hundred times the width, it must be able to be converted into yarn, and it must be strong enough to withstand mechanical action during production. So, a textile fiber must have at least 5 mm of length so that it will be supple, flexible, and strong enough to be spun. Other properties like elasticity, fineness, uniformity, durability, luster, and crimp should also be possessed by a textile fiber.

4.2 Fiber Classification

Fibers for textiles are classified by many systems. In 1960, the Textile Fiber Products Identification Act became effective. One of the basic ways to classify fiber is by its origin, and this is indeed the most commonly employed method. Figure 4.1 gives a general overview of fiber classification.

There are various types of fibers used in the textile industry, each having their unique properties. These characteristics are largely dependent upon their origins. Natural fibers are obtained from nature, where the source could be a plant, an animal, or a mineral. Regarding plants, we obtain fibers from seeds (cotton, coir), from leaves (sisal), and from stems (jute, flax, ramie, etc.). From animals we get wool and silk and from minerals we obtain asbestos. With the increasing population, the demand for textiles is ever increasing and to meet these demands mankind has started to develop fibers commonly classified as manmade fibers. Manmade fibers are produced from polymer sources, either from nature (regenerated fibers) or from synthetic polymers.

4.3 Fiber Identification Methods

4.3.1 Physical

Microscope analysis can be easily employed for natural fibers. Optical microscopes with a magnification power of several hundred can be employed for the analysis. The analyzer must know what the fibers look like under

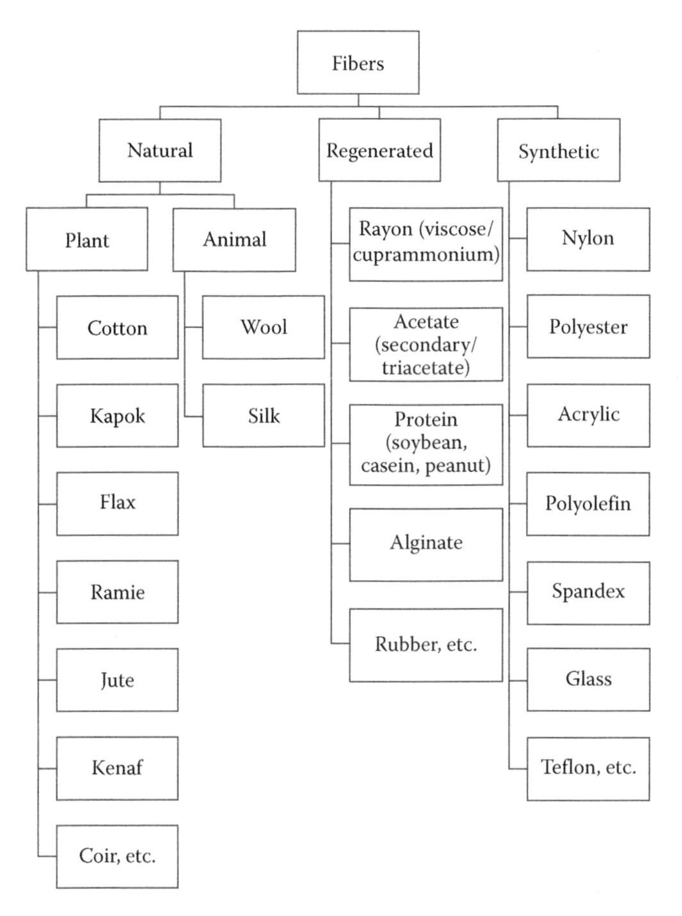

FIGURE 4.1
Classification of fibers. (From Cook, J.G., *Handbook of Textile Fibres, Natural Fibres*, 5th edn., Vol. 1, Woodhead Publishers, Cambridge, England, 2001.)

a microscope, as many processes change the appearance of the fibers, such as mercerizing and delustering. Dark colored fibers cannot be tested under a microscope as light cannot pass through them. For these fibers chemical analysis is the most suitable method, though not for synthetic fiber, as many of them have a similar appearance. Natural fibers have their characteristic markings that aid identifying them in a blend.

The following are examples of the appearance of natural fibers under an optical microscope:

- Flax has a sharp polygon cylindrical shape with well defined edges. The fiber filament shows cross-markings at intervals.

- Ramie shows a rounded polygon structure with an irregular lumen structure when viewed cross-sectionally.
- Cotton shows a spirally twisted ribbon-like tube with a rough surface. However, mercerized cotton does not have these natural twists due to swelling.
- Wool has a multicellular rough cylindrical shape. Under a microscope three basic structures are visible: the epidermis, medulla, and cortex. Characteristic scales can be seen on the surface of the wool fiber.
- Silk has an elliptical shape composed of two filaments. The single filament has a triangular shape.

Manmade fibers are difficult to identify under a microscope as they are available on the market in multiple variants such as trilobal, multilobal, channeled, and circular cross sections.

For microscopic cross-sectional examination, a parallel bundle of fibers or yarns is obtained from the bulk. This is then adjusted by a copper thread looped through a stainless steel plate so that the fiber bundle is perpendicular to the plate. With a sharp razor blade the fibers are then cut on both side of the plate which is now put under the microscope for examination. The plate can be covered with mineral oils under a cover slip for evaluation. If a microtome is used instead of the steel plate, then the bundle of fibers or yarns is inserted into the slot, where it is compressed with the help of a metallic tongue. The fiber is then cut from both sides before a drop of collodion solution is applied to one of the cut faces. After the solution has shown through at the other side the fiber is left to dry thoroughly. Excess collodion and fiber is removed with the razor blade. Finally, the fiber is transferred to a glass slide and a drop of mineral oil is added, before the cover slip is positioned and examination can begin.

4.3.2 Chemical

Chemical analysis is of two types: (a) the stain method and (b) the solvent method. The solvent method is the most suitable for quantitative analysis and will be discussed in detail in this section.

For this method a preweighed sample is treated with various solvents to distinguish one type of fiber from another. It is extremely difficult to use the same solvent for chemically similar fibers. For analysis, the blended fiber sample has to be thoroughly cleaned and any finishes applied have to be removed. Table 4.1 provides the type of solvents used to identify particular fibers.

TABLE 4.1

Solubility of Fibers

	Acetic Acid	Acetone	Sodium Hypochloride	Hydrochloric Acid	Formic Acid	1,4 Dioxane	m-Xylene	Cyclo Hexanone	Dimethyl Formamide	Sulfuric Acid	Sulfuric Acid	m-Cresol	Hydrofluoric Acid
Concentration (%)	100	100	5	20	85	100	100	100	100	59.5	70	100	50
Temperature (°C)	20	20	20	20	20	101	139	156	90	20	38	139	50
Time (min)	5	5	20	10	5	5	5	5	10	20	20	5	20
Acetate	S	S	I	I	S	S	I	S	S	S	S	S	
Acrylic	I	I	I	I	I	I	I	I	S	I	I	P	I
Anidex	I	I	I	I	I	I	I	I	I	I	I	I	
Aramid	I	I	I	I	I	I	I	I	I	I	I	I	I
Azlon	I	I	S										
Cotton and flax	I	I	I	I	I	I	I	I	I	I	S	I	I
Glass	I	I	I	I	I	I	I	I	I	I	I	I	S
Modacrylic	I		I	I	I	SP	I	S	SP	I	I	P	
Novoloid	I	I	I	I	I	I	I	I	I	I	I	I	I
Nylon	I	I	I	S	S	I	I	I	N	S	S	S	
Nytril	I	I	I	I	I	I	I	S	S	I	I	SP	
Olefin	I	I	I	I	I	I	S	S	I	I	I	I	
Polyester	I	I	I	I	I	I	I	I	I	I	I	S	I
Rayon	I	I	I	I	I	I	I	I	I	S	S	I	I
Saran	I	I	I	I	I	S	S	S	S	I	I	I	
Silk	I	I	S	I	I	I	I	I	I	S	S	I	
Spandex	I	I	I	I	I	I	I	I	S	SP	SP	SP	
Teflon	I	I	I	I	I	I	I	I	I	I	I	I	I
Vinal				S	S	I	I	I	I	S	S	I	
Vinyon	I	S	I	I	I	S	S	S	S	I	I	S	
Wool	I	I	S	I	I	I	I	I	I	I	I	I	

Note: S, soluble; I, insoluble; P, form plastic; SP, soluble or form plastic; N, nylon 6.

4.4 Physical Properties

4.4.1 Length and Length Uniformity

The length of fiber is an important aspect of raw material classification. There are two types of fibers: filament and staple. Staple fibers are shorter in length as compared to filament fibers. Natural fibers do not have the same length in a single tuft and are distributed according to their length. The classification of fibers according to their length is called the length distribution which is used to define different practical parameters for specific uses. The sample fibers are sorted into different length groups and placed onto a velvet board to obtain a sample staple length distribution. A graphical representation of fiber length distribution and defined parameters is shown in Figure 4.2.

The figure shows the fiber length on the y-axis and the number of fibers on the x-axis. The calculated points described in the figure are obtained as follows:

$$OQ = \frac{1}{2}OA$$

$$OK = \frac{1}{4}OP$$

$$KS = \frac{1}{2}KK'$$

$$OL = \frac{1}{4}OR$$

$$\text{Short fiber percentage} = \left(\frac{RB}{OB}\right) \times 100$$

$$LL' - MM' = NL' = \text{inter quartile range}$$

$$\text{Dispersion percent} = \frac{NL'}{LL'}$$

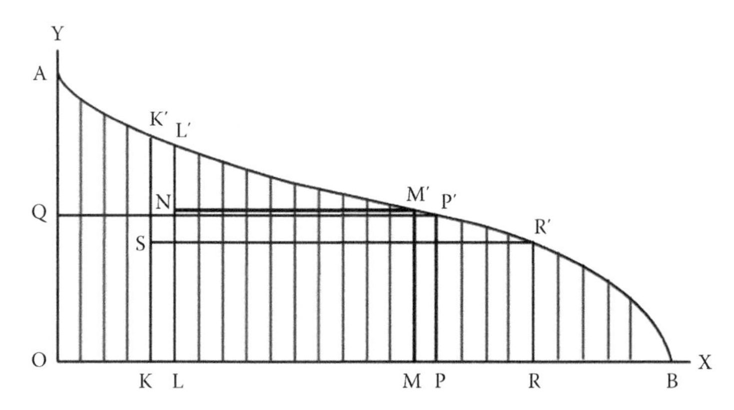

FIGURE 4.2
Comb-sorter diagram of staple length distribution.

Fiber length can be classified as staple length (STPL), mean length (ML), upper quartile length (UQL), effective length (EL), modal length (ML), span length (SL), and upper half-mean length (UHML) [3]. The other parameters that can be used to estimate fiber length variations include the uniformity index (UI), uniformity ratio (UR), short fiber content (SFC), and floating fiber index (FFI).

Fiber length was measured by classers up until 30–35 years ago. The fibers were aligned parallel to each other and measured against a scale. Subjective analysis was done only by hand and eye judgment and is the length of a specific portion of fibers in comparison with prescribed standards [4]. A representative portion of parallel fibers is achieved by the classifier by pulling a tuft of fibers from the sample and pulling, lapping, and discarding to compare with different standards. The staple length is defined by USDA as "the normal length of typical portion of staple fibers having relative humidity of 65% and temperature of 70°F without regard to value or quality." The classification of cotton according to staple length is given in Table 4.2.

The mean length of the fibers is defined as "the average length of all fibers in the test specimen based on weight–length data" [5]. It can also be calculated by number–length data as an alternative. The number–length data emphasize short fibers in the fiber sample, while weight–length data tend to hide them. The mean lengths by mass–length data (ML_w) and number–length data (ML_n) are calculated using Equations 4.1 and 4.2, respectively.

$$ML_w = \int_0^\infty f_w(l)\,dl \tag{4.1}$$

$$ML_n = \int_0^\infty f_n(l)\,dl \tag{4.2}$$

TABLE 4.2

Classification of Cotton Fibers according to Their Staple Length

Staple Length		
(mm)	**(in.)**	**Description**
<20.6	<26/32	Short
20.6–25.4	26/32–32/32	Medium
26.2–27.8	33/32–35/32	Medium long
28.6–33.3	36/32–42/32	Long
>34.9	>44/32	Extra long

TABLE 4.3

Classification of Cotton according to Upper Quartile Length

Upper Quartile Length (mm)	Classification
<27.9	Short
27.9–31.5	Medium
31.8–35.3	Long
>35.3	Extra long

According to American Society for Testing and Materials (ASTM) international standards, upper quartile length (UPL) is defined as "the fiber length which is exceeded by 25% of fibers by weight in test specimen" [5]. The classification of cotton based on UQL is given in Table 4.3 [4]. The mathematical equations to find the UPL by weight and by number are given in Equations 4.3 and 4.4, respectively [6].

$$\int_{UQL_w}^{\infty} f_w(t)dt = q_w(UQL_w) = 0.25 \tag{4.3}$$

$$\int_{UQL_n}^{\infty} f_n(t)dt = q_n(UQL_n) = 0.25 \tag{4.4}$$

The effective length is longer than the average length and is a measure of the length of the majority of longer fibers in a sample [4]. The effective length is described statistically as the upper quartile of the fiber length distribution obtained by ignoring short fibers whose length is less than half of the effective length of fibers. The length of fibers in Figure 4.2 observed at LL′ is the effective length obtained by a comb-sorter diagram. The effective length is more independent of the tail of short fibers and is therefore used for machinery settings, particularly for the distances between the nips of successive pairs of drafting rollers. The effective length is about 1.1 times the staple length [4].

The modal length is the staple length of the high frequency fiber in the sample obtained from the fiber length frequency diagram. The modal length of long staple cotton is higher than the mean staple length due to the progressive increase in skewness with increasing staple length in the fiber length distribution.

The span length is the length of fibers at a distance spanned by a specific percentage of fibers (it can be by number or by weight) in the test beard, considering the reading as 100% at the starting point of scanning. Different span length reference points are considered. The most common span lengths (t) used commercially are 2.5% and 50% span length (S.P). The 2.5% span length is defined as the length of fiber at which only 2.5% of long fibers are excluded.

It provides the reference length for roller drafting ratchet settings to be adjusted so that few, if any, fibers are broken. The 2.5% span length agrees best with the upper half-mean length for U.S. upland cotton. The 50% span length is more valuable as a potential measure of yarn quality and spinning performance [4]. It can be calculated from a fibrogram [6]:

$$p_w\left(SL_{t\%w}\right) = \frac{t}{100} \tag{4.5}$$

$$p_n\left(SL_{t\%n}\right) = \frac{t}{100} \tag{4.6}$$

The upper half-mean length (UHML) as defined by ASTM standards is the average length by number (respectively by weight) of one-half of the longest fibers when they are divided on a weight basis (respectively number basis) [6].

$$UHML_n = \frac{1}{q_n(ME)} \int_{ME}^{\infty} tf_n(t) \tag{4.7}$$

$$UHML_w = \frac{1}{q_w(ME)} \int_{ME}^{\infty} tf_w(t) = 2\int_{ME}^{\infty} tf_w(t) \tag{4.8}$$

where ME is the median length that exceeds 50% of fibers by weight, therefore q_w (ME) = 0.5.

The uniformity index (UI%) is the ratio of the mean length to the upper half-mean length. It is a measure of the uniformity of fiber lengths as a percentage of the sample. The uniformity index of the U.S. upland cotton classification is given in Table 4.4 [4].

$$UI\%_w = \frac{ML_w}{UHML_w} \times 100 \tag{4.9}$$

TABLE 4.4

Uniformity Index of U.S. Upland Cotton Classification

$UI\%_w$	Description
Above 85	Very high
83–85	High
80–82	Average
77–79	Low
Below 77	Very low

TABLE 4.5

Classification of Uniformity Ratio of U.S. Upland Cotton

Uniformity Ratio	Description
Above 48	Very high
47–48	High
44–45	Average
41–43	Low
Below 41	Very low

$$UI\%_n = \frac{ML_n}{UHML_n} \times 100 \tag{4.10}$$

The uniformity ratio (UR%) is the ratio of the 50% span length to the 2.5% span length. It is a smaller value than the uniformity index by a factor close to 1.8 [6]. The uniformity ratio of U.S. upland cotton with classification is reported in Table 4.5 [4].

$$UR_w\% = \frac{SL_{50\%w}}{SL_{2.5\%w}} \times 100 \tag{4.11}$$

$$UR_n\% = \frac{SL_{50\%n}}{SL_{2.5\%n}} \times 100 \tag{4.12}$$

Short fiber content (SFC%) is the percentage by weight of fibers having a length less than half of an inch [6]. It can also be measured in percentage by number of fibers having a length less than half of an inch, with the respective short fiber content percentage in number. The presence of short fibers in the cotton increases the cost of processing waste and also contributes to weaker yarn and less efficient spinning. Long length fibers are mostly preferred due to the reduced number of fiber ends with a higher yarn strength in the same length.

$$SFC_w\% = 100 \times \int_0^{127} f_w(t)dt = 100 \times \left(1 - q_w(127)\right) \tag{4.13}$$

$$SFC_n\% = 100 \times \int_0^{127} f_n(t)dt = 100 \times \left(1 - q_n(127)\right) \tag{4.14}$$

Short fiber content can be calculated by measuring two span lengths from the high volume instrument using Equation 4.15 [4]:

$$\text{SFC\%} = 50.01 - 0.766 \times (2.5\%\text{SL}) - 81.48 \times (50\%\text{SL}) \tag{4.15}$$

The FFI is an alternative to short fiber content. It explains the number of short fibers which are not clamped between the nips of a pair of rollers in the drafting system. These fibers are floated on long fibers to pass through the drafting zone without the influence of the applied drafting mechanism. The index is calculated using Equation 4.16 [4]:

$$\text{FFI} = \left(\frac{\text{UQL}}{\text{ML}} - 1\right) \times 100 \tag{4.16}$$

It can also be calculated from a fibrogram:

$$\text{FFI} = \left(\frac{\text{UHML}}{\text{ML}} - 1\right) \times 100 \tag{4.17}$$

The fiber length is measured by a number of techniques including the Suter-Webb fiber array [7], the fibrograph, Pyer, advance fiber investigation system (AFIS), and high volume instrument (HVI) systems. Each technique has its own advantages. Commercial fiber testing has been done using an HVI instrument which is a fast measurement technique but with less accuracy as compared to Suter-Webb.

The commercial testing of fiber length distribution is done using an HVI. The fiber beard is developed using needles of a comb/clamp to pick up specimen fibers through holes of an HVI fibrosampler. The collected fiber sample is in the shape of a tapered beard. The loose fibers and fiber crimps are removed by the brushing and combing principle. The sample beard is analyzed using scanning light attenuation at each length to determine the fiber mass at every point length of the sample. The mass–length curve obtained by this tapered beard is called a fibrogram. Much work has been carried out to determine fiber length parameters using a fibrogram by different researchers. A number of equations for calculations have also been proposed in the literature [3]. In the HVI method, a part of the fiber which is in clamp is a hidden portion for the fibrogram. This instrument can measure the length, fineness, strength, elongation, maturity index, color, trash, and moisture of a fiber, as well as the short fiber index and the number of neps. The true fiber length distribution scanning is done using AFIS, where each fiber is scanned individually, but which is a slow and costly process. Image processing is considered to be an accurate and precise method for the measurement of fiber length [8].

TABLE 4.6
Correlation Coefficients between Different Length Parameters

Measures	Correlation Coefficient
SL(n) vs. ML(n)	0.974
SL(w) vs. ML(w)	0.985
STPL vs. 2.5% SL	0.32
STPL vs. UQL	0.42
UQL vs. 2.5% SL	0.72
STPL vs. UHML	0.42
STPL vs. UQL	0.62
STPL vs. 3.0% SL	0.99
UQL vs. 3.1% SL	0.98
UHML vs. 3.1% SL	0.88
UQL vs. ML	0.89
UHML vs. UI	0.52
UHML vs. ML	0.97
ML vs. UI	0.72
SFC(n) vs. UHML	0.59
SFC(n) vs. ML	0.66
SFC(n) vs. UI	0.75
SFC(w) vs. UHML	0.71
SFC(w) vs. ML	0.77
SFC(w) vs. UI	0.38
SFC(w) vs. UR	0.328
SFC(n) vs. UR	0.425
SFC(w) vs. HVI-UI	0.955
SFC(w) vs. Fib.-UI	0.535
ML vs. 2.5% SL	0.980
SFC(w) vs. 2.5% SL	0.523

The correlation coefficients found in the literature between various length parameters are given in Table 4.6 [4].

4.4.2 Moisture Regain

The amount of moisture (water) present in a textile sample is referred to either by its regain or its moisture content. These two terms are often confused with each other. Moisture regain is expressed as the percentage of water in a sample compared to its oven dry weight, also referred to as its bone dry weight. Moisture content is expressed as a percentage of the total weight of the sample. The standard test method ASTM D2495-07 is most commonly employed in the textile industry to measure the regain and moisture content. The test is a simple one and can be easily performed. A sample of fiber is collected and weighed, before being oven dried at 105°C until it maintains a constant weight. The difference between the original mass before drying and the oven dried mass is calculated as a percentage, and is denoted either as moisture content or moisture regain. Moisture regain and moisture content can be measured using the following equations.

$$R = \frac{W}{D} \times 100 \tag{4.18}$$

$$C = \frac{W}{D + W} \times 100 \tag{4.19}$$

where
- R is the moisture regain
- C is the moisture content
- W is the weight of water
- D is the oven dry weight

4.4.3 Trash Content

The presence of undesirable material in the fiber is considered to be trash. Other synonyms include contamination and nonlint matter. It is comprised of fragments of leaves, stalks, grasses, seeds, and dust. It also includes feathers; pieces of plastic, rugs, and cloths; foreign fibrous material other than the desired fiber (like polyester or jute in cotton); and immature fibers. The immature fibers of the same desired fiber are also considered to be trash as they are not wanted in the final product. A Shirley analyzer is used to determine trash content in the fibrous tuft. It comprises a pair of rollers for gripped feeding into a sawtooth beater, rotating at a high surface speed.

4.4.4 Cross-Sectional Shape of the Fiber

Shape affects the physical and mechanical properties of textile fiber. There are many properties which are changed by the shape of the fiber's cross section, like flexural rigidity, fabric softness, drape, crispness, and stiffness. Different natural fibers have different types of shapes, while the shape of man-made fibers depends upon the shape of the spinneret from which they are extruded, as shown in Figure 4.3. Like silk it has a triangular cross section. Cotton fibers are kidney shaped. Wool fibers are round or oval in shape.

4.4.5 Fiber Color

The color of the fiber is an important aspect in regard to aesthetic sense and dye shade. Every fiber type has its particular color regarding its natural or synthetic origin. In natural fibers, cotton is found as white to yellowish in color, wool fiber has whitish to blackish color grades, silk fiber is found in a lustrous

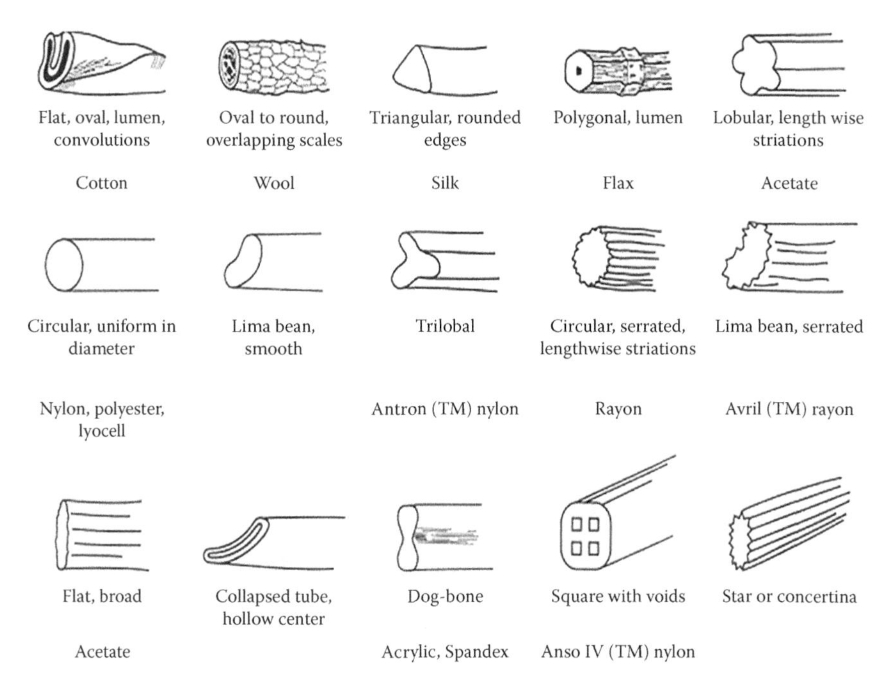

FIGURE 4.3
Schematic illustration of different cross-sectional shapes of fibers.

white color, and jute fiber is brown. Regenerated rayon fibers are transparent in color unless dulled by pigments. In synthetic fibers, acrylic ones are white to off-white, nylon ones are off-white, polyester ones are white, para-aramid ones are dull yellowish, and carbon ones are black. Some fibers can be decolored to introduce new colors by the dying process, while other fibers have a permanent color which can't be removed, as in the case of Kevlar and carbon fibers.

4.4.6 Fiber Fineness

Fineness is one of the major aspects of fiber characteristics and explains cross-sectional thickness. A fine fiber can be used to spin fine yarns. As the linear density of yarn decreases, the number of fibers also decreases by yarn diameter. The presence or absence of a single fiber shows longitudinal unevenness and variation in diameter. The decrease of fiber diameter will increase the number of fibers in a cross section of yarn and hence better yarn evenness can be achieved. Fiber fineness has great influence on the properties of yarn and fabric. The evenness of the yarn is improved by the use of fine fibers. In addition, fine fibers need less twist and have less stiffness than coarser fibers. The increase in fiber surface due to a decrease in fiber diameter contributes to a cohesion of fibers to achieve the same strength with less twist than coarser fibers [9]. These characteristics contribute to the hand feel of the products developed from them.

Transversal imaging of a fiber is not an easy task, and not even feasible for commercial testing. The natural fibers have their distinctive particular cross-sectional shape. These cross sections are not circular, which makes it difficult to determine their fineness. Cotton has a flattened tube, wool has a circular tube, while silk fiber has a triangular cross-sectional shape. Man-made fibers can be manufactured in any shape, in which star, trilobal, and hollow structures are common. Natural fibers have variability in their length, diameter, and cross-sectional shape. Therefore, it would not be easy to determine fiber fineness using the transversal magnified imaging method at a commercial scale of testing. In this case, two methods are commonly used: the gravimetric and the air flow method. In the gravimetric method, the mass of the given fiber length is used as a measure of its fineness:

$$\text{Mass of a fiber} = \text{length} \times \text{density} \times \text{cross-sectional area} \tag{4.20}$$

Where the length and density of the given fiber are known, the mass will be directly proportional to its cross-sectional area. The primary measurement unit is a tex (g/1000 m), but other conversion units are also used including decitex, militex, and denier, where

$$\text{decitex} = \frac{\text{mass in grams}}{10{,}000\,\text{m of fiber}}$$

$$\text{militex} = \frac{\text{mass in miligrams}}{1{,}000\,\text{m of fiber}}$$

$$\text{denier} = \frac{\text{mass in grams}}{9{,}000\,\text{m of fiber}}$$

These measurement units are like those for the linear density measurement of yarn. The diameter of circular cross-sectional fibers can be measured using the following equations [9]:

$$d = \sqrt{\frac{\text{Decitex} \times 10^3}{7.85 \times \rho}} \tag{4.21}$$

$$d = \sqrt{\frac{\text{Denier} \times 10^3}{7.07 \times \rho}} \tag{4.22}$$

where

d is the fiber diameter in micrometers

ρ is the fiber density in g/cm^3

The cross-sectional area of fibers other than those that are circular shaped is difficult to determine, resulting in complex relationships to find their diameter.

The fiber fineness by airflow method is measured using an ASTM standard test method that employs an air-flow instrument [10]. This measurement method is particular to cotton fiber testing due to its high bulk consumption and commercial testing point of view. It is an indirect measurement method of fiber fineness in which air is blown at a given pressure difference through a uniformly distributed mass of fibers to determine the total surface area of the fibers. The coarser fibers provide more spacing between themselves and hence provide less resistance to air flow, while finer fibers compact closely to each other leaving minimal spacing to allow air to flow between them. Therefore, a high pressure difference indicates a large surface area, resulting in a lower fiber diameter. A scale calibrated with a micronaire reading attached to an air regulator and tube is used to take readings directly. The sample is conditioned first as per the standard conditioning provided in ASTM before the start of the test [11]. The weighted fiber sample plug is inserted into the tube end which pressurizes the sample and allows air to pass through it. The pressure difference observed is calibrated and recorded as a micronaire reading of the fiber.

4.4.7 Fiber Crimp

The waviness in a fiber is known as crimp. It is measured as the number of crimps or waves per unit length or the percentage increase in the extent of the fiber on removal of the crimp. Crimps also govern the capacity of fibers to cohere under light pressure. The bi-component structure of wool increases the crimp in it. Cotton has a low crimp. Crimp allows the scattering of light due to its wavy structure and provides a dull appearance on developed products. Synthetic fibers are lustrous in structure, which can be reduced by the introduction of crimps in them. Crimps increase the thickness and enhance the bulky aspect of products.

4.5 Mechanical Properties

4.5.1 Fiber Strength

The strength and elongation of a cotton fiber can be measured by a single fiber or by the bundle method. The bundle fiber strength can be measured using an ASTM standard test procedure that employs a fiber bundle tensile testing machine [12]. These machines are commercially available in pendulum and inclined plane mechanisms. A fiber sample is conditioned as per the standard conditioning procedure [11]. A parallel bundle of fiber ends is clamped in a Pressley type vise weighing 60–80 mg. The fiber bundle is spread to a ¼ in. wide and placed at the center of the open clamps. The clamps are tightened to grip the fiber bundle straight and parallel between them. The clamps are then removed from the vise and inserted into the tensile testing machine. The locking mechanism is released to allow the specimen to break in accordance with the instructions of the manufacturers for the particular instrument. The breaking force is recorded to the nearest scale reading after the breakage of the fiber bundle. In the case of irregular breakage, where less force than the minimum required for the instrument is realized, a new sample must be tested. The gauge length can be zero or 1/8 in. The parameters are calculated using Equations 4.23 and 4.24 at zero gauge with a bundle length of 0.465 in.

$$B = 5.36 \times \frac{F}{m} \tag{4.23}$$

$$T = 10.81 \times \frac{F}{m} \tag{4.24}$$

where
- B is the breaking tenacity (gf/tex)
- F is the breaking force (lbf)
- T is the tensile strength (1000 psi)
- m is the bundle mass (mg)

The fiber strength contribution to yarn strength is found to be 20% as shown in Figure 4.4 [13].

4.5.2 Tenacity

Tenacity is the measure of the breaking strength of a textile fiber. It is also defined as ultimate breaking strength and is the maximum force a fiber can bear without breakage. The tenacity value for individual fibers is the value of load applied at breakage. For comparison purposes between different fibers, specific stress at break is used as given in Table 4.7. The specific stress is the ratio of load to linear density and is measured in units of g/denier, cN/tex, and MPa.

4.6 Chemical Properties

4.6.1 Blend Ratio

Blending is the easiest way to obtain synergistic effects of two different materials. In the textile industry the blending of a fiber is a common practice to obtain the desired functional and aesthetic properties. Blends can

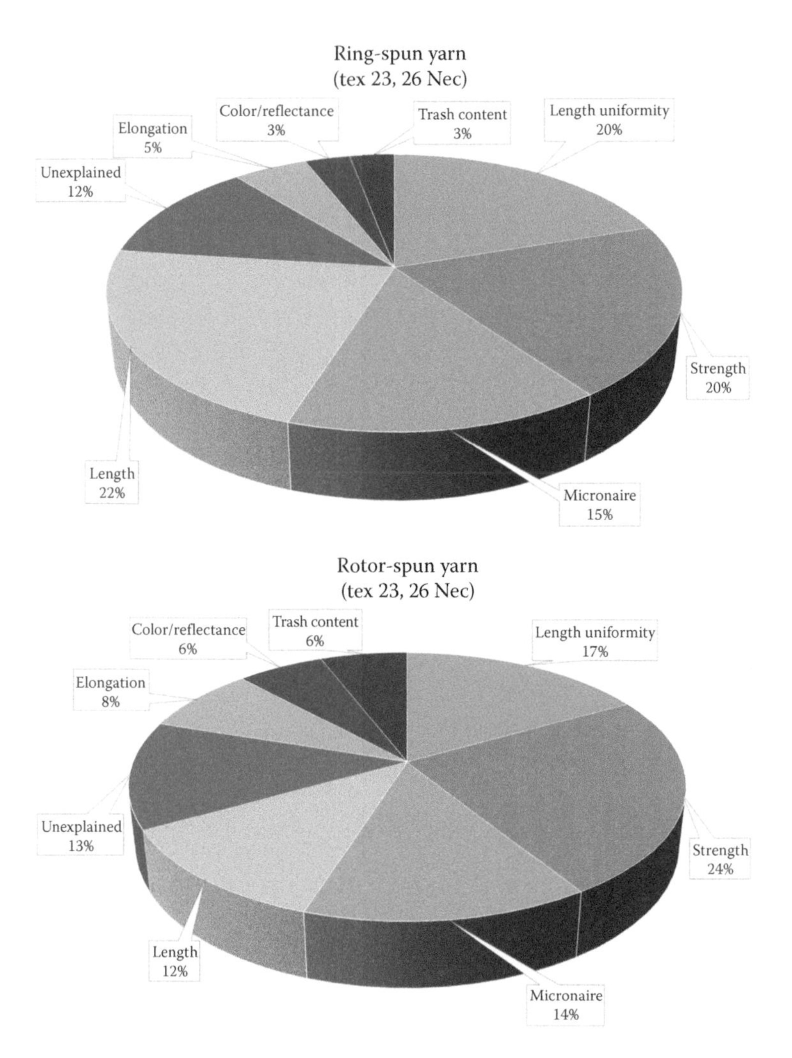

FIGURE 4.4
Influence of various fiber properties in yarn strength.

be identified either qualitatively or quantitatively. Furthermore, the testing methods for the identification of fibers in a blend can be technical or nontechnical. The nontechnical tests include the feeling test or the burning test; these tests are for qualitative assessment.

The technical test for finding fibers in a blend includes optical analysis (qualitative) and chemical testing (quantitative). These tests are far more reliable than the nontechnical tests. In this section, we focus on these two testing methods.

TABLE 4.7

Tenacity Values for Different Fibers

Fiber		Tenacity (Dry) (cN/tex)	Tenacity (Wet) (cN/tex)
Cotton		24.3–36	31.5–49.5
Wool		16.2–18	7.2–14.4
Silk		21.6–45	18–36
Acrylic		17.66–31.79	14.12–25.61
Polyester	RT	35–44	—
	HT	56–70	—
Nylon 6	RT	40–50	36–45
	HT	66–73	45–62
Nylon 6,6	RT	40–50	35–45
	HT	80	68
PAN	RT	17–31	14–23
	HT	29–37	26–31
Polypropylene	RT	26–44	—
	HT	80	—
Polyethylene		500	—
Carbon	HT	385.79	—
	HM	104.76	—
Kavlar	HT	206	—
	HM	209	—
Twaron	HT	220	—
	HM	210	—
Nomax		48.5	—
Vectron		207–252	—
PBO (Zylon)		352.56	—
Steel cord		33	—
Ceramic fiber		84.4–126.6	—
Glass fiber		118	—
Melamine formaldehyde (Basofil)		20–40	—
Nooloid (Kynol)		12–16	—
Polyimide (P84)		35	—
Polyphenylene sulfide		43	—
FR viscose		24	—
Oxidizing acrylic (Panox)		14.8–22.2	—
PBI		24	—
Polyamide-imide (Kermal)		24.5–58.8	—

Note: RT, regular tenacity; HT, high tenacity; HM, high modulus.

Before performing these tests, some basic sample preparation must be done. The first thing is to obtain a clean fiber content for analysis. To do this the following instructions must be followed.

Take a fiber sample of minimum 5 g and dry it until it maintains a constant weight—to remove any moisture content present in it—in an oven at 105°C–110°C. Record the weight as the sample dry weight.

Perform the following treatments if the nonfibrous content is not known:

- Freon treatment for removal of oils, fats, waxes, etc.
- Alcohol treatment for removal of soap, cationic finishes, etc.
- Aqueous treatments for removal of water soluble materials
- Enzyme treatment for removal of sizes, e.g., starch
- Acid treatment for removal of amino resins

After treatment, rinse the materials thoroughly and dry at 105°C–110°C until a constant weight is maintained.

The calculations for nonfibrous materials are

$$N = \frac{C - D}{C} \times 100$$

where

N is the nonfibrous material percentage

C is the dry weight of specimen before treatment

D is the dry weight of specimen after treatment

Calculations for fibrous content are

$$F = \frac{D}{C} \times 100$$

where F is the percentage of clean fiber content.

For colored fibers, stripping or bleaching is performed accordingly before subjecting the samples to analysis.

The fiber weight percentage is calculated as

$$F = \frac{B - A}{B} \times 100$$

where

F is the fiber weight percentage in blend

B is the weight of fiber sample before treatment

A is the weight of fiber sample after treatment

The blend ratio can be calculated as

$$BR = \frac{F_A}{F_B}$$

where
BR is the blend ratio
F_A is the weight percentage of fiber A
F_B is the weight percentage of fiber B

4.6.2 Maturity Ratio

The maturity of the cotton fiber is analyzed using the ASTM standard test procedure by employing polarized light or the sodium hydroxide swelling technique [14]. A solution of 18% concentration of sodium hydroxide is used to swell the cotton fibers by soaking them. The fibers are then laid parallel on the microscope slide and covered with glass and viewed at a magnification of 400× to distinguish between immature and mature fibers. The mature fibers swell to become almost round in cross-sectional shape. This method is not considered acceptable for commercial testing due to the poor precision of the results between different laboratories.

4.6.3 Chemical Composition

The chemical composition of fibers depends on their origin. Natural fibers obtained from plants are made up of cellulosic structures like cotton, jute, and hemp fibers. The fibers obtained from animals are composed of amino acids like wool and silk. Regenerated fibers have the same chemical composition as those obtained from their parent origin, like viscose rayon which is cellulosic in nature. The chemical composition of synthetic fibers is based on the nature of their raw materials and the chemical reactions that occurred.

References

1. Cook, J.G., *Handbook of Textile Fibres, Natural Fibres*, 5th edn., Vol. 1, Woodhead Publishers, Cambridge, England, 2001.
2. Karmakar, S.R., *Chemical Technology in the Pre-Treatment Processes of Textiles*, 1st edn., Vol. 12, Textile Science and Technology, Elsevier Science Amsterdam, the Netherlands, 1999.
3. Cui, X. et al., Obtaining cotton fiber length distributions from the beard test method. Part 1-Theoretical distributions related to the beard method. *Journal of Cotton Science*, 2009. **13**(4): 265–273.

4. Ikiz, Y., Fiber length measurement by image processing. PhD, NC State University, Raleigh, NC, 2000.
5. ASTM, Standard terminology relating to textiles. D123-15be1. ASTM International, West Conshohocken, PA, 2012.
6. Azzouz, B., Hassen, M.B., and Sakli, F., Adjustment of cotton fiber length by the statistical normal distribution: Application to binary blends. *Journal of Engineered Fibers and Fabrics*, 2008. **3**(3): 35–46.
7. ASTM, Standard test method for length and length distribution of cotton fibers (Array Method). ASTM D1440-07. ASTM International, West Conshohocken, PA, 2012.
8. Ikiz, Y. et al., Fiber length measurement by image processing. *Textile Research Journal*, 2001. **71**(10): 905–910.
9. Saville, B.P. (ed.), Fibre dimensions, in *Physical Testing of Textiles*. Woodhead Publishing Ltd., Cambridge, England, 1999, pp. 44–76.
10. ASTM, Standard test method for micronaire reading of cotton fibers. ASTM D1448-11. ASTM International, West Conshohocken, PA, 2011.
11. ASTM, Standard practice for conditioning and testing textiles. ASTM D1776/D1776M-16. ASTM International, West Conshohocken, PA, 2016.
12. ASTM, Standard test method for breaking strength and elongation of cotton fibers (Flat Bundle Method). ASTM D1445/D1445M-12. ASTM International, West Conshohocken, PA, 2012.
13. Furter, R., *USTER High Volume Instrument*, USTER Technologies, Uster, Switzerland, 2005.
14. ASTM, Standard test method for maturity of cotton fibers (Sodium Hydroxide Swelling and Polarized Light Procedures). ASTM D1442-06. ASTM International, West Conshohocken, PA, 2012.

5

Textile Yarns

Khurram Shehzad Akhtar, Fiaz Hussain, and Faheem Ahmad

CONTENTS

5.1 Introduction

Yarn has an intermediary position in the manufacturing of woven or knitted fabric, which is produced from different types of yarn. The properties of yarn are therefore important for appreciating the quality of its raw materials and the quality control of the manufactured product. The significant properties of yarn are

- Twist
- Linear density
- Single yarn and lea strength
- Elongation
- Evenness
- Hairiness

Most of the fibers used for the manufacturing of yarns are hygroscopic in nature and their characteristics change considerably with the change of moisture in the air. Moisture affects yarn strength, elongation, and the imperfection index. Therefore, yarn conditioning and yarn testing are carried out in standard conditions of temperature and relative humidity. The standard atmospheric conditions in textile testing requires a temperature of 20°C ± 2°C and a relative humidity of 65% ± 4% [1]. These standard conditions have a vital impact on the precision and accuracy of test results.

5.1.1 Sampling

It is not necessary to perform test on the complete material, which is practically impossible anyway, due to time and cost factors. Also some tests are of a destructive type, in which all the material is lost after performing the test. Due to this we use model/symbolic samples from bulk material for testing. A major objective of sampling is to produce an unbiased sample representing the whole. The following terms are used in sampling:

- *Consignment.* This can be described as the quantity of material that is transported at the same time. Each consignment may consist of single or multiple lots.
- *Test lot.* All the containers of a material having one specific type and quality that are delivered to the customer are considered as a test lot. It is presumed that the whole material is uniform and all the parameters of the material have been checked.
- *Laboratory sample.* Samples of the material that are used to perform testing in the laboratory are known as laboratory samples. Random sampling procedures are adopted to derive laboratory samples from the test lot.
- *Test specimen.* This is the portion of the material that will actually be utilized for testing, which is taken from the laboratory sample. Several specimens are used to obtain reliable test results.
- *Package.* Elementary units within each container in the consignment is known as a package, which might include bobbins, cones, and fabric rolls [2].

5.2 Yarn Classification

A yarn is normally classified as a fine fiber strand of adequate length, which may comprise either continuous filaments laid in parallel or staple fibers twisted together. Both types have the ability to be interlaced into woven fabric or interloped to manufacture a knitted fabric. So, yarns have two major classes: staple spun yarns and continuous filament yarns. Staple spun yarns are produced from natural or man-made fibers, using production processes like mixing, opening, cleaning, carding, drawing, combing, and then spinning to orient the fibers and convert them into yarn by real or false twisting and several other means.

- On the basis of their structural complexity, staple spun yarns are categorized into single, plied, or cabled yarn.

- On the basis of spinning method, staple spun yarns are categorized into ring-spun, air jet, rotor-spun, vortex, and friction-spun yarns.
- On the basis of the fiber preparation method, staple spun yarns are categorized into carded, combed, worsted, and woolen yarns.

A continuous filament yarn is made up of more than one filament and aligned parallel to a very small amount of twist. Molten polymer or a solution of polymer is drawn through the spinneret and drafted as per the requirement defined prior to solidification. Monofilament yarn comprises one filament, and multi-filament consists of more than one filament. Different types of processes could be applied to obtain the desired properties, for example, texturizing to obtain stretchy or bulky yarns [3].

5.3 Linear Density of Yarn

Among the other parameters of a yarn, its diameter is a significant factor. But determining the diameter is impossible by any means due to the fact that it varies significantly as the yarn is squeezed. Besides the optical technique, all other methods involve compressing the yarn during testing. Due to this compressive character of yarn, the measured diameter varies with the pressure applied. Optical techniques for determining diameter have the difficulty of specifying where the peripheral edge of the yarn lies as its surface can be unclear or rough due to hairiness on it. That is why the determination of the yarn's edges is subject to the operator's understanding. Due to these problems a system must be designed to ascertain the delicacy of a yarn by weighing its predefined length. This quantity is called the linear density and it can be determined with accuracy if the tested amount of yarn is sufficient. There are two main systems for assessing the linear density of yarn: the direct and the indirect.

5.3.1 Direct System

In this system we determine the weight per unit length. The most well known direct systems are

- *Tex*: This is the number of grams in 1000 m of yarn.
- *Decitex*: This is the number of grams in 10,000 m of yarn.
- *Denier*: This is the number of grams in 9000 m of yarn.

5.3.2 Indirect System

In this system we determine the length per unit weight. This linear density is also called a count, due to the fact that it is established by determining the hanks

	Tex	dtex (Decitex)	Deneir	English Cotton Count (Nec)	Metric Count (Nm)
Tex		dtex/10	Denier/9	591/Nec	1,000/Nm
dtex (Decitex)	Tex × 10		Denier/0.9	5910/Nec	10,000/Nm
Deneir	Tex × 9	dtex × 0.9		5319/Nec	9,000/Nm
English cotton count (Nec)	591/Tex	5,910/dtex	5319/denier		Nm × 0.59
Metric count (Nm)	1000/Tex	10,000/dtex	9000/denier	Nec × 1.693	

FIGURE 5.1
Conversion table.

of a specified length. It is the most widely used system of measuring a yarn's linear density [3]. A conversion table for different systems is shown in Figure 5.1.

The most well-known direct systems in use are

- *Yorkshire skeins woollen* (*Ny*) *Count*: This is the number of hanks per pound (where one hank = 256 yards).
- *Worsted count* (*New*) *Count*: This is the number of hanks per pound (where one hank = 560 yards).
- *Cotton count* (*Nec*)* *Count*: This is the number of hanks per pound (where one hank = 840 yards).
- *Metric count* (*Nm*) *Count*: This is the number of kilometers per kilogram [4].

5.3.3 Measurement of Linear Density

Yarn linear density is often measured using standard test method ASTM D1907. Specified lengths of yarn are wound on reels as skein, and then weighed. From this the linear density is calculated from the weight and length of the skein. In some options, the skein is scoured before weighing, or the mass of the skein may be determined after oven drying or after conditioning [4].

A reel having a length between 1.0 and 2.5 m or between 1.5 and 3.0 yards may be used with a tolerance of ±0.25%. The tension on the reel is adjusted to

* In these systems the fineness and count of a yarn are directly proportional to each other.

0.5 cN/tex with an adjustable tension device. The yarn sample is preconditioned for a minimum of 3 h before drying in an oven. A ventilated oven with the ability to control the temperature at 105°C ± 3°C is required to dry the skeins. After oven drying, the weight of the sample is measured by a weight balance having a sensitivity of 1 part in 1000 to measure the linear density. This sample is then conditioned for a minimum of 24 h at a standard condition of 20°C ± 2°C and 65% ± 4% relative humidity then weighed to measure the linear density at the standard moisture regain, from which the linear density is calculated [4].

$$\text{Weight at correct condition} = \text{Dry weight} \times \frac{(100 + \text{Standard regain})}{100}$$

5.4 Tensile Properties of Yarn

A yarn's tensile property can be defined as the maximum force/load that is required to break the material. It is a parameter of vital importance regarding the fabrication of yarn because it directly influences the strength of the developed fabrics. Two different approaches are used to measure yarn strength. In the first approach a single yarn strength is determined. Normally Newton (N) and cN units are used. The amount of force required by an object having mass of one kilogram to accelerate it to one meter per second square is called as one Newton force. A single yarn strength provides information about the warping machine and loom efficiency. In order to calculate the combined strength effect of the yarn, the count lea strength product (CLSP) of the yarn is calculated. A lea of 120 yards is made using the wrapping reel, and the weight of the lea is determined in order to calculate the yarn count by using the formula: number of hanks/pound. The lea strength is determined by using a lea strength machine that has two jaws, one fixed and the other attached to the load. Using the constant rate of loading principle, the tensile strength of the lea can be determined.

5.4.1 Types of Tensile Strength Testing Machines

On the basis of the working principle, tensile strength testing machines can be categorized into three major categories.

5.4.1.1 Constant Rate of Extension (CRE)

The machines have a constant rate of elongation of the specimen; as the load increases there is negligible movement of the measuring mechanism. The working principle of the Tensorapid-4, which is used to evaluate the tensile strength of a single yarn, is a constant rate of elongation.

5.4.1.2 Constant Rate of Loading (CRL)

The machines apply the load on the test sample, which is increased constantly with time. The specimen is free to elongate and its extension depends on its properties for any applied load [2]. The working principle of the lea strength machine is of this category.

5.4.1.3 Constant Rate of Traverse (CRT)

In this type of machine two pulling clamps are used to evaluate the tensile strength of the sample. One clamp moves with constant speed and application of the load is done by the second clamp, which is responsible for the activation of a load measuring mechanism. Normally old machines use this mechanism, such as the old fabric tensile strength testing machine.

5.4.2 Tensile Strength of Yarn by Skein Method

This method involves the reeling of the yarn onto a skein more commonly known as a lea through a wrapping reel, which is used for measuring linear density, the two loose ends being tied together. This lea is mounted on two jaws of a tensile strength tester. After that the lea is subjected to increasing extension and the force applied is recorded. As one portion of the lea is broken from a point in the weakest region, the maximum force applied is noted in kilograms or pounds. The strength of at least 10 leas of the same count is measured by using the above method from which the mean is calculated. The British Standard determines a hank or lea of 100 wraps of 1 m distance across. This is tested at a certain rate, namely, up to the point at which it breaks within 20 ± 3 s. Alternatively, a consistent velocity of 300 mm/min may be used. On the off chance that the yarn is spun on the cotton, worsted frameworks of 10 skeins ought to be tested with 20 leas for woolens. The strategy is not utilized for persistent fiber yarns.

5.4.2.1 Count Strength Product or CLSP

The check quality item (CLSP) is a term utilized for staple spun yarns of cotton and the lea (hank) quality. It depends on determining the quality of the lea made on a wrapping reel having a circumference of 1.5 yards; a 80 turns lea has a total length of 120 yards. The quality is commonly noted in pounds force (lbf). This strength, measured in pounds, is then multiplied with the English cotton count of that yarn to get the CLSP.

$$\text{CLSP}(\text{Count Lea Strength Product}) = \text{Count}(\text{Nec}) \times \text{Leastrength}(\text{in pounds})$$

5.4.3 Tensile Strength by Single Yarn Method

The tensile testing of yarns is commonly done in accordance with ISO 2062 and ASTM D2256. These tests are used to determine the breaking force, elongation, and toughness properties of the yarn. The breaking tenacity, a ratio of the breaking force to yarn linear density, is also a common property for evaluating the strength of a yarn's material and for comparison and validation purposes. It is necessary to clamp the yarn test specimen so that the machine loading axis is aligned with the specimen axis. This alignment is most easily achieved and repeatable using capstan style grips. Sharp edges or changes in the path can cause specimen failure to occur outside the gauge section and far below the actual strength of the yarn. Capstan grips also help to avoid this, as the yarn never encounters sharp changes in geometry. Since elongation properties are important for the product application of yarn materials, it is necessary to prevent slippage of the yarn during testing, which is accomplished through the even distribution of the load over the capstan instead of using only a set of clamp jaws [5].

Mostly the test standards are similar. In order to obtain more precision in results, the tests are performed many times so that accurate results are obtained. According to British standards the following number of tests should be performed:

- For a single yarn
  For continuous filament yarns perform 20 tests
  For spun yarns perform 50 tests
- For cabled and plied yarn perform 20 tests

5.4.3.1 Test Procedure

Before the start of the test, the atmospheric conditions of the laboratory should be maintained according to standard. The settings of the machine should also be accurate and meet the demands of the standard. Mostly the USTER TENSORAPID/USTER TENSOJET testing machine is used for this purpose. The gauge length for the test is 500 mm and pretension is set to 0.5 cN/Tex.

First of all the conditioned yarn is fixed into the USTER TENSORAPID/ USTER TENSOJET and is adjusted between the two jaws of the machine, one of which is movable and the other stationary. It works at a speed of 5000 mm/s and the gauge between the two jaws is 500 mm. The machine is turned on and the test is started. The tests are performed automatically and stop after 20 have been completed. After completion the result is printed, which gives the value of the tensile strength and its coefficient of variation [5].

We require a large number of tests to be performed in less time and with higher efficiencies and accuracy levels. To meet this requirement USTER Technologies produces the USTER TENSORAPID/USTER TENSOJET testing machine, which is frequently used for measuring the single yarn strength. USP or USTER statistics enable us to compare the results of single yarn strength, whether they fall within an acceptable range or not. Mostly, the mean strength is not so important, though the frequency of any weak place is. Due to these weak places, yarn breakage occurs during subsequent processes of weaving and causes low production efficiencies or fabric faults that must be avoided to obtain high quality and more production. In high speed production the weak places still cause problems even if they occur after hundreds of meters. Therefore, in such cases, the coefficient of the strength of a single yarn is of greater importance than the mean value.

In order to check greater lengths of yarn the speeds of the machine are kept higher, otherwise the tests would take longer if the standard test time of 20 s was used. The greater the number of tests, the better will be the statistical prediction of weak spots and more precise results of tensile strength will be obtained.

Parameter	Unit	Description
B-Force	cN	The maximum tensile force measured, called the breaking force.
Elongation	%	Elongation at maximum force, called the breaking elongation.
Tenacity	cN/tex	Breaking force divided by the linear density of the yarn.
B-Work	cN × cm	Work done to break the yarn.
Max values		Maximum values denote the maximum value of force, elongation, tenacity, or work within one test series.
Min values		Maximum values denote the minimum value of force, elongation, tenacity, or work within one test series.
Percentile values, e.g., P. 0.01		0.01%, 0.05%, 0.1%, 0.5%, and 1.0% of all measurements are below the reported value.

5.5 Evaluation of Yarn Twist

"Twist" could be elaborated as the spiral disposition of the components of a thread as a result of the relative rotation of the two ends. The presence of twist in a yarn binds the fibers together and helps to keep them in their corresponding positions. It provides coherence between the fibers and adequate strength to the yarn. Twist is also imparted to create different effects that are highly visible when fabric is manufactured from this yarn. The effects are attained by combining yarns with different twist levels and twist directions in the fabric. Twist is usually expressed as the number of turns per unit length of yarn, for example, turns per inch or turns per meter.

5.5.1 Effect of Twist on Strength of Yarn

5.5.1.1 Short Staple Yarn

Twist is imparted into a staple yarn to bind the fibers together and provide coherence to give strength to the yarn. Firstly, with the increment of twist, the lateral force holding the fibers also increases, which allows more fibers to contribute to the yarn *strength*. But secondly, when twist increases from the optimum level, the angle of fibers made with the axis of the yarn also increases, which makes the fibers contribute less to the yarn strength. The influence of increasing twist on the strength of a staple spun yarn is shown in Figure 5.2.

5.5.1.2 Continuous Filament Yarns

In the case of filament yarns, a small amount of twist is needed to hold the filament fibers together. Therefore increase in twist will decrease the strength of filament yarn. This is due to the fact that the filaments are stronger than

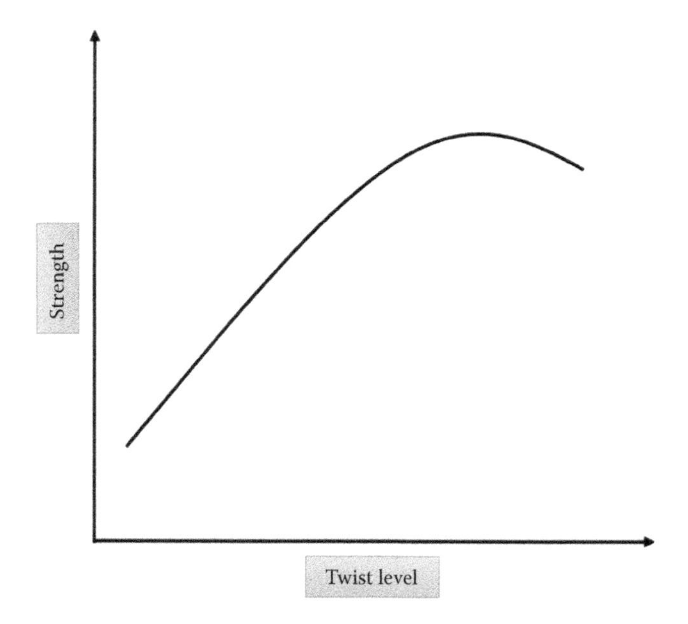

FIGURE 5.2
Strength of yarn versus twist level (staple spun yarn).

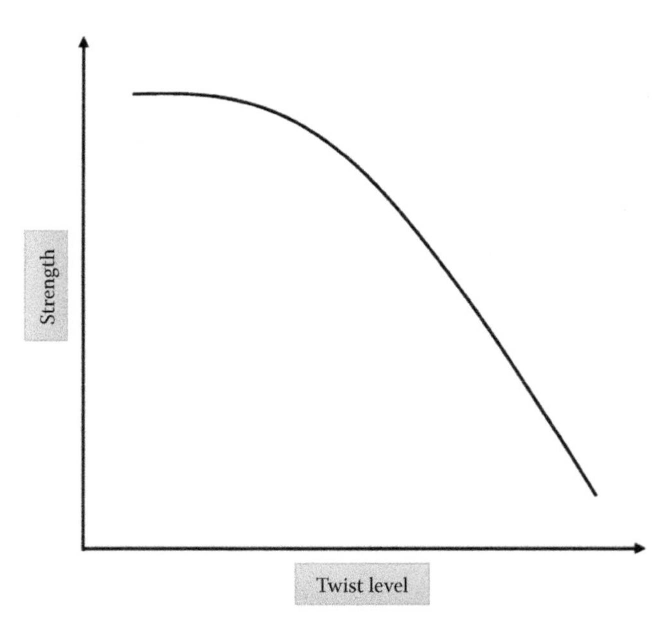

FIGURE 5.3
Strength of yarn versus twist level (filament yarn).

short staple fibers, so less twist is needed to impart strength in the case of filament yarns. Theoretically the maximum strength of a continuous filament yarn is possible when the filaments are oriented parallel to the axis of the yarn. Since the filaments are of variable strength, so the primary function of twist is to provide support to weaker filaments thus resulting in an increase in the yarn strength. The influence of increasing twist on a continuous filament yarn is shown in Figure 5.3.

5.5.2 Effect of Twist on Yarn and Fabric Properties

The twist level has an effect on the properties of yarn as well as fabric. The following parameters are affected by twist:

- Hand feel
- Moisture absorption
- Wearing properties
- Aesthetic effects
- Moisture wicking
- Air permeability
- Luster

5.5.3 Twist Directions

5.5.3.1 Z-Twist

When a twisted yarn is held vertically and the individual filaments appear as the diagonal in the letter "Z," then it is called a "Z–twist." Similarly, when several yarns are combined and given a Z-twist then the individual yarns appear as the diagonal in the letter "Z."

5.5.3.2 S-Twist

When a twisted yarn is held vertically and the individual filaments appear as the diagonal in the letter "S," then it is called an "S–twist." Similarly, when several yarns are combined and given an S-twist, then the individual yarns appear as the diagonal in the letter "S." S and Z twisted yarns are shown in Figure 5.4.

5.5.4 Types of Fabrics with Respect to Different Types of Twist

Soft surfaced fabrics have slack twist.

- Smooth surfaced fabrics have optimum twist. This stimulates strength, smoothness, and elasticity.
- Crepe fabrics have the maximum number of twists.
- Poplins have two single yarns that are Z-twisted individually and plied together using an S-twist.
- Sewing thread has three S-twisted single yarns, which are then Z-twisted together. Tear-resistance of this thread will be higher.

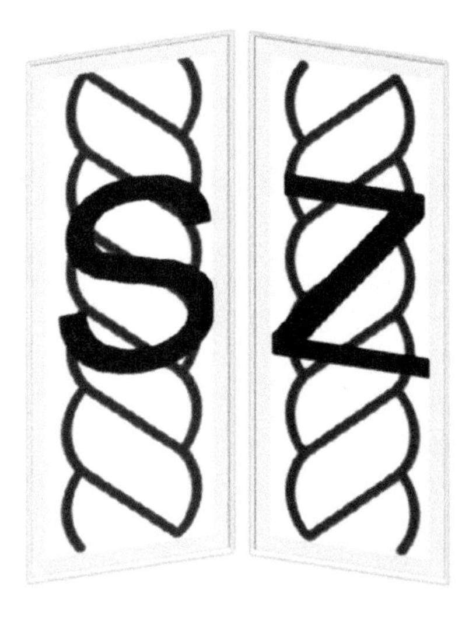

FIGURE 5.4
Illustration of Z and S twist directions.

5.5.5 Twist Measurement

5.5.5.1 Simple/Direct Counting Method

This test is performed at a temperature of 20°C ± 2°C and a relative humidity of 65% ± 4% [1].

5.5.5.1.1 Sampling

Twenty-five meters of yarn is discarded from every package and each sample is withdrawn in the direction normally used. The most preferable method is the side end withdrawal because the over end withdrawal inserts a twist in the yarn. Twenty-five specimens of spun yarn and eight specimens of filament yarn are referred to in ASTM D1423-99 [6].

5.5.5.1.2 Spun Single Yarn

A moveable clamp gauge length is set as long as convenient but should be less than the staple length of the fiber. The counter is set at zero and a tension of 0.25 ± 0.05 cN/tex is maintained. Twist is removed completely by turning the rotatable clamp until the fibers in the yarn become parallel. The number of rotations for untwisting a yarn gives the twist in a specific length of clamp. The number of turns is counted and the turns per unit length are calculated.

5.5.5.1.3 Filament Single Yarns

The clamp gauge length is adjusted to 250 ± 0.5 mm or 10 ± 0.02 in. The counter is set at zero and the specimen is mounted in the clamp at a tension of 0.25 ± 0.05 cN/tex and from both free ends. Twist is removed completely by rotating the clamp until the mono filaments in the strand become parallel and the parallelism is ensured by passing the needle through the strand. The number of rotations for untwisting the yarn gives the twist in the specific length of the clamp. The number of turns is counted and the turns per unit length are calculated.

5.5.5.1.4 Folded Yarn

Five samples are referred to in ASTM D1423-99 for folded yarn twist measurement. The clamp gauge length is adjusted to 250 ± 0.5 mm or 10 ± 0.02 in. The counter is set at zero and the specimen is mounted in the clamp at a tension of 0.25 ± 0.05 cN/tex and from both free ends. A specified length of specimen is mounted in a twist device [6]. One end is rotated until all the strands are free of twist. The number of turns is counted and the turns per unit length are calculated.

5.5.5.2 Continuous Twist Tester

The continuous twist tester is designed to increase the number of tests performed per unit time. Yarn passed from a rotating jaw end is wrapped on a rotatable drum. Twist is measured by untwisting and twisting a specific length but, after removal of the twist, it is imparted back onto the yarn. By this method an instrument measures the twist per unit length of yarn.

5.5.5.3 Untwist and Retwist Method

When the level of twist increases, the length of yarn is contracted, and when the twist is removed, the length is increased; if all the twist is removed then the length reaches its maximum value. This method is used on equipment in which one end of the yarn is attached to a counter and the other is attached to a weight-pointer. When the yarn is untwisted, the pointer identifies the slight change in length. When the yarn is untwisting, the length of yarn is increased and the pointer moves from right to left; and when all the twist is removed then the length of yarn is at the maximum and the pointer does not move further from right to left but the rotating jaw continuously rotates in the same direction; further rotation causes length contraction due to twist insertion, at that point, before length contraction, the untwisted twist is the yarn twist. The level of twist is indicated by the instrument.

5.5.5.4 Automatic Twist Tester

An automatic twist tester (Zweigle D302 and USTER ZWEIGLE TWIST TESTER-5) makes the largest number of tests for measuring the level of twist. These automatic twist testers are also governed by untwist–twist methods for measuring the twist levels with a special tensioning system.

5.6 Evaluation of Evenness of Yarn

This could be defined as the variation in a yarn's weight or thickness per unit length. Evenness of yarn is measured by the following methods.

5.6.1 Visual Examination

In this method, yarn evenness is checked by wrapping it onto a black board in uniformly spaced turns to reduce the effect of optical illusions caused by irregularity. These boards are then checked under proper lighting using a uniform and unidirectional light. Normally visual examination is done without any comparison with a standard, though comparison could also be made with the ASTM standard if it is available. Nowadays more uniformly spaced yarn boards are prepared with the help of motorized wrapping machines. By way of these wrapping reels the yarn moves gradually along the tapered black board as it is revolved. Tapered black boards are preferred for evaluating or determining periodic faults. If there are periodic faults in the yarn they produce a woody pattern, which is clearly visible. This visibility of the yarn faults on the tapered boards is due to the equal spacing of the yarns on the board [2].

5.6.2 Cut and Weigh Method

For determining the mass variation of yarn, the cut and weigh method is considered to be the simplest. In this method consecutive lengths of yarn are cut and weighed. For this testing, we need a precise way of cutting the yarn as all the lengths should be same. The small error in cutting the lengths of yarn results in wrong and inaccurate measurements. To avoid this problem, the yarn is wrapped around a grooved rod with a circumference of exactly 2.5 cm. Then yarn with equal lengths of 2.5 cm is cut by running a razor blade along the groove. These lengths of yarn are then weighed on a sensitive weighing balance. By plotting the mass of each length, we produce the graph shown in Figure 5.5.

The visual indication of unevenness of yarn can be found by plotting the line showing mean value and thus comparing the scatter of individual readings. A mathematical measurement is also necessary for this unevenness of yarn, which can be expressed in the following two ways.

The first term is used by USTER Technologies to designate U%. The average value for all the deviations from the mean is calculated and then expressed as a percentage of the overall mean, which is known as the percentage mean deviation (PMD). The value of the standard deviation is calculated by squaring the deviations from the mean, which is then expressed as a percentage of the overall mean. The deviations having a normal distribution about the mean are correlated as [2]:

$$\mathrm{CV} = 1.25\,\mathrm{PMD}$$

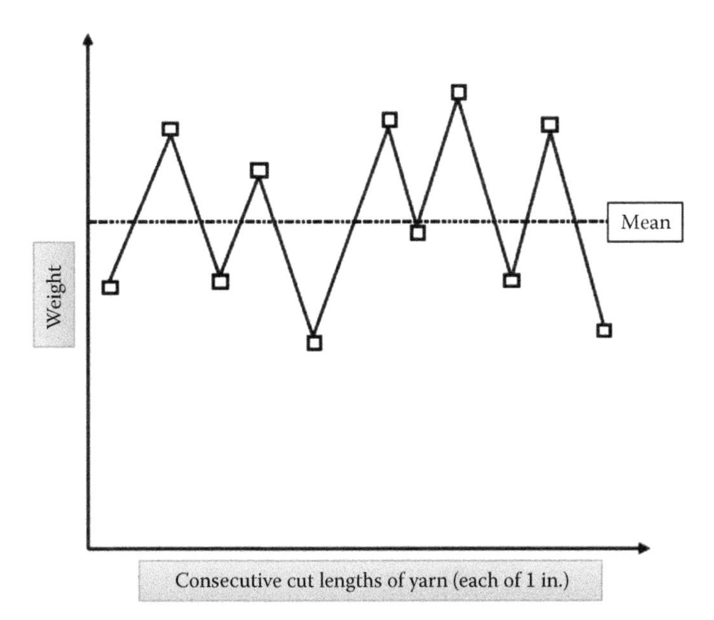

FIGURE 5.5
Variation in mass per unit length of a yarn.

5.6.3 USTER Evenness Tester

The evenness tester of USTER Technologies finds the variations in thickness of a yarn by using capacitive techniques. The yarn to be examined is passed through a pair of parallel plates of a capacitor whose capacitance is continuously measured electronically. The presence of yarn between the capacitor's plates continuously changes the capacitance of the system. The capacitance depends on the mass of yarn between the plates and the type of raw material used. For the same dielectric constant, the signals are directly related to the mass of yarn present between the capacitor's plates. To get the same relative permittivity for yarn, it should be made up of the same type of fiber and it must have uniform moisture content throughout its length. The varying moisture content, or an uneven blend of two or more fibers, will vary the dielectric constant in different parts of the yarn and this variation will be signaled as unevenness. The readings made by the USTER tester are equivalent to weighing successive 1 cm lengths of the yarn [2,7].

The following can be possible reasons for yarn unevenness:

- The number of fibers in the yarn's cross-section is not constant but varies widely depending upon the fiber parameters. This is the most significant reason of yarn unevenness.
- The staple spun yarn is made up of natural fibers having a variable fineness. This variation leads to a difference in yarn thickness even when the number of fibers in the cross-section remains the same.

Parameter	Unit	Description
Um	%	The mean linear irregularity, but now CVm is more preferable.
CVm	%	The coefficient of variation of the yarn mass.
CVm (L)	%	The coefficient of variation of the yarn mass at cut lengths of 1, 3, 10, 50, and 100 m.
Mass deviation		m(max) = maximum mass m(min) = minimum mass cut lengths for the calculation are 1, 3, 10 m, or Half Inert.
Index		The ratio between the ideal and actual evenness of staple fiber strands.
Imperfections (IPI)		Summation of thin places, thick places, and neps at certain sensitivity levels. Generally, thin places: –50%, thick places: +50%, neps: +200%.
Rel. count	%	Count deviation relating to the length of yarn tested, the mean corresponds to 100%.
Abs. count	e.g., tex, Ne	Linear density of the yarn unit length (yarn count).

(*Continued*)

Parameter	Unit	Description
Hairiness		The hairiness H corresponds to the total length of protruding fibers divided by the length of the sensor of 1 cm. The hairiness is, therefore, a unit-less quantity.
sh		Standard deviation of hairiness.
sh (L)		Standard deviation of hairiness at cut lengths of 1, 3, 10, 50, 100 m.
Hairiness deviation		h(max) = maximum hairiness h(min) = minimum hairiness cut lengths for the calculation are 1, 3, 10 m
2DØ	mm	Mean value of the two-dimensional diameter over the measured yarn length.
s2D8 mm	mm	Standard deviation of the diameter over the reference length of 8 mm.
CV2D8 mm	%	Coefficient of variation of the diameter over the reference length of 8 mm.
CV2D0.3 mm	%	Coefficient of variation of the diameter over the reference length of 0.3 mm.
CV FS	%	Coefficient of variation of the fine structure, assessment of short-wave variations.
CV1D0.3 mm	%	Coefficient of variation of the one-dimensional yarn diameter, related to 0.3 mm.
Shape		Nondimensional value between 0 and 1, which describes the roundness of a yarn (1 = circular, 0.5 = elliptical).
D	g/cm^3	Mean yarn density related to the nominal count.
Trash count	Cnt/km or yd	Trash particles per km or yard > 500 μm.
Trash count spec.	Cnt/g	Trash particles per gram > 500 μm.
Dust count	Cnt/km or yd	Dust particles per km or yard > 100–500 μm.
Dust count spec.	Cnt/g	Dust particles per gram > 100–500 μm.
Trash size	μm	Mean trash particle size.
Dust size	μm	Mean dust particle size.

5.7 Classification of Yarn Faults

Broadly we can classify yarn faults into the following categories:

- Imperfection index (IPI)
- Coefficient of variation of mass (CV%)
- Hairiness index
- Periodic faults
- Classimate faults

5.7.1 Imperfection Index (IPI)

Broadly there are three basic subcategories of the imperfection index that are given here:

1. Thick place (+50%)
2. Thin place (−50%)
3. Neps (+200%)

These are the faults of the yarn, and excessive numbers of thick places, thin places, or neps cause the different types of faults that affect the fabric quality and appearance. Loom efficiency is also decreased as IPI causes excessive breakage during the warping process. These faults are classified on the basis of yarn diameter, which can be calculated by the formula:

$$\text{Yarn diameter} = \frac{1}{\sqrt{28}\,\text{count}}\ \text{inch}$$

Yarn diameter can also be calculated by optical or capacitive method. The USTER Tester 5 uses the capacitive method to determine the yarn diameter [8]. If the diameter of any place along the length of the yarn is 50% higher than the actual diameter of the yarn then it is considered as a thick place, and if the diameter is 50% less than the yarn diameter then it is counted as a thin place. If the diameter of any place is so high that it becomes 200% of the yarn diameter then such a place is considered to be a nep of yarn. The IPI tells us about the evenness of the yarn; the greater the value of the IPI, the greater will be the evenness. The IPI of the yarn is determined by following the standard test method ASTM D 1425 [7]. The USTER Tester is used to determine the unevenness of yarn and is based on the capacitive method. In this method, the yarn is passed through two capacitive plates and on the basis of capacitance changes the diameter of the yarn passing through the plates is analyzed [7].

5.7.2 Coefficient of Variation of Mass (CV%)

CV percentage can be defined as the standard deviation expressed as a percentage of average:

$$\text{CV\%} = \frac{\text{SD}}{\text{Average}} * 100$$

$$SD = \sqrt{\frac{\Sigma(x - \text{average})^2}{n - 1}}$$

where
x are the individual values of mass of specific length over which unevenness is being measured
n is the number of readings

ASTM D 1425 also determines the CV% of the yarn.

5.7.3 Hairiness Index (H)

This is the accumulated length of hairs in centimeters in the unit length. It is denoted by H and has no unit because it is the ratio of two lengths:

$$\text{Hairiness index (H)} = \frac{\text{Total length of all protruding fibers (cm)}}{\text{one cm of the yarn}}$$

The hairiness range for the traditional yarn is 2–12. In the case of finer yarn, there are less fibers per unit cross-section and normally longer fibers are used so they have less tendency to move outward, which results in low hairiness. Normally ring spun yarn has a higher H value than air vortex and rotor yarn because of the friction that occurs between the traveler and the yarn and also the higher mobility that is observed. There are many techniques by which hairiness can be measured, such as optical, capacitive, and image analyses as well as the theoretical method. In most cases, hairiness is measured by the optical method, normally an USTER or Zweigle Hairiness Tester [9,10].

5.7.4 Periodic Faults

These are the faults that arise periodically and appear as a wood pattern or as stripes or streaks in the fabric. The appearance of periodic faults in yarn is shown in Figure 5.6. After dyeing such faults result in shade variation.

5.7.5 Classimate Faults

On the base of the USTER Classimate, yarn faults are classified into the following categories, most of which can be removed with the help of any yarn clearer in the auto cone department during the winding process.

- Neps
- Short thick
- Long thick or double yarn
- Long thin

FIGURE 5.6
Appearance of the yarn periodic faults in the fabric. (Courtesy of Uster Technologies AG, Uster, Switzerland.)

Normally neps have a length of 0.1–1 cm and a diameter of about +420%. A short thick place appears as a very small fault and covers the length of 1–4 cm with a diameter range of +150% to +400%. Long thick places have a diameter range of about +50% to +200% with a length of 8–38 cm. Double yarn is also produced by doubling the material at any stage during yarn manufacturing as a doubling of roving or sliver. This fault is known as a double yarn. A long thin place and a long single yarn are also produced. A long single yarn is the yarn whose diameter is half of the main yarn. The cross-section of the thin yarn is –30%, –45%, and –75%.

5.7.6 Yarn Slub Test

Slub yarn or fancy yarn have a variety of applications in textiles. All the slubs are calculated on the base of the reference level, which is on the level of the base yarn. The mass increase is the amount of mass that is increased from the base level/reference level, which is 0%. The maximum length of the slub at the bottom is named the slub length and the distance between the two consecutive slubs is known as the slub distance. The ratio T/P (top length to bottom length) is also a very importance factor and it tells us about the steepness of the slubs present in the yarn. We can produced a fancy yarn with two or more populations, this means that there are at least two different slub sizes regarding for example different mass increases or slub lengths or a combination of both in a slub yarn. A mass decrease of a slub is an important quality parameter because mass decrease before and after a slub produces weak places in the yarn, so it is very important to maintain the setting of the spinning machine or the slub yarn device. The count of the fancy yarn is normally given as the overall nominal count that describes the weight per unit length. It is also possible to describe the base yarn count and slub count separately. A yarn slub test is performed by the USTER Tester 5 and works on the capacitive principle in which the diameter of the yarn passing through the dielectric plates is determined by the change in the capacitance [8,11].

5.8 Yarn Friction

The frictional properties of textiles are of enormous interest to both manufacturer and end user. Yarn friction is interlinked with the friction between fibers. There are number of stages in yarn manufacturing where yarn comes under the influence of friction, which includes drafting and twisting, yarn clearing and winding. The physical (touch and feel) and mechanical (strength and elongation) properties of textiles are largely affected by yarn friction.

Yarn friction normally depends on many variables, the major ones being the type of fibers, level of twist, and methods of yarn spinning. Yarn friction is also changed by yarn tension and the speed of yarn.

Mostly yarn to yarn friction and yarn to other surface friction, like rubber and metal, are active during the formation of yarn. By reducing the friction in the above mentioned situations, we can reduce the breakage of yarn [12].

5.8.1 Measurement of Yarn Friction

The friction of any object is mainly governed by two parameters: the force acting on the object, and the contact force between the object and the other surface. When yarn comes into contact with another surface, the force that opposes the movement of the yarn is frictional force. Both static and dynamic friction are associated with yarn.

The Zweigle Friction Tester 5 by USTER is used to measure yarn friction. It works on the principle of classical friction, according to which the coefficient of friction can be calculated by sliding an object over another surface [8,12].

Consider A1 and A2, the two surfaces shown in Figure 5.7. The coefficient of friction of these depends upon their surface roughness and can be calculated as

$$F2 = \mu F1$$

where

F1 is the perpendicular force on the object (N)
F2 is the frictional force (N)
μ is the coefficient of friction

In the Zweigle Friction Tester 5, yarn is subjected to move through a disk tensioner horizontally, as shown in Figure 5.8. The disk consists of two plates. Force is applied on the yarn by pressing the upper plate of the disk. So the coefficient of friction between the yarn and the other surface, which is metal here, can be calculated by the above expression.

This friction tester is also capable of measuring yarn friction by the principle of tensioner friction. In the tensioner principle, the tester has a sensor with two rollers. Firstly, the zero force area is established between these two rollers. Yarn is allowed to pass in one direction and then the other with a speed of 200 m/min as shown in Figure 5.9.

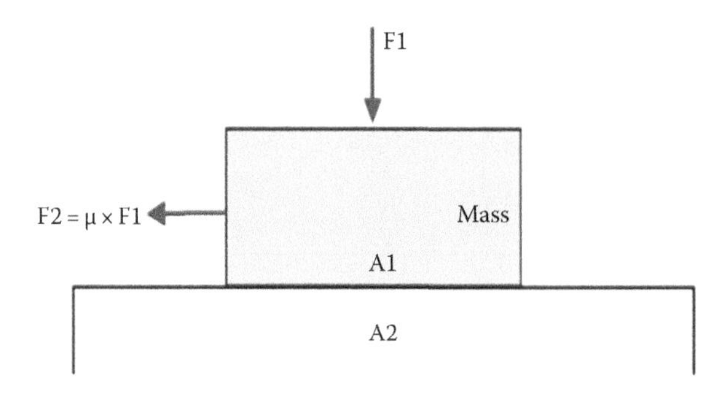

FIGURE 5.7
Classical friction principle. (Courtesy of Uster Technologies AG, Uster, Switzerland.)

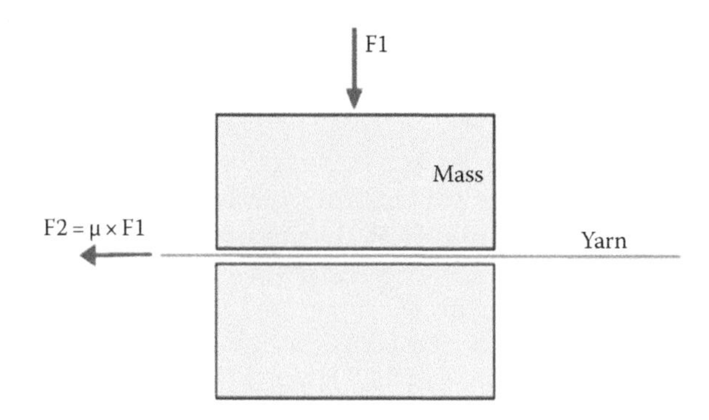

FIGURE 5.8
Friction measurement system by USTER Zweigle friction tester 5. (Courtesy of Uster Technologies AG, Uster, Switzerland.)

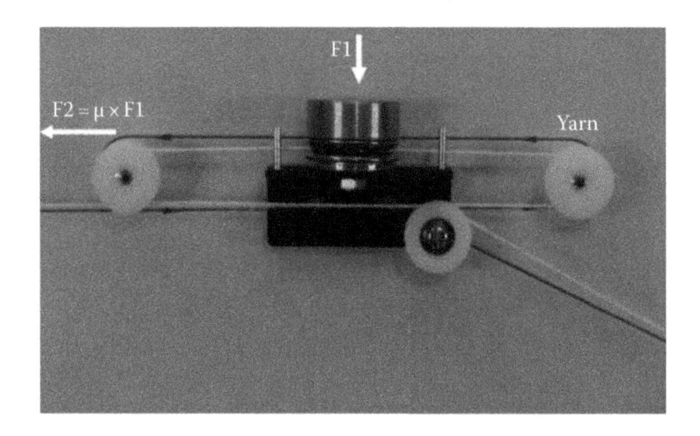

FIGURE 5.9
Friction tensioner measuring system by USTER Zweigle friction tester 5. (Courtesy of Uster Technologies AG, Uster, Switzerland.)

The yarn tensioner is loaded with a 20 CN load to apply the force on the yarn. The rollers are made sensitive to measure each lateral movement of the yarn. Rollers are connected with the sensor that measures the variation of force applied to the yarn. As F1 applies a known force on the yarn, and F2 is measured from the sensor, so the coefficient of friction can be calculated from the above expression [13].

5.9 Yarn Diameter

The dimensional stability of textiles is significantly affected by yarn diameter from which these are weaved or knitted. The spacing between the warp and weft or the course and wales is changed due to the variation in yarn diameter. Thermal conductivity, air permeability, moisture permeability, and the cover factor of fabric is greatly influenced by yarn diameter [14].

Yarn diameter can be altered by the number of fibers, fiber fineness, the density, and the amount of twist given to yarn. Yarn diameter is mainly also the function of the manufacturing technique used to form it. For the same linear density, the diameter of yarn is different for different yarn spinning methods.

5.9.1 Measurement of Yarn Diameter

The USTER tester 5 has an optical measuring module to measure the diameter of yarn. Yarn diameter is measured more accurately by optical means in this module. Yarn is made to pass through two parallel beams of light, as shown in Figure 5.10. The yarn travels through an optical field so as to be

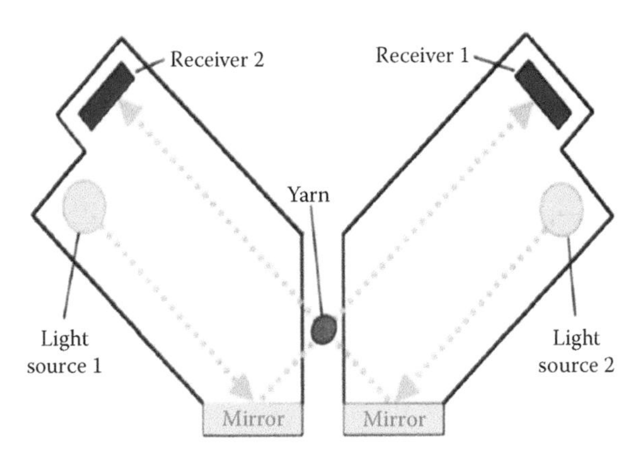

FIGURE 5.10
USTER tester 5 OM sensor. (Courtesy of Uster Technologies AG, Uster, Switzerland.)

illuminated by two parallel beams of light. These beams are adjusted to be perpendicular to each other. So the diameter of the yarn is calculated from the fluctuation of the light intensity after passing through a cross-section of yarn. As the measurement is done without pressing the yarn so optical measurement has more accuracy than other methods. It also can be used for all types of conductive and non-conductive materials [8,14].

References

1. ISO 139:2005, Textiles—Standard atmospheres for conditioning and testing. International Organization for Standardization, Geneva, Switzerland.
2. B.P. Saville, *Physical Testing of Textiles*. Woodhead Publishing Limited, Cambridge, England, 2000.
3. P. Schwartz, *Structure and Mechanics of Textile Fibre Assemblies*. Woodhead Publishing Limited, Cambridge, England, 2008.
4. American Society for Testing and Materials ASTM D 1907-01, Standard test method for linear density of yarn (yarn number) by the skein method. ASTM, West Conshohocken, PA, Vol. 7, pp. 1–9, 2000.
5. American Society for Testing and Materials ASTM D 2256-97, Standard test method for tensile properties of yarns by the single-strand method. ASTM, West Conshohocken, PA, Vol. 7, 1997.
6. American Society for Testing and Materials ASTM D 1423-99, Standard test method for twist in yarns by direct-counting. ASTM, West Conshohocken, PA, Vol. 7, pp. 1–7, 2000.
7. American Society for Testing and Materials ASTM D 1425-96, Standard test method for unevenness of textile strands using capacitance testing equipment. ASTM, West Conshohocken, PA, Vol. 7, pp. 1–5, 1996.
8. S.D. Kretzschmar and R. Furter, USTER® Tester 5-S800 application report the measurement of the yarn diameter, density and shape of yarn, 2009. Available: https://www.uster.com/en/service/download-center/. Accessed June 13, 2016.
9. X.X. Huang, X.M. Tao, and B.G. Xu, A theoretical model of maximum hairiness of staple ring-spun yarns, *Text. Res. J.*, 84(11), 1121–1137, 2014.
10. D. Yuvaraj and R.C. Nayar, A simple yarn hairiness measurement setup using image processing techniques, *Indian J. Fibre Text. Res.*, 37, 331–336, 2012.
11. S. Edalat-pour, USTER® Tester 5-S800 application report measurement of slub yarn part-1/basics, 2007. Available: https://www.uster.com/en/service/download-center/. Accessed June 13, 2016.
12. S.D. Kretzschmar and R. Furter, USTER® Zweigle friction tester 5 application report friction measurement, 2010. Available: https://www.uster.com/en/service/download-center/. Accessed June 13, 2016.
13. H.G. Howell, K.W. Mieszkis, and D. Tabor, *Friction in Textiles*. Butterworths, London, U.K., 1959.
14. J. Matukonis, A. Kauzoniene, and S. Gajauskaite, Frictional interaction between textile yarns, *Mater. Sci.*, 4, 50–52, 1999.

6

Textile Greige Fabrics (Woven and Knitted)

Muhammd Umair, Muhammad Umar Nazir, and Sheraz Ahmad

CONTENTS

6.1 Introduction

Textile fabric is a flexible assembly of fibers or yarns, either natural or man-made. It may be produced by a number of techniques, the most common of which are weaving and knitting. Conventional fabrics (woven, knitted) are produced in such a way that the fibers are first converted into yarn before being converted into fabric. These fabrics are used for a variety of applications from clothing to technical purposes. The primary purpose of textile testing and analysis is to assess textile product performance and to use test results to make predictions about product performance. Product performance must be considered in conjunction with end use; therefore, tests are performed based on the ultimate end use. Different types of testing is done to check the performance of greige fabrics, for example, GSM, tensile strength, tear strength, drapability, and stiffness.

6.2 Classification

6.2.1 Woven Fabric

Weaving is a chain process of textile products that involves the interlacement of two sets of yarn. The interlacing is carried out at right angles to each set. The set of yarn that runs along the length of the product or in a longitudinal direction is termed a warp. The other set that runs in the cross-direction to the warp or at a right angle is termed a weft. The two sets are usually called ends (longitudinal yarn) and picks (transversal yarn). The interlacement method of these two sets will govern the characteristics of the resultant fabric. The type of interlacement of warp and weft yarn is termed the weave design, or weave.

Weaving motions can be divided into three types:

1. Primary motions
2. Secondary motions
3. Tertiary motions

The primary motions are

- *Shedding*. The separation of ends is carried out with the help of a heald frame, which are raised or lowered in a certain pattern so that the pick can pass through and the required weave design could be obtained.
- *Picking*. This is the weft mechanism in which the yarn is inserted in a cross-direction to the warp yarn. This could be achieved with the help of a projecting force, such as solid media (shuttle, projectile, rapier) and air or water.
- *Beating-up*. The newly inserted weft is beaten up with the already beaten up weft yarn so as to weave the fabric.

The secondary motions are

- *Let off motion*. The let off or warp yarns are released at a uniform speed from the beam for continuous weaving of the fabric.
- *Take up motion*. The woven fabric is drawn forward in a uniform way so that the density of pick insertion is properly regulated.

Then come the stop motions, which are also known as tertiary motions. These are of two types:

- Warp stop motion
- Weft stop motion

6.2.2 Basic Types of Woven Fabrics

6.2.2.1 Plain

The *plain weave design* is also called a homespun taffeta or weave design. For this type of weave design, there is a right angle between each warp and weft yarn, and it can be manufactured by two distinct methods: as

1 end × 1 pick and 2 ends × 2 picks. The major portion of the total woven fabric production of the world is composed of plain weave fabric. It is the simplest form of woven fabric and can be produced by the simplest arrangement at the weaving machine. The number of interlacements between warp and weft yarn are the highest in this type of weave design, hence the fabric produced in this way will have the most compact structure. The derivatives of this weave include rib structure (warp or weft) and basket weave structure.

6.2.2.2 Twill

There is distinct difference between the front and back of *twill weave design*. There should be a proper balance between the rows so that the final structure gives a unique slanting appearance. A common example of twill weave design is denim fabric.

6.2.2.3 Satin and Sateen

This type of *weave design* is considered the most popular for woven fabrics. This weave is expensive as compared to other designs as it has some glimmer to it. The satin weave is characterized by four or more fill or weft yarns floating over a warp yarn or vice versa. In warp-faced (satin), warp yarn lies on top of the weft yarn, while in weft-faced (sateen), weft yarn lies on top of the wrap yarns. It shows good luster and excellent extensibility. Some of its derivatives include slipper satin and crepe back satin.

6.2.3 Knitted Fabrics

The second most common method of fabric manufacturing after weaving is knitting. This involves the interlooping of one set of yarn. In terms of production, knitting is the most widely used fabric after weaving. A single set of yarn is provided to a set of needles, and the fabric is obtained by interlooping the vertical set of loops. The term "knitting" is derived from the Saxon word *cnyttan*, which was itself derived from the primeval Sanskrit word *nahyat*. This type of fabric manufacturing can be classified into two types, according to the loop formation direction with respect to the fabric formation direction (Figure 6.1) and the comparison of properties is given in Table 6.1. The global annual production of knitted items is estimated at 7 million tons per annum. The diversity of knitted fabric and knitting techniques results in different shapes and fitting which has made this method very common as well as important. The end use of knitted fabric ranges from apparel to domestic and industrial uses [1].

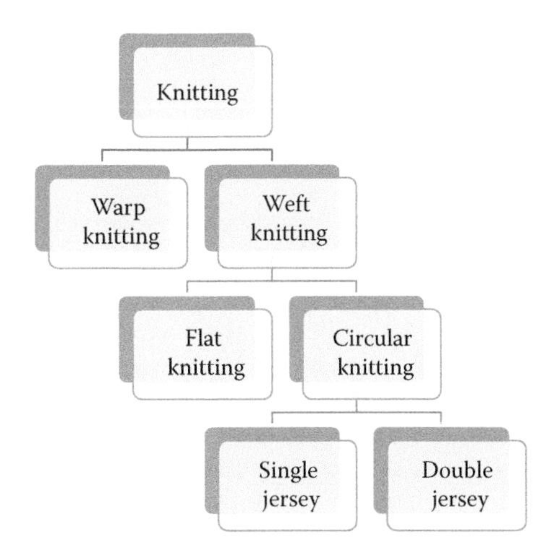

FIGURE 6.1
Types of knitted fabrics.

TABLE 6.1
Comparison of Warp and Weft Knitting

Weft Knitting	Warp Knitting
Plain, rib, interlock, purl	Tricot, raschel, milanese, crochet etc.
Loops are produced along the width of fabric	Loops are produced along the length of fabric
Less production speed	More production speed
Not necessarily each needle has its own thread	Each needle has its own thread
Yarn is supplied from a cone held on creel	Yarn is supplied from a beam
Knitting process can be done from a single yarn	Large number of yarns are required for knitting a fabric
Staple yarn is preferable but filament is also used	Filament yarns are preferable but staple yarns are also used
Less preparatory processes are required	More preparatory processes are required
Latch needle are used in all machines	Bearded needle is mostly used but latch needle can also use in some cases
Less variety of structure can be made	Wide variety of structure can be made
Fabric has less aesthetic value	Fabric has more aesthetic value
Fabric has good stretch ability in both directions, higher in width direction	Fabric has low stretch ability in both directions, higher in width direction
Dimension stability of fabrics is lower	Dimension stability of fabrics is higher
Machines may be flat or circular	Warp knitting machines are generally flat
Width wise more elastic	Length wise more elastic
More shrinkage	Less shrinkage
It may ravel from edges	Does not ravel from edges
Easy snagging	Less snagging

6.2.4 Types of Knitting

6.2.4.1 *Weft Knitting*

In this type of knitting, the direction of loop formation is at right angles to the direction of fabric formation. Normally the fabric is formed vertically and the loops are formed horizontally. It is the most common fabric formation technique for knitted fabric. It is usually knitted with one piece of yarn, and can be made either by hand or using a knitting machine. Weft knitting is the most common form of knitting as it is simpler than warp knitting, the other form of knitting. There are four basic weft knitted fabric structures: interlock, purl, plain, and rib. The action of the needle during loop formation produces all these distinct weft knitted structures. On the basis of the type of weft knitting machine, the weft knitted fabric can be classified as single jersey or double jersey [2].

6.2.4.2 *Warp Knitting*

The second knitting method is termed *warp knitting*, though its share in the production of knitted fabric is low compared to weft knitting but it is used in technical areas. In warp knitting, the yarn runs zigzag along the length of the fabric, i.e., following adjacent columns ("wales") of knitting, rather than a single row ("course"). It requires the preparation of a warp sheet for further use on machine. The most common warp knitted designs or structures are raschel and tricot [3].

The purpose of woven and knitted fabric testing is to check the raw materials, to separate out faulty materials, to monitor production at the different steps, to assess the final product, and to facilitate research and the development of products.

6.3 Fabric Aerial Density/Grams per Square Meter

6.3.1 Introduction

"Fabric aerial density" refers to the mass per unit area of the fabric. It is expressed as GSM (grams per square meter) or ounce per square yard, and grams per linear meter (ounce per linear yard). Fabric aerial density can also be expressed inversely as length per unit weight when the width of fabric is given. For light weight quality, the unit of GSM is preferred, and for heavy weight quality, ounces per square yard are preferred. This is one of the important parameters of fabric quality that categorizes the fabric into light weight, medium weight, or heavy weight quality [4]. The weight of the fabric is directly affected by yarn types and density. The weight of the fabric can be expressed as

$$\frac{g}{m^2} = \frac{10^3 M}{LW} \tag{6.1}$$

$$\frac{g}{m} = \frac{10^3 M}{L} \tag{6.2}$$

$$\frac{m}{kg} = \frac{L}{M} \tag{6.3}$$

where

M is the fabric mass in kilograms
L is the fabric length in meters
W is the fabric width in meters

Or alternatively

$$\text{Mass, } \frac{g}{m^2} = \frac{oz}{yd^2} \times 33.906 \tag{6.4}$$

$$\text{Mass, } \frac{g}{m} = \frac{oz}{yd} \times 31.000 \tag{6.5}$$

$$\frac{m}{kg} = \frac{yd}{lb} \times 2.016 \tag{6.6}$$

6.3.2 Measurement of GSM

The circular cutter die and weighing scale shown in Figure 6.2 is used for the measurement of aerial density.

ASTM D 3776-07 is used to measure the mass per unit area (weight) of the fabric (woven/knitted).

The conditioning of the sample is very important, as it will help to maintain the proper weight of the fabric. The samples are conditioned at a temperature of 21°C ± 1°C and at a relative humidity of 65% ± 2% following the standard ASTM D 1776. To analyze the GSM, a specimen of 13.3 cm diameter is taken [5].

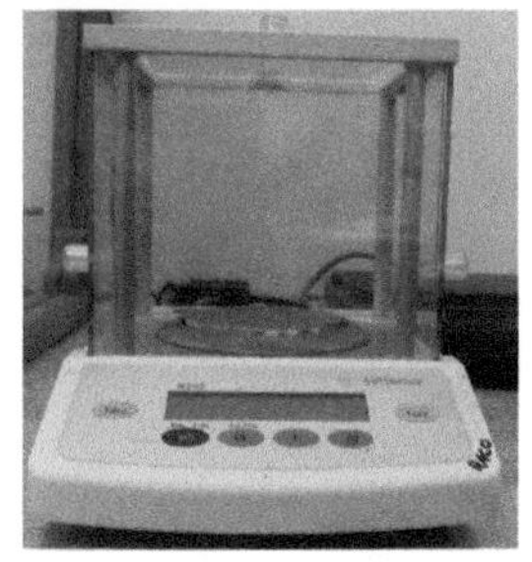

FIGURE 6.2
GSM cutter and weight balance.

The fabric aerial density can be expressed in two ways, either as the "weight per unit area" commonly known as GSM (Gram per square meter) or the "weight per unit length" commonly known as GLM (Gram per linear meter); the former is self-explanatory but the later requires a little explanation because the unit length in the case of knitted fabric will be affected by stitch length and stitch density.

The sample is cut to the desired dimension and placed on the weighing scale. The reading is noted and Equation 6.7 gives the formula used to calculate the GSM of the fabric. There should be 10 repetitions with the average of all the results [6].

$$\text{GSM} = \text{Weight of sample}(\text{grams}) \times 100 \tag{6.7}$$

If there is no cutter, then the GSM of any sample of any size can be found by

$$\text{Weight per square meter}(\text{in gram}) = \frac{\text{Weight of the sample in gram} \times 10,000}{\text{Area of sample in cm}^2} \tag{6.8}$$

6.4 Fabric Tensile Properties

The tensile property is one of the important factors that determines the quality of the fabric. There are different methods employed for the determination of tensile strength of woven and knitted fabric testing which are discussed below.

6.4.1 Tensile Strength of Woven Fabric

The ability of a material to bear an axial load or force is its tensile strength. It is normally expressed as force per cross-sectional area. The tensile property or stress–strain curve is the most commonly measured mechanical property. This type of measurement shows how a material will behave when it is subjected to a tensile pull or force. The breaking load and the elongation at break are the most important information derived from a tensile test. The measurement of tensile properties are usually considered as arbitrary rather than absolute. The results will depend on the geometry, the structure of the fabric as well as the type of fiber used in its construction. The values of tensile strength are more important in the case of brittle material as compared to ductile materials.

6.4.1.1 Tensile Breaks

Tensile breaks can be classified into two types: sharp and percentage.

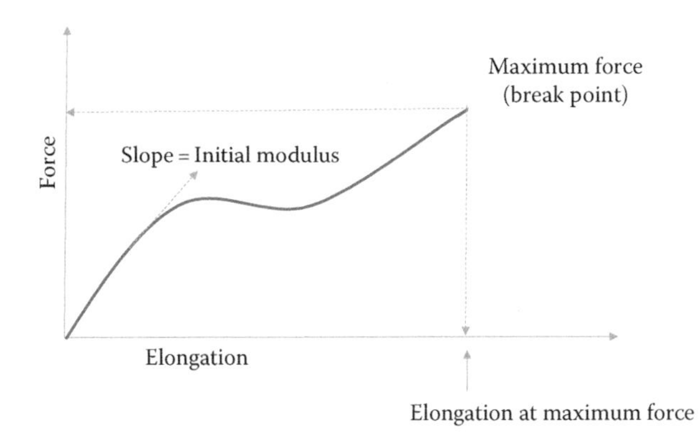

FIGURE 6.3
Sharp break.

6.4.1.1.1 Sharp Break

If the drop in the load is sudden then it will be termed a *sharp break*. This test is normally called a "pull to break." This is shown in Figure 6.3.

6.4.1.1.2 Percentage Break

The gradual reduction in the load from its maximum value as the further extension is achieved is termed the *percentage break*. A percentage drop from the maximum load is often used to define an end point or break point. This is usually termed "the pull to yield" and the same testing setup or mechanism can be employed as in the case of the "pull to break" method. The majority of the test methods will yield both types of information, that is the maximum load as well as the load at break. It is important to note that the breaking strength will not always be the maximum strength of that material, particularly in the case of elastic and soft fabrics. This is shown in Figure 6.4.

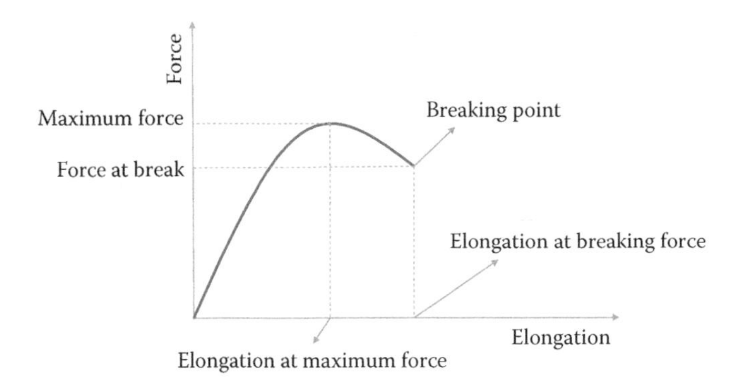

FIGURE 6.4
Percentage break.

6.4.1.1.3 Initial Modulus (IM)

The ratio of tensile stress and tensile strain is termed "Young's modulus" or the "initial modulus" (IM). The materials having a higher value of modulus are termed "stiff" materials. The deformation will be very little as a small amount of stress is applied to those materials. Those having a lower value of modulus are termed "soft." In the case of fabric, the IM is related to the fabric handle. A higher IM will be related to the stiffer or harsher fabric handle or feel, whereas a lower IM will give softer fabric handle or feel.

6.4.1.2 Methods

There are two methods for testing tensile strength, that is, grab test and strip test which are shown in Figure 6.5. These tests are further classified into subtypes:

1. Grab test
   - Grab test
   - Modified grab test
2. Strip test
   - Raveled strip test
   - Cut strip test

In the grab test, the width of the jaws is less than the width of the specimen. When the fabric has a higher thread density or where the pulling of threads

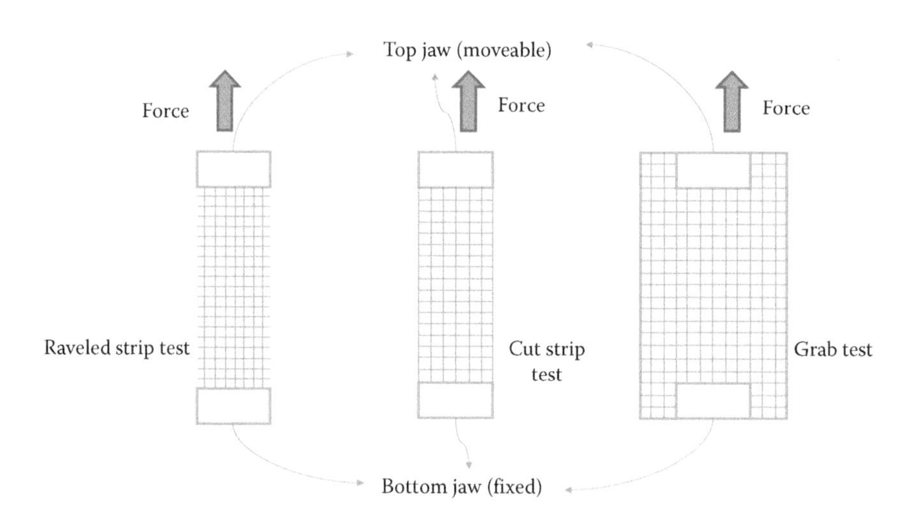

FIGURE 6.5
Strip and grab test methods.

from the edges is not easy, the grab test is used. In order to measure the effective fabric strength, the grab test is the most widely used.

In the modified grab test, lateral slits are made in the specimen to serve all yarn bordering the portion to be strength tested, reducing to a minimum the fabric resistance inherent in the grab method.

A fabric test specimen of a specified dimension is extended at a constant rate until it ruptures. The maximum force and the elongation at maximum force and, if required, the force at rupture and the elongation at rupture are recorded. The atmosphere for conditioning includes a temperature of 21°C ± 1°C and a relative humidity of 65% ± 2% according to ASTM D 1776. It is recommended that samples be conditioned for at least 24 h in the relaxed state. The width of each test specimen should be 50 ± 5 mm (excluding any fringe) and its length should be long enough to allow a gauge length of 200 mm (60 mm × 300 mm). Sample size varies according to the standard being used. Five samples are taken along both sides of the warp as well as the weft for this test.

According to the British Standard adopted for the woven fabric tensile strength, the breaking load and the breaking extension is measured as the fabric specimen (strip) is extended to its breaking point by the use of a suitable mechanical pulling mechanism (see Figure 6.6). The same test methods

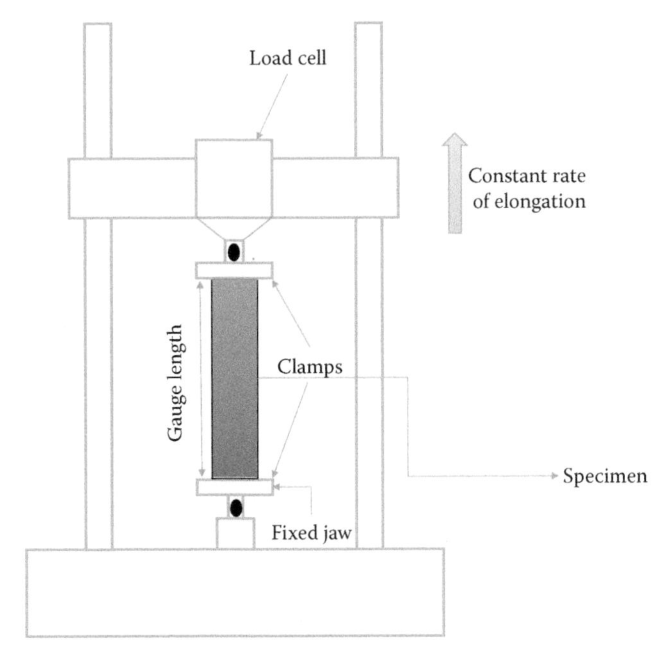

FIGURE 6.6
Apparatus for fabric tensile test.

are repeated five times for warp direction and five times for weft direction. The samples are chosen in such a way that each of them has a distinct set of yarns. The dimensions of the test specimens are 60 mm × 300 mm. The cut specimens are then equally frayed widthwise at both sides to obtain a sample of exactly 50 mm width. This ensures the test specimen will have the threads that will run the length of the specimen and will contribute to the measured strength of the specimen as well as having accurate width. The testing machine should have a constant rate of extension of 20 mm/min. The gauge length between the jaws before the test should be fixed at 200 mm. A pre-tension of 1% will be used before the breaking load is applied. The breaks occurring within 5 mm of the jaws should be rejected, the samples having slippage of 2 mm should also be rejected, and those having improper values should be retested [1].

The strip test is desirable for high-strength fabric. The raveled strip test is only suitable for woven fabric and is so-called because the specimen is prepared by raveling the threads from each side of the specimen until the required width is obtained. The cut strip test is employed for those fabrics from which the threads cannot raveled from the sides, such as knits, nonwoven, felts, and coated fabrics. The test specimens should be prepared accurately by the precise cutting of the sample.

A fabric test specimen is gripped at its center by jaws of specified dimensions and is extended at a constant rate until it ruptures. The maximum force is recorded. The atmosphere for conditioning includes a temperature of 21°C ± 2°C and a relative humidity of 65% ± 2%. It is recommended that samples be conditioned for at least 24 h in the relaxed state.

The width of each test specimen should be 100 ± 2 mm. It should have enough length that the gauge length of 100 mm can easily be achieved. Five samples are taken along both sides of the warp as well as the weft for this test.

According to the British Standard adopted for fabric tensile strength, the test should be repeated five times for the warp direction and five times for the weft direction. The samples should be prepared in such a way that each of them has a distinct set of yarns. This ensures the specimen will have threads that run along its length and contribute to its measured strength as well as having accurate width. The tester should have a constant rate of extension of 50 mm/min. The gauge length between the jaws before the test should be fixed at 75 mm.

The test specimen should be mounted centrally in such a way that the longitudinal center line of the specimen passes through the center point of the front edges of the jaws, is perpendicular to the edges of the jaws, and the line drawn on the test specimen coincides with one edge of the jaws. After closing the upper jaw, pre-tension needs to be avoided when adjusting the specimen along the guide line in the lower jaw so that the fabric hangs under its own weight when the lower clamp is closed. Any rupture occurring within 5 mm of the jaws should be rejected and the result reported as a jaw break; test results that are significantly lower than the average value should be retested [2].

6.4.2 Tear Strength of Woven Fabric

It is a property of textile material to withstand the effect of tearing or resistance to propagation of tear after it has been initiated. Tear strength depends majorly on two factors. The first is the number of yarns that are going to bear the load; as the number of yarns per inch increases, tear strength decreases, and vice versa. The second factor that affects the tear strength of a fabric is the amount of yarn slippage in the fabric structure; this is related to the weave design: fabric with a longer float tends to have a higher tear strength as compared to fabric with a shorter float. For instance, plain weave, which has the maximum amount of interlacement, has the least tear strength because, when the tear force is applied, only one yarn bears the load, whereas in twill weave there is a greater float and when force is applied the yarn starts to slip and work as a unit to bear the load, hence greater force is born by the fabric. Tear strength is also affected by fiber type, yarn type, the GSM of the fabric, and the type of finish applied to the fabric.

6.4.2.1 Measurement of Tear Strength

The apparatus to perform this consists of a fixed jaw, a moveable jaw, a pointer, and a pendulum. This apparatus as a whole is known as Elmendorf Tear Tester. A sample of a specific dimension is placed between the fixed jaw and the moveable jaw. An initial slit along the sample is created either by a cutter mounted on the fixed jaw or by using a manual cutter. The sample is torn between the fixed distances and resistance to tearing is noted. Factors like the scale reading of the instrument and the capacity of the pendulum will affect the tearing resistance of the specimen. The Elmendorf tear tester is shown in Figure 6.7.

This test method is based on the standard ASTM D1424.

The dimension of the testing sample is 100 × 63.5 mm. The pendulum and pointer are brought to their starting position and the sample is placed in the jaws. A slit of 20 mm is created along the width of the fabric. The pendulum is released and the reading is noted. The experiment is performed five times and the average of the five readings is taken [3].

Another method involves the use of a simple tensile testing machine, using the principle of constant rate of extension. The sample is conditioned for 24 h. The testing speed is adjusted to 50 ± 2 mm/min or as described by the standard association. The sample of 75 × 200 mm is cut. The specimens for machine direction measurement will have a longer dimension parallel to the cross-machine direction. A slit of 75 mm is put in the middle of the sample along the machine direction and the tear strength measured. The average of five samples is taken in order to get the best results [7]. The apparatus used for this testing is shown in Figure 6.8. This test method is based on ASTM D 2261.

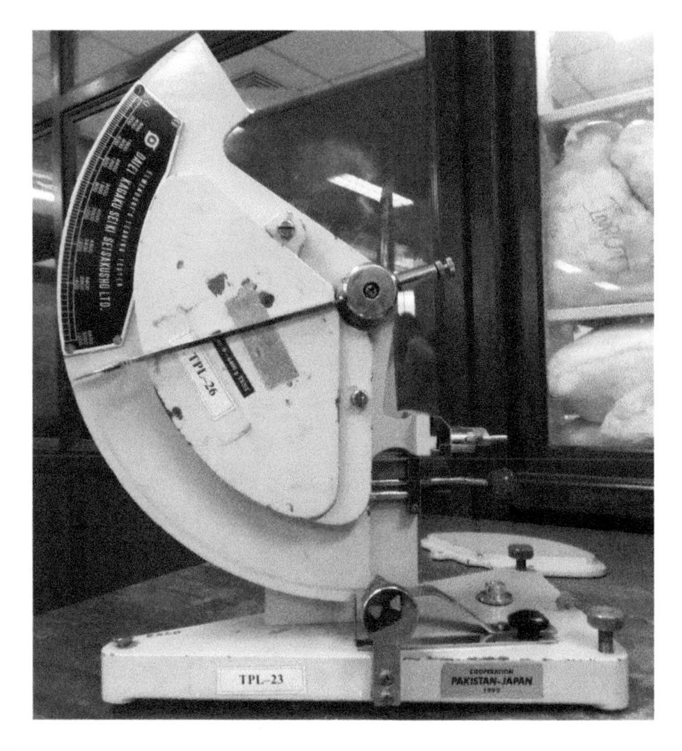

FIGURE 6.7
Elmendorf tear tester.

6.4.3 Bursting Strength of Knitted Fabric

The standard test method employed for knitted fabric bursting strength was issued by the American Standard for Testing and Materials standard, with the last revision made in 2007. This standard was issued under the fixed designation D 6797. There are some other standards too for the measurement of the bursting strength of fabrics. The principle of measurement varies across these different standards, which are set out in the following.

This test method describes the measurement of the bursting strength of woven and knitted fabrics. The measured values are expressed in SI units. The tensile tester is so adjusted that the test is performed according to the standard test method. A test specimen is clamped in a secure manner in such a way that there is no tension on the ball burst attachment of the constant rate of extension (CRE) machine. A force is applied on the specimen with the help of a polished, hardened steel ball until the rupture occurs.

Fabric rolls, pieces of fabric, or cartons of fabric may be used as the primary sampling unit, as applicable. For the laboratory sample at least one full-width piece of fabric that is 1 m in length along the selvage (machine direction) will be taken from the primary sampling unit. The laboratory sampling will be carried out after the removal of the first 1 m from the primary

FIGURE 6.8
Tongue (single rip) tear tester.

sampling unit. A band having a width of 300 mm will be cut in the case of circular knit fabrics. When applicable, the entire fabric component of the fabricated system will be used.

There will be five repetitions for each laboratory fabric sample. The preparation of the specimen can be carried out without the standard atmosphere for testing. For the end-use fabric component of fabricated systems, the specimens should be taken from different areas. That is, if the product is a garment worn on the upper body, then take specimens from the shoulder, shirt tail, shirt back and front, and sleeve. For fabric widths of 125 mm or more, no specimen should be taken closer than 25 mm from the selvage edge. For fabric widths less than 125 mm the entire width of specimens is used. The specimens should be cut to represent a broad distribution diagonally across the width of the laboratory sampling unit. The lengthwise specimens should be taken from different positions across the width of the fabric. The widthwise specimens should be taken from different positions along the length of the fabric.

Ensure specimens are free of folds, creases, or wrinkles. Avoid getting oil, water, grease, and so on on the specimen when handling. If the fabric has a pattern, ensure that the specimens are a representative sampling of the pattern. For test specimens from each laboratory sampling unit, proceed as follows. For fabrics, cut five specimens at least 125 × 125 mm.

The specimens (or laboratory samples) should be brought from the prevailing atmosphere to moisture equilibrium before the testing of these textiles. This can be achieved by keeping the samples in a standard atmosphere for testing as prescribed in ASTM D 1776.

The specimen should be placed in the ring clamp, carefully avoiding any tension. A constant speed of 305 ± 13 mm/min should be maintained on the CRE machine. This speed should be maintained until the specimen is ruptured. The bursting strength of the specimen should be noted to the nearest 5 N (1.0 lbf).

This method is employed to measure the load needed to burst the textile fabric with the help of a steel ball that is forced through the fabric with a constant-rate-of-extension tensile tester. In the case of significant differences in the results of two laboratories, a relative test should be conducted to conclude if there is a statistical bias between them, with the help of competent statistical assistance. The samples used for such a test should be as homogeneous as possible, drawn from the same lot of material as the samples that resulted in disparate results during initial testing, and randomly assigned in equal numbers to each laboratory.

The second test method reveals a measurement of bursting strength with the help of a ball burst strength tester. The measured values are expressed in SI units. The standard test method employed for this type of testing is noted as ASTM D 3787.

A test specimen is clamped between grooved circular plates securely in such a way that there is no tension on the movable clamp of constant rate of traverse testing machine. A force is applied to the specimen with the help of a polished, hardened steel ball that is attached to the fixed clamp of the machine until the rupture occurs.

A constant-rate-of-traverse (CRT) tensile testing machine, with a ball burst attachment replacing the clamp assembly, is used. The movement of the ring clamp pushes the fabric into the clamp against the steel ball, the diameter of which should be 25.400 ± 0.005 mm and spherical to within 0.005 mm. The ring clamp should have an internal diameter of 44.450 ± 0.025 mm.

A lot sample should be taken as per the directions available with the material specification. If there are no such directions, then the rolls should be selected randomly as shown in Table 6.2.

TABLE 6.2

Lot Samples Specifications

Number of Rolls or Pieces in Lot	Inclusive Number of Rolls or Pieces in Lot Sample
1–3	All
4–24	4
25–50	5
Over 50	10% or a maximum of 10 of the rolls or pieces

The fabric rolls, pieces of fabric, or cartons of fabric may be considered as the primary sampling unit, as applicable. For the laboratory sample at least one full-width piece of fabric that is 1 m in length along the selvage will be taken from the primary sampling unit. The laboratory sampling will be carried out after the removal of the first 1 m from the primary sampling unit. A band having a width of 300 mm will be cut in the case of circular knit fabrics. When applicable, use the entire fabric component of the fabricated systems.

There will be five repetitions from each laboratory fabric sample. The dimensions of the specimen should be 125 mm^2 or in the case of a circular sample it should have a diameter of 125 mm. Each sample should consist of a distinct set of warp and weft yarn so that it will be a true representative of the whole fabric width. The samples should not be cut near the selvage but to about one-tenth of the fabric width. This limitation is not applicable to circular knitted fabrics.

The specimens (or laboratory samples) should be brought from the prevailing atmosphere to moisture equilibrium before the testing of these textiles. This can be achieved by keeping the samples in a standard atmosphere for testing as prescribed in ASTM D 1776.

The specimen should be placed in the ring clamp carefully and secured with the help of a lever or screw device. A constant speed of 305 ± 13 mm/min should be maintained on the CRT machine. This speed should be maintained until the specimen is ruptured. The bursting strength of the specimen should be noted to the nearest 5 N (1.0 lbf).

The third method to measure the fabric bursting strength involves the use of a hydraulic or pneumatic diaphragm bursting tester. A wide variety of textile products are tested by this method. This test can also be used for stretch woven and woven industrial fabrics such as inflatable restraints. A specimen is clamped over an expandable diaphragm, which is expanded by fluid pressure to the point of rupture. The difference between the total pressure required to rupture the specimen and the pressure required to inflate the diaphragm is reported as the bursting strength.

A sufficient amount of pressure is used to minimize the slippage of the specimen. The upper and lower clamping surfaces should have a circular coaxial aperture of 31 ± 0.75 mm in diameter and should be durable with any edge, which might cause a cutting action being rounded to a radius of not more than 0.4 mm. The lower clamp should be integral with the chamber in which a pressure medium inflates the rubber diaphragm. This can be achieved by two methods: hydraulic and pneumatic.

The fabric rolls, pieces of fabric, and cartons of fabric may be considered as the primary sampling unit, as applicable. For the laboratory sample at least one full-width piece of fabric that is 1 m in length along the selvage will be taken from the primary sampling unit. The laboratory sampling will be carried out after the removal of the first 1 m from the primary sampling unit. A band having a width of 305 mm will be cut in the case of circular knit fabrics.

There will be 10 repetitions for each laboratory fabric sample. The dimensions of the specimen should be 125 mm^2.

The specimens (or laboratory samples) should be brought from the prevailing atmosphere to moisture equilibrium before the testing of these textiles. This can be achieved by keeping the samples in a standard atmosphere for testing as prescribed in ASTM D 1776.

This method for the determination of the diaphragm bursting strength of knitted, nonwoven, and woven fabrics is used by the textile industry for the evaluation of a wide variety of end usages.

This standard describes a hydraulic method for the determination of bursting strength and bursting distension of textile fabrics. In this part, a hydraulic pressure is applied using a device that delivers a constant rate of pumping. The method is applicable to knitted, woven, nonwoven, and laminated fabrics. It may be suitable for fabrics produced by other techniques. The test is suitable for specimens in the conditioned or wet state.

A test specimen is clamped over an expansive diaphragm by means of a circular clamping ring. Increasing fluid pressure is applied to the underside of the diaphragm, causing distension of the diaphragm and the fabric. The volume of fluid is increased at a constant rate per unit time until the test specimen bursts when the bursting strength and bursting distension are determined.

Avoid areas that are folded or creased, selvages, and areas not representative of the fabric as shown in Figure 6.9. The system of clamping used generally permits tests to be applied without cutting out specimens.

Set a constant rate of increase in volume of between 100 and 500 cm^3/min depending on the test area and fabric requirements. Or adjust a time to distend a test specimen to burst of 20 ± 5 s using preliminary trials, if a constant rate of increase in volume is not applicable. Place the test specimen over the diaphragm so that it lies in a flat tensionless condition, avoiding distortion in its own plane. Clamp it securely in the circular holder, avoiding jaw damage,

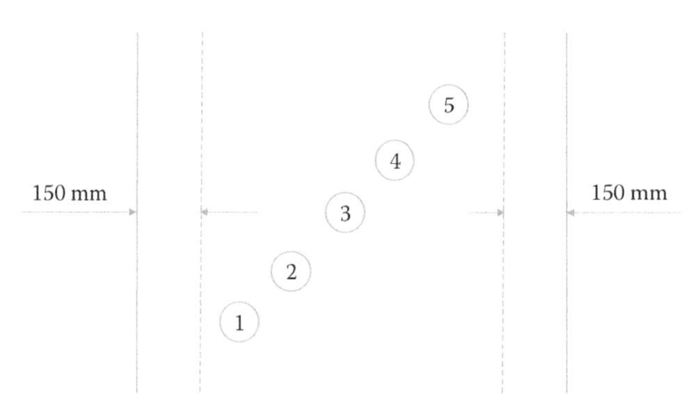

FIGURE 6.9
Width of fabric.

to prevent slippage during the test. Place the distension recording device into the measuring position and adjust it to the zero position. Fasten the safety cover in position according to machine requirements. Apply pressure to the test specimen until the fabric bursts.

Immediately after the burst, reverse the apparatus to its starting position. Note the bursting pressure and height at burst and/or bursting volume. If the test specimen bursts close to the edge of the clamping device, report this fact. Reject jaw breaks occurring within 2 mm of the clamping line.

For tests in the wet condition, immerse the test specimen for a period of 1 h in grade 3 water in accordance with EN ISO 3696:1995 at a temperature of 20°C ± 2°C. For tropical regions, a temperature in accord with ISO 139:1973 may be used. An aqueous solution containing not more than 1 g of a non-ionic wetting agent per liter may be used instead of water. Immediately after removal of a test specimen from the liquid, briefly place it on blotting paper to remove excess water.

This standard describes a pneumatic pressure method for the determination of bursting strength and bursting distension of textile fabrics. The method is applicable to knitted, woven, nonwoven, and laminated fabrics. It may be suitable for fabrics produced by other techniques.

A test specimen is clamped over an expansive diaphragm by means of a circular clamping ring. An increasing compressed air pressure is applied to the underside of the diaphragm, causing distension of the diaphragm and the fabric. The pressure is increased smoothly until the test specimen bursts. The bursting strength and bursting distension are determined.

Prior to testing the sample should be conditioned in the relaxed state. Adjust the control valve of the bursting tester so that the mean time to distend a test specimen to burst falls within 20 ± 5 s. Place the test specimen over the diaphragm so that it lies in a flat tensionless condition, avoiding distortion in its own plane. Clamp it securely in the circular holder, avoiding jaw damage, to prevent slippage during the test. Place the distension recording device into the measuring position and adjust it to the zero position. Fasten the safety cover in position according to machine requirements. Apply pressure to the test specimen until the fabric bursts.

Immediately after burst, close the main air valve. Note the bursting pressure and height at burst. If the specimen bursts close to the edge of the clamping device, record this fact. Reject jaw breaks occurring within 2 mm of the clamping line.

For tests in the wet condition, immerse the test specimen for a period of 1 h in grade 3 water in accordance with EN ISO 3696 at a temperature of 20°C ± 2°C. For tropical regions a temperature in accord with ISO 139:1973 may be used. An aqueous solution containing not more than 1 g/L of a non-ionic wetting agent may be used instead of water. Immediately after removal of a test specimen from the liquid and briefly placing it on blotting paper to remove excess water, perform the test according to the method described in the procedure.

6.5 Fabric Stiffness

6.5.1 Introduction

Stiffness is the tendency of fabric to continue to stand without any support; it is the resistance to bending. Stiffness of fabric is one of its handling properties, which are those that can be assessed by touch or feel, such as smoothness, drape, and luster. Stiffness is considered the basic feature for checking fabric suitability or drape ability. Stiffness is a very important parameter of fabric with respect to its end use. It affects the aesthetic as well as the comfort properties. Fabric with more stiffness is required in the case of upholstery, but in clothing less stiffness is required to meet comfort properties. This depends upon the fiber type, yarn twist per inch, yarn stiffness, diameter, and fabric structure [8]. Fabric with more thickness will be stiffer [9]. To measure the stiffness, the bending length of the fabric is determined by using the Shirley stiffness tester. The bending length is the length of fabric when it bends or falls due to its own weight for a specific length having attained a specific angle. Stiffer fabric has a high bending length, and a lack of drape and flexibility. When a load is applied to a material it bends or deflects and the measure of the amount to which this occurs is known as its stiffness. It can also be found by the modulus "E." Material that deflects a lot under a load has a low stiffness, for example, rubber, and material whose deflection is low under a load and breaks without any more deflection has a high stiffness, for example, glass.

6.5.2 Cantilever Test Method

The cantilever test method is mostly preferred as its procedure is very simple. The cantilever test apparatus as shown in Figure 6.10 is used to determine fabric stiffness under the ASTM standard D 1388-07 [10].

The woven fabric sample of 25 × 200 ± 1 mm is required to perform the test. For accurate measurement three or four samples should be tested and the average bending value taken. The specimens are preconditioned under ASTM D 1776 in the standard atmosphere of a temperature of 21°C ± 1°C and a relative humidity of 65% ± 2%. The bending angle indicator is set at 41.5° as marked on the scale. The moveable slide is removed and the sample is placed on the horizontal platform, keeping the length of the sample parallel to the edge of the platform. The face side of the sample should be up. The moveable slide should be placed on the sample carefully so as not to disturb its position. Move the clamped sample by hand at approximately 120 mm/min so that its edge does not touch the knife edge. Note the overhang length as well.

FIGURE 6.10
Cantilever test apparatus.

Four readings are measured by testing the face, back, and both ends of the sample. Calculate the bending length of the sample for each testing direction by using the formula

$$C = \frac{O}{2} \tag{6.9}$$

where
C is the bending length in cm
O is the length of the overhang in cm

6.5.3 Stiffness of Fabric by the Circular Bend Procedure

Fabric stiffness can also be checked by using the circular bend procedure. This test under ASTM standard D 4032 can be used to determine the stiffness of all types of fabric such as woven, knitted, and nonwoven. This test method is used for quality control testing and commercial shipment in the trade field. In this method, a multidirectional (circular) deformation of a double ply fabric sample takes place and a force value related to fabric stiffness in all directions shows the average fabric stiffness [11].

To check stiffness, a fabric sample of 8-inches length along the warp and a 4-inch width along the weft is required. First identify the warp and weft direction of the sample and mark it by placing a template (8 in. × 4 in.) on it. Cut the marked sample and precondition it under ASTM D 1776 in the

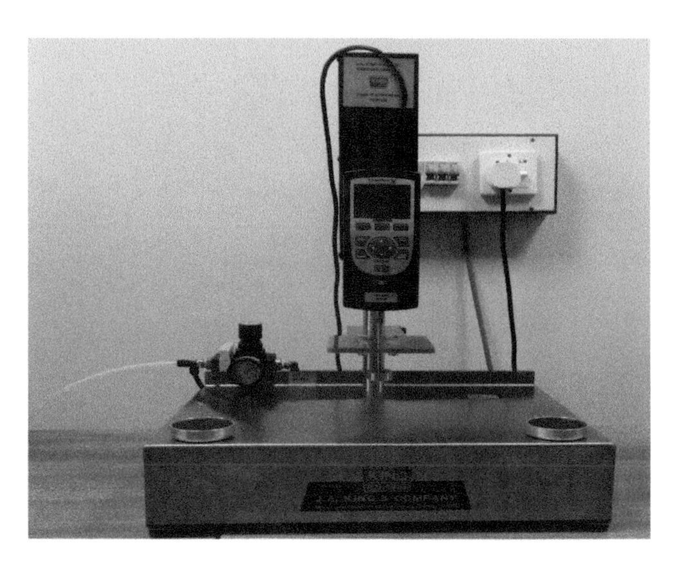

FIGURE 6.11
King air operated digital model.

standard atmosphere of a temperature of 21°C ± 1°C and a relative humidity of 65% ± 2%. The circular bend stiffness tester used in the circular bend procedure is shown in Figure 6.11.

The tester is placed on a flat surface and air pressure should be 47 psi to provide a plunger speed of 1.7 ± 0.15 s. The bottom of the plunger should be 3 mm above the top surface of the orifice plate. The fabric sample is folded along the warp direction to obtain a double ply fabric sample of 4 in. × 4 in. After folding the sample place the template on the folded fabric and apply hand pressure to the template to remove the creases. Then place the sample on the orifice plate with minimum handling below the plunger of the tester. Press both pressure buttons at the same time to turn on the machine. The plunger moves downward and passes through the orifice by pressing the sample. The flattened specimen is converted into a concave shape on one side and convex shape on the other side. The screen displays the stiffness value of the tested fabric.

6.6 Fabric Drape

6.6.1 Introduction

Fabric drape is the shape or fall of fabric under its own weight [12]. From an aesthetic point of view, fabric drape is a very important factor. It is the ability of a fabric to deform when suspended under its own weight under specified conditions [13].

Fabric drape is an important factor that defines the look of a garment along with color, luster, and texture. The drape quality of a fabric affects the final look and appearance of the garment, so it is important to the designer. Fabric drape defines how a garment adapts to the shape of the human body. Drape quality with respect to the end use of a product is different, so its values are not classified as good or bad. Woven fabrics are stiffer, having less drape ability as compared to the knitted fabrics, therefore they are used in tailored clothing where fabric hangs away from the body. Curtains, tablecloths, and women's clothing need to exhibit good drape shape and appearance. Fabric with good drape leads to its fitting on the surface or structure without undesired creases or wrinkles. Fabric drapability is related to the bending stiffness of fabric and depends upon various factors such as fiber content, yarn type, fabric structure, or design and fabric finishing.

6.6.2 Method for Testing Fabric Drapability

Fabric drape can be checked by different methods, such as the fabric research liberating method (FRL) drape meter, Peirce's cantilever method, a 3D body scanner, and the CUSICK drape test method. The most used and simplest method to measure the fabric drape is the CUSICK drape test.

There are three standard sizes of circular fabric diameter (sample) used for different fabrics to measure the fabric drape [13]:

- For limp fabrics the circular sample diameter should be 24 cm. Drape coefficient of such fabric is below 30% with a sample size of 30 cm.
- For medium fabrics the sample diameter should be 30 cm.
- For stiffer fabrics the circular fabric diameter should be 36 cm. Such fabrics have a drape coefficient above 85% with a sample size of 30 cm.

The samples are cut according to the above dimensions and are placed on a circular rigid disk of the CUSICK drape tester. The diameter of the disk is 18 cm. The fabric deforms, resulting in a number of folds around the disk. The draped sample makes a shadow on a paper ring placed on a glass screen of the tester. The drape coefficient is calculated by using a very simple method described by Cusick in which a circular paper of radius R is placed on the screen of the tester. The 10 R indicates the radius of the circular paper and r is the radius of the support disk, as shown in Figure 6.12 [14].

The perimeter of the shadow of the draped fabric is drawn on the paper. The weight of the circular paper is indicated by W_1. After that the paper is cut down according to the perimeter of the draped fabric shadow and weighed as W_2. The drape coefficient is calculated as the ratio of W_1 and W_2.

$$\text{Drape coefficient}(F) = \frac{W_2}{W_1} \times 100 \tag{6.10}$$

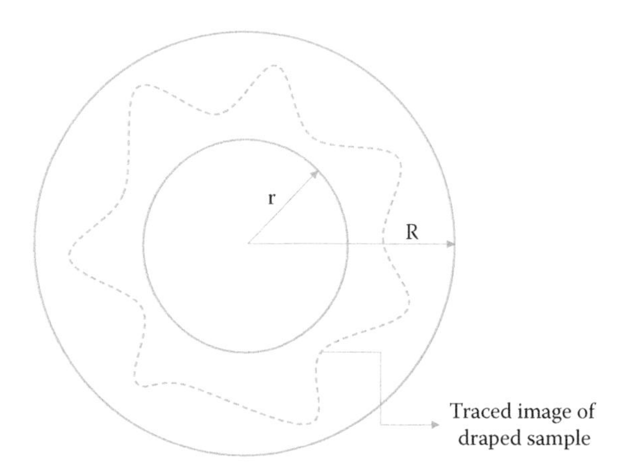

FIGURE 6.12
Image tracing on paper.

The drape coefficient can also be calculated by the formula

$$F = \frac{\text{Area under the draped sample} - \text{Area of supported disk}}{\text{Area of the specimen} - \text{Area of the supported disk}} \times 100 \qquad (6.11)$$

The drape coefficient F, the number of folds or nodes developed by draping the circular sample of fabric on the pedestal, and the node dimensions are the factors used to describe the drape quality of the fabric. The drape coefficient F is the main parameter for describing the fabric drape.

A lower value of drape coefficient means that the fabric is softer, having better drapability, and a higher value indicates the fabric is stiffer. The drape coefficient describes the fabric drapability but it is not sufficient for characterizing drape formation or drape shapes. Fabrics with the same F may have different drape shapes, so the other parameters are also important.

6.7 Fabric Skew and Bow

Bow is a fabric condition resulting when weft yarns or knitting courses are displaced from a line perpendicular to the selvages and form one or more arcs across the width of the fabric; skewness is a fabric condition resulting when weft yarns or knitted courses are angularly displaced from a line perpendicular to the edge or side of the fabric.

6.7.1 Introduction

Woven fabric is made by the interlacement of warp and weft yarn at a right angle. During the process of fabric manufacturing when weft yarn is angularly placed instead of being straight, then bow and skew are produced in a fabric. There is little difference between bowed and skew in a fabric. A bow in the fabric is when the filling yarns are even at two edges but slightly curve in the middle. If the filling yarn is straight from one side to the other side of the fabric, but makes some angle in the whole width of the fabric, such fabric is known as skewed fabric. During the garment manufacturing process, such as in sewing and tailoring, skew in the fabric creates difficulties. Also skewed fabric loses its shape, and as a result of this it shows different behavior on each part of the body. Skew and bow is more prominent in colorful patterns, which distort the appearance of the fabric [15].

6.7.2 Reasons for Skew and Bow

It was seen during manufacturing of fabric, skew and bow in the fabric can be produced. On loom, when the weaver beam begins to unwind at different tensions, or when there is uneven take up of the fabric, skew is produced. Uneven tension across the warp sheet or along the whole fabric width may also lead to manufacturing of faulty fabric having skew or bow. During the finishing process, when the fabric is subjected to drying, it is placed in a heated oven and their edges are attached to tendering frames on a chain driven machine. If there is a speed difference in the movement of chains on the fabric attached to the frames in such a way that the weft yarns are pulled off at a certain angle, then skew in the fabric may be produced. Skewness of fabric also depends upon the weave structure of the fabric. If there is a weave that has less interlacement in one repeat, such as a 3/1 or 2/2 twill, then skew in the fabric will be produced because more float in the weave will favor the yarn to move freely after cutting the fabric from the loom or after washing. In 1/1 weave, there is no chance of skew due to the greater interlacement in one repeat. So skew in the fabric depends upon the weave design of the fabric [16]. This comparison of weave design is made on the base that all other weaving parameters are kept constant. Also this skewness in the fabric will be produced in later processes on the woven fabrics. To avoid skewness in the fabric, maintain proper tension across whole width of the warp sheet/fabric during the production and finishing stages.

6.7.3 Method for Measuring Fabric Bow and Skew

Bow and skew are measured under ASTM D3882. To measure them in the fabric, a measuring scale, a flat surface for opening fabric rolls, a rigid straight edge, and an inspection table will be required.

Three readings are taken randomly at some distance about one-fifth of the way from the first reading on each sample of the fabric. It was observed that if a roll of fabric is not properly conditioned then there is a variation in results, which do not match with those test results taken under standard conditions. So conditioning of a sample is necessary to get accurate results. This conditioning is done under the standard ASTM D1776.

6.7.3.1 Bow Measuring Procedure

First of all, the sample is placed on the inspection table by smooth handling. The random readings are taken from three places at some distance about one-fifth of the way from the first reading. After this, mark the weft yarn along the width of the fabric. To measure the bow in the fabric, place the sample on the smooth surface inspection table with minimum handling to get the most accurate result. Take care that the fabric sample is not under tension in any direction. Measure the bow of the fabric from at least three places having a minimum distance of 1 m or more along the fabric length between each sample. After placing the sample on the inspection table, select one filling yarn across the full width of the fabric. If colored filling yarns are used in the fabric construction, then select the color filling yarn for easy measurement. Take a rigid straight edge and place it across the fabric width in such a way that its two edges meet at the edges of the selected filling yarn. Measure the base line distance BL, the distance between two selvages of the fabric sample along the rigid straight edge, as shown in Figure 6.13.

Measure the maximum distance between the rigid straight edge and the selected filling yarn. This distance is shown by D. Calculate the bow percentage of the individual sample by using the formula

$$\text{Bow\%} = \frac{\text{D}}{\text{BL}} \times 100 \tag{6.12}$$

This method also allows the calculation of bow when a double bow, a double reverse bow, a double hooked bow, or a hooked bow (as shown in Figure 6.14) is present in the fabric.

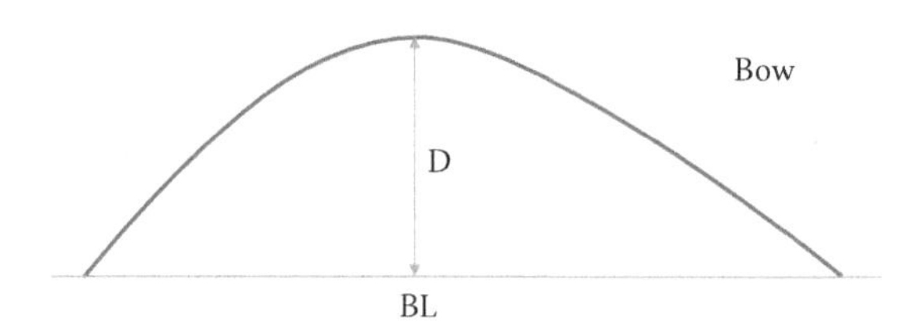

FIGURE 6.13
Measurement of bow in fabric.

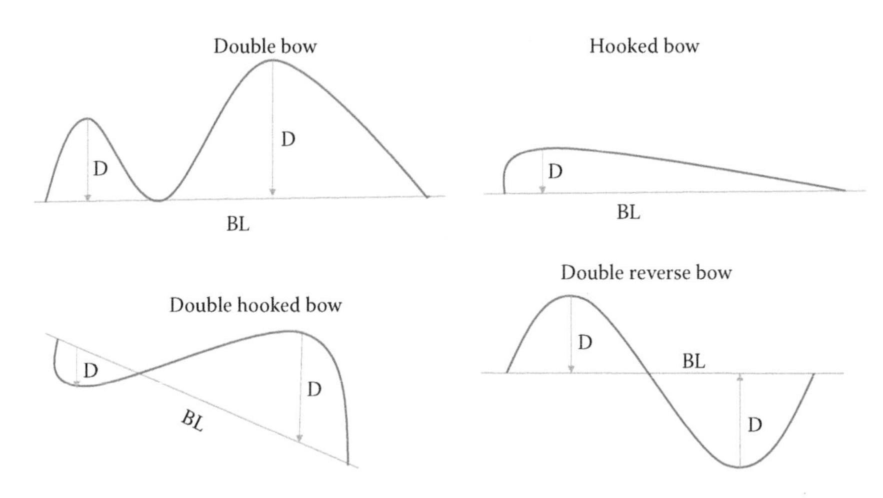

FIGURE 6.14
Different shapes of bow.

6.7.3.2 Skew Measuring Procedure

To measure the skew in the fabric sample, place the sample on the smooth surface inspection table with minimum handling. The sample should be tension free from all sides of the fabric to get the most accurate result. Measure the skew of the fabric from at least three places having a minimum distance of 1 m or more along the fabric length between each place. Select one filling yarn or knitting course across the whole width of the fabric and mark it using a marker or soft pencil, as line AC or line AD as shown in Figure 6.15.

Then place the rigid straight edge along the width of the fabric in such a way that its one edge meets with the edge of the selected marked filling yarn (line BC) as shown in Figure 6.15. Measure the distance BC that is the width

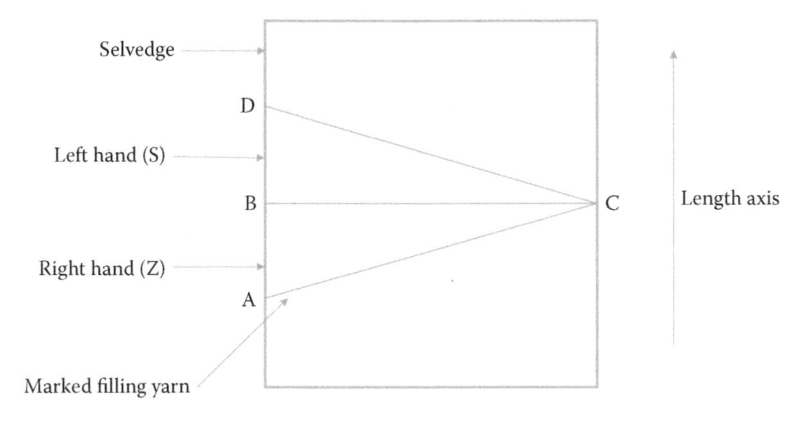

FIGURE 6.15
Skew measuring in the fabric.

of the fabric. Measure AB in the case of right hand skew "Z" or BD in the case of left hand skew "S" and calculate the skew in the sample using

$$\text{Right hand skew\%} = \frac{AB}{BC} \times 100 \tag{6.13}$$

$$\text{Left hand skew\%} = \frac{BD}{BC} \times 100 \tag{6.14}$$

6.8 Fabric Inspection

6.8.1 Significance of Inspection

There are many reasons for inspecting the fabric for its quality level. A producer does it so that he will know he is delivering an acceptable product in conformance with a sale contact. There are many secondary reasons as well for doing it. The producer may want to know the quality level of the given lot of fabric.

In trade relationships between sellers of fabrics and their customers, a system of defect analysis or defect grading must be agreed upon and be in use. At the core of any grading system is simplicity and accuracy. The system must be clearly stated and easy to execute. Most producers don't want to produce low quality fabric. Many of them go to great lengths to monitor the quality and efficiency of every loom. Fabric inspection is frequently used to identify weaving and yarn faults, which are then identified and corrective action taken. The reason why buyers inspect the fabric is to assure themselves that they are receiving the quality level they have ordered. Buyers will also use fabric inspection to judge the relative fabric quality they are receiving from competing suppliers. Also a seller should have a fabric inspection made by an independent experienced examiner if customers make a quality complaint. The fabric for inspection must be selected by the independent examiner. The buyer or the seller might want another party to see only the best or worst fabrics while the examiner must see a representative sample, if only a portion and not 100% of the fabric to be inspected. So it can be said that the prime reason for fabric inspection is to find out the quality of the product. Fabric inspection is carried out at inspection frames and rolling machines, and inspection is done before delivering the product to the customer.

6.8.2 Fabric Rejection Parameters

Some of the standard parameters are

- Ends per centimeter
- Picks per centimeter

- Warp and weft count
- Weave of the fabric
- Fabric width
- Piece length
- Fabric weight
- Quality assurance

Among all the above, the first five have characteristics that, once set on the loom, will not vary as the fabric is produced. But the remaining do have the tendency to vary from loom to loom; the same aimed for quality is often produced on different looms in order to complete the fabric in a given time.

The fabric width is dependent on the weft crimp; more the weft crimp less will be the width of the fabric. Fabric weight is largely influenced by the size percentage on the warp. Hence the last parameters are under the close observation until the order is dispatched. But the quality assurance is the most important.

6.8.3 Classification of Faults

The classifications of defects are based upon the judgment and experience of fabric inspectors. It must be conceded that all "point able" defects do not result in the determination of the quality of product (the end item). Further, it must be understood that certain small or minor defects may be acceptable in certain areas of an end item (garment or home product) while being unacceptable in others, and that a large percentage of small or minor defects are lost in the cutting and fabrication of end items. For these reasons, it seems logical that the quality of a fabric, represented by a point value, should be more reliable and correlative to end item quality and fabric utilization if these facts are taken into account.

It is known that across the industry almost all conditions exist, so that a minor defect to one end use may be a major defect to another, and vice versa, but if the nature of the fabric and the demands of an end use are understood, there should be little problem relating quality determined by the major/minor concept. Generally, the quality of the fabric is determined against worldwide quality standards. There are two most commonly used systems for woven fabrics:

- American system (4 point system)
- Japanese system (10 point system)

Both are used in the mill according to the buyer's requirements.

TABLE 6.3

American Grading System

Size of Faults	Number of Penalty Points
3 or <3 in.	1
Between 3 and 6 in.	2
Between 6 and 9 in.	3
Over 9 in.	4

6.8.3.1 American System

The rating of faults is determined through the use of this four-point system was developed by the American Society for Quality Control (see Table 6.3).

Not more than four points are awarded to a single yard. Total points are normally awarded on the basis of 100 sq. yds. Generally, not more than 40 points per 100 yards are acceptable. Points can be determined by the formula

$$\frac{(\text{Fault points assigned})(100)}{\text{Yardage of piece}} = \text{Points per 100 yards}$$

6.8.3.2 Japanese System

This system is also known as the 10-points system, and it is the oldest. The standards explain the penalties assigned to faults in the fabrics. The frequency of the imperfection is important because it will reflect the potential quantity of fabric with faults. The faults rating in this system are determined as in Table 6.4.

6.8.4 Grading Method

A piece is graded "A" if the total penalty points do not exceed the total yardage of the piece. A piece is graded "B" if the total penalty points exceed the total yardage of the piece. The 10-point system is used when major faults are important and small faults are not critical, while the 4-point system is used

TABLE 6.4

Japanese Grading System

Size of Faults	Number of Penalty Points
Up to 2.5 cm	1
Between 2.5 and 12.5 cm	3
Between 12.5 and 25 cm	5
Greater than 25 cm	10

when dealing with a certain special use. Faults are calculated by the square yard when faults are primarily small ones and the fabric is wide, while linear yard calculations are used when faults are full width.

- Latent faults can be recognized by only testing or by regular processing of the fabric.
- Patent faults can be recognized by reasonable inspection.
- Imperfections are faults that can be prevented under normal conditions or with reasonable care.
- Irregularity concerns faults beyond the reasonable control of the manufacturer or which are natural to any particular quality.

6.8.5 Fabric Faults and Quality Assurance

The quality assurance as well as the economic viability of the weaving plant is significantly influenced by the extent of its success in eliminating cloth faults. The close monitoring of the fabric width, piece length, and fabric weight is useless if the quality is not assured. Fault detection and mending continue to create extra cost. The occurrence of fabric faults not only reduces the quality of the fabric but also reduces the production of the mill. Because of its important nature, special attention will be given to defining each and every fault clearly, along with the technical remedial action needed.

The term "fabric faults" covers all the faults appearing on the loom-state fabric resulting from all previous stages of processing.

One of the problems faced when inspecting fabric is the conditions under which the inspection is being carried out. The correct way is to use an examining machine with an overhead light. The object of using a light that illuminates the face of the fabric is to allow the inspector to observe those faults in the fabric, which would normally be visible in a garment or other end use product. However, there are some faults that are not visible readily under this form of lighting. Some of the weaving and yarn faults become apparent only when the fabric is observed with backlighting. Holes, missing yarns, or shifted yarn become quite apparent as light shines through open spaces in the fabric. Slub or heavy yarn appears as dark spots or lines. Mixed yarn and some types of bare yarn can be seen because of variation in the shade of fabric.

The standard lighting for fabric is 240 W daylight fluorescent bulbs, placed parallel to each other and perpendicular to the direction of the moving fabric. There should be 4 ft of fabric visible for inspection. The light should be directly above the viewing area, so they will need to be at about 60° to the fabric.

6.8.6 Classification of Fabric Faults

Fabric faults can be classified according to their origin:

- Spinning faults
- Warping faults
- Sizing faults
- Looming faults

6.8.6.1 Spinning Faults

- *Count variation.* For a certain count, if the CV% of the yarn is more than 1% then the yarn is said to have count variation. This is the long term variation of the yarn appearing regularly in both warp and weft direction. This spoils the appearance of the fabric.
- *Shade variation.* If different shades are found on the surface of the fabric, even if all the threads are of the same count, then this fault is called a shade variation. The result is the fluorescent effect, which also comes out as a non-uniform dyeing of the fabric.
- *Black end.* If the yarn has, up to some length, a black color while the rest of the portion has a white color, then this fault is called a black end.
- *Polypropylene.* If white, blue, orange or red colored material appears in the yarn, then this fault is termed polypropylene.
- *Slub.* Part of the yarn whose cross-section is more than 120% of the original cross-section of a certain count, and the length is more than 2 cm, is termed a slub. Thick places in the yarn of about one inch are also slubs.
- *Yarn hairiness.* When there are a number of protruding fibers on the surface of the yarn then this is termed hairiness. Originally it may have been a spinning fault or it may occur during the process due to friction.

6.8.6.2 Warping Faults

- *Loose end.* A loose-end fault occurs when one or more warp ends coming from the weaver beam have less tension in them than the others. Loose ends are crimpier and give an uneven look when seen on the folding frame.
- *Knot.* This gives a bead-like appearance to the fabric.
- *Extra end.* This fault occurs on a warping machine when a thread breaks and the operator fails to find that end on the warper beam to knot with the broken end from the creel. The operator knots this end to the other end. This is a wrong practice for broken warp ends. In this fault two ends are passed through the same healed eye and behave similarly. Here the normal and the extra end increases the warp density.

6.8.6.3 Sizing Faults

- *Sizing beads*. On the surface of the fabric small balls appear due to the entanglement of the protruding fibers.
- *Sizing stain*. Sizing machine is covered with a hood, it gets dirty with the passage of time and when steam (from drying cylinders) interacts with this dirt and fell on warp sheet and produce stain on it. It can be avoided by occasional cleaning of the hood.
- *Sizing hole*. When the yarn sheet is oversized, then small holes appear in the fabric surface. This is termed oversize or a size hole. It makes the fabric surface rough and also causes sticky ends. Due to the sizing hole, warp breakage occurs in the sizing comb.

6.8.6.4 Looming Faults

- *Miss pick*. Miss pick represents the absence of the weft thread in the fabric. The weft thread can be missing from the whole width of the fabric or for a part of the fabric edge. In plain weave, the miss pick appears as the double pick with low weft density. In other weaves, a miss pick disturbs the weave design.
- *Double pick*. Insertion of two picks in the same shed is called a double pick. A double pick may occur across the whole width of the fabric or from the selvage up to the center of the fabric width. A double pick may not cause a distortion to the design but instead appear as a coarser weft that spoils the smooth and uniform appearance of the fabric. The density of the weft increases because the take up motion on the loom moves the cloth after every pick. Now if the pick is inserted and the other is repeated in the same shed, the cloth is taken up for one pick while two picks are inserted in the same length of warp threads, resulting in increased weft density.
- *Hanging thread*. This occurs when a long warp thread protrudes from the face of the fabric, with one end gripped in the fabric and the other end freely protruding. When the warp thread breaks after reed, the operator knots the broken thread with extra end and brings the extended length toward the fell of the cloth. He or she aligns the yarn with the cloth fell then starts the loom. After a few picks the residual yarn becomes firmly gripped in the body of the fabric while the other end protrudes freely. This extra length is broken during looming or in the folding department, leaving no fault on the fabric.
- *Jala*. Jala represents the area where picks are not inserted. Only the warp thread is present.
- *Wrong drawing*. The recurring of incorrect warp end interlacement is termed wrong drawing. It is basically the incorrect sequence of the drawing of the warp end through the eyes of the heald frame.

- *Wrong denting*. This fault appears as a narrow transparent stripe or crack running in the warp direction adjacent to the warp ends that are too densely spaced.
- *Count mix*. This fault occurs when the weft of the different counts is used in the same quality of fabric, causing a stripe of a different cover factor in the weft direction.
- *Broken end*. This fault occurs as a narrow transparent stripe, that is, a crack with insufficient warp density.
- *Starting mark*. This is a sort of stripe along the weft and appears as the variation in the pick spacing. It causes variation in pick spacing.

  Dense mark. This fault occurs when the loom starts after a certain stoppage. Because of the creep movement of the fell of the cloth toward the back of the loom, at a new start the beat up occurs more powerfully, resulting in a closure of the pick spacing known as a dense mark. The width of a starting mark may vary from a few picks to several picks to an inch.

  Crack. A crack is a term employed to insufficient weft density. It represents the increased pick spacing.
- *Repping mark*. In the case of a repping mark there is no variation in the pick spacing but there is a difference in the warp tension. So alternate lines are seen on the fabric in the weft direction. This fault may occur at loom stop or during the running position of the loom. A repping mark cannot be seen on the inspection frame with the light below the fabric. It is visible with upper light.
- *Fluff*. Fluff in the fabric looks like slub in the yarn. The main difference between slub and fluff is that the slub is twisted in the yarn while fluff is not. It is easily removed in the inspection department.
- *Nozzle mark*. This appears as the crack in the warp direction and occurs only on the air jet looms.
- *Temple cut*. If there are cuts or small holes near the selvage that extends in for about four inches from the edge, it is termed a temple cut. Temples are used on the loom to hold the cloth at its proper width, gripping it from the reed for a distance of about four inches on either side.
- *Let off mark*. Continuous variation in pick spacing is called a let off mark. It is also known as a weaving mark. It is much more frequent than a starting mark. This fault occurs during the running of the machine. It is observed on the surface of fabric in the form of stripes.
- *Shadow*. When the tuck in threads are woven into fabric that has the same weave as that of the body then it will result in shadow.
- *Lashing in*. This occurs when the tuck in thread comes out of the upper side of the body of the fabric.

- *Floats.* Floats occur when a thread in operation does not take part in shedding at one or a number of picks, so that the yarn remains free on the surface or at the back of the fabric, that is, it is not woven with the design.
- *Reed cut.* Tearing of the fabric due to reed is termed a reed cut. This fault appears in the warp direction.
- *Tails.* These are the abnormally protruding threads at the outer edge of the selvage.
- *Oil stains.* These are oil spots observed on the surface of the fabric. They mostly occur on a Sulzer loom because it involves mechanical components.
- *Snarling.* This is the form of loose weft that occurs when the yarn becomes twisted with itself and forms a loop. It mostly occurs on air jet looms.

References

1. BS EN ISO, Textiles—Tensile properties of fabrics—Part 1: Determination of maximum force and elongation at maximum force using the strip method, British Standards Institution (BSI), Chiswick High Road, London, U.K., 1999.
2. BS EN ISO, Textiles—Tensile properties of fabrics—Part 2: Determination of maximum force using the grab method, British Standards Institution (BSI), Chiswick High Road, London, U.K., 1999.
3. Standard ASTM D1424-96, Tearing strength of fabrics by falling-pendulum type (elmendorf) apparatus, ASTM International (American Society for Testing and Materials International), West Conshohocken, PA, 1996.
4. Standard ASTM D3776-96, Mass per unit area (Weight) of fabric, ASTM International (American Society for Testing and Materials International), West Conshohocken, PA, 1996.
5. M. I. Kiron, Determination of GSM (Gram per Square Meter) of woven and knitted fabrics, http://textilelearner.blogspot.com/2012/02/determination-of-gsm-gram-per-square.html, 2012. Accessed September 15, 2016.
6. K. Anderson, It's not just an Aesthetic Decision: Choosing the right weave design, Report for [TC]^2, www.techexchange.com, 2007.
7. Standard ASTM D2261-96, Tearing strength of fabrics by the tongue (single rip) procedure (CRE tensile testing machine), ASTM International (American Society for Testing and Materials International), West Conshohocken, PA, 1996.
8. M. E. Yüksekkaya, T. Howard, and S. Adanur, Influence of the fabric properties on fabric stiffness for the industrial fabrics, *Tekstil ve Konfeksiyon*, 18(4), 263–267, 2008.
9. H. M. Elder, S. Fisher, K. Armstrong and G. Hutchinson, *The Journal of The Textile Institute*, Fabric stiffness, handle, and flexion, 75(2), 99–106, 2008.

10. Standard Test Method for Stiffness of Fabrics ASTM D 1388, ASTM International (American Society for Testing and Materials International), http://www.astm.org/Standards/D1388.htm. Accessed on September 15, 2016.
11. Standard test method for stiffness of fabric by the circular bend, ASTM D4032–08 (Reapproved 2016), ASTM International (American Society for Testing and Materials International), West Conshohocken, PA, pp. 1–5.
12. B. P. Saville, *Physical Texting of Textiles*, Woodhead Publishing Series in Textiles, Elsevier, Cambridge, U.K., 1999.
13. X. Wang, X. Liu, and C. Hurren, Physical and mechanical testing of textiles, in Hu, J., *Fabric Testing*, Woodhead Publishing in Textiles, London, England, pp. 90–123, 2008.
14. Fabric and garment drape measurement—Part 1, *J. Fiber Bioeng. Informat.* University of Leeds, Wood house lane, Leeds, Yorkshire LS2 9JT, 5(4), 341–358, 2012.
15. Standard ASTM D3882-08, Standard test method for bow and skew in woven and knitted fabrics American Society for Testing and Materials International (ASTM), West Conshohocken, PA, Reapproved 2016, pp. 1–5.
16. A. Alamdar-Yazdi, Weave structure and the skewness of woven fabric, *Res. J. Textile Apparel*, 8(2), 28–33, 2004, doi: 10.1108/RJTA-08-02-2004-B004.

7

Textile Dyed and Finished Fabric

Muhammad Mohsin

CONTENTS

A range of chemicals as dyes, finishes, and auxiliaries are used to enhance the performance of the desired fabric. However, these chemical treatments need to be assessed on a regular basis to provide the maximum benefit. Using doses of chemicals below and above the optimum levels, and using non-optimum application conditions, can lead to the rejection of the fabric and wastage of the expensive chemicals. The selection of the appropriate testing methods and conditions is as important as the successful application of the required dyes and chemicals to the textile. A range of test methods has been developed for the testing of a single performance. However, it is critical to use a testing method that correlates with the actual conditions and that is less costly, easily adaptable, reproducible, precise, and accurate. There are numerous test methods for dyed and finished fabric; however, only the most frequently used methods are discussed below. The most common standard test methods in the textile dyed and finished area are those of the International Organization for Standardization (ISO, 2016), the British Standard (BS, 2016), and the American Association of Textile Chemist and Colorist (AATCC, 2016).

7.1 Color Fastness Properties

One of the major sources of consumer complaints relating to dyed fabric is their variable and nonacceptable fastness. Dyed fabrics behave differently when they are in contact with various conditions like rubbing, washing, perspiration, and exposure to light during the fabric/garment's life. Fabric can be fast to one condition but can easily exhibit poor fastness in another condition. Therefore, it is necessary to test the fabric as per the required end use, and that test method needs to be mentioned clearly to avoid any complications. Typically, the class of dye, subtype of dye, color of dye, shade depth, and dyeing process impart significant effects on the final fastness rating of the fabric.

It is also very important to mention how the fastness rating is given. Typically, there are two ways to do this. One is the subjective assessment of the observer by using the standard gray scales under standard light conditions as shown in Figures 7.1 and 7.2. The rating is given by assessing the difference on the gray scale rating and comparing it with the obtained similar difference on the tested and nontested fabric.

The final rating is given on the basis of the similar level of difference on the gray scale. However, different ratings can be given by different observers on the same tested fabric.

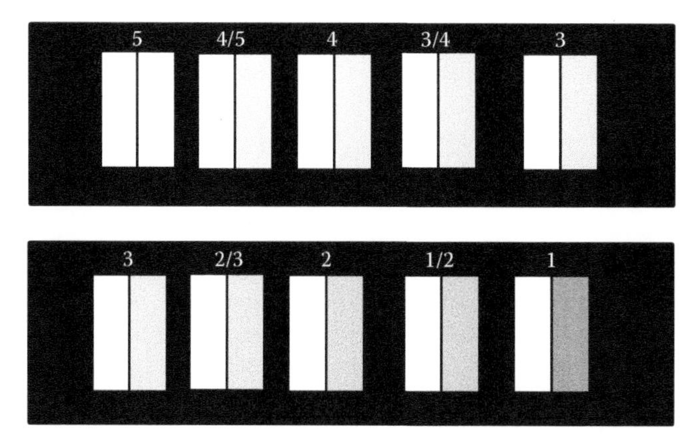

FIGURE 7.1
Standard gray scale for staining.

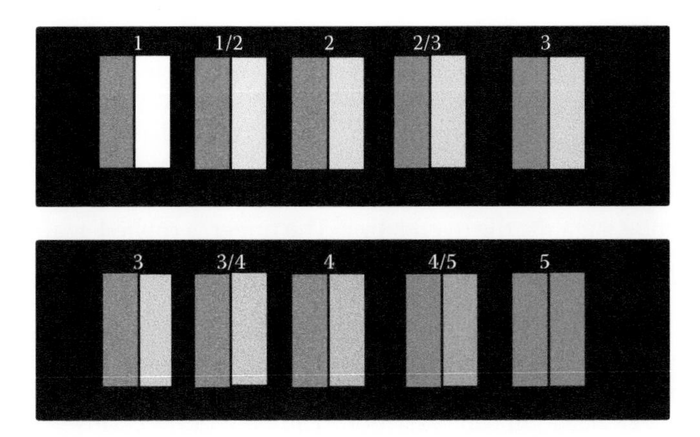

FIGURE 7.2
Standard gray scale for coloring.

TABLE 7.1
Gray Scale Rating of Spectrophotometer Based on ΔE

ΔE		Gray Scale
≥0	<0.4	5
≥0.4	<1.25	4–5
≥1.25	<2.10	4
≥2.10	<2.95	3–4
≥2.95	<4.10	3
≥4.10	<5.80	2–3
≥5.80	<8.20	2
≥8.20	<11.60	1–2
≥11.60		1

A second rating can be given by the use of an instrument called a spectrophotometer, which is based on ΔE as shown in Table 7.1. The rating is still given on the basis of the standard gray scale, which is fed into the computer system, but it is given by using ΔE and so human error and bias can easily be avoided.

7.1.1 Color Fastness to Rubbing

This test exhibits the amount of color that is transferred from the dyed sample onto the white fabric under specific conditions of rubbing. It is performed by using crock meter equipment, which can be operated manually or motorized. There are two main types of test methods that are used to determine the effect of rubbing, one is BS EN ISO-105-X12, while the second is the AATCC crock meter method, test method 8. With the BS EN ISO-105-X12 test method, dyed fabric dimensions should not be less than 50 mm × 140 mm. However, with the AATCC-8 specimen, dimensions should be at least 50 mm × 130 mm. The AATCC also developed another method, the rotary vertical crock meter AATCC-116, which is used for smaller samples as well as for printing fabrics. A dyed fabric sample is locked onto the base of the crock meter, while white bleached fabric, whose dimensions are 5 cm × 5 cm, is mounted onto the finger and rubbed against the dyed fabric. The finger is rubbed against the dyed fabric at the speed of one turn per second (10 × 10 s). Then the white finger cloth is removed and evaluated by using the gray scale shown in Figure 7.1. A rating is given on a scale of 1–5. The procedure for wet rubbing fastness is exactly similar to that of the dry one, apart from the fact that the finger cloth is wet prior to rubbing. It is very important to keep the pick-up of the fabric at 65%, otherwise significant variations can be reported. This test method is equally good at determining the rubbing rating for a range of solely dyed fabrics as well as for after-treated dyed fabrics (Mohsin et al., 2013b, 2014a).

7.1.2 Color Fastness to Washing

This method is used for assessing the resistance of the color of the dyed fabric to all kinds of wash in water with soap and detergent. There are numerous methods and submethods for assessing dyed fabric fastness to washing as washing conditions vary from country to country as well as from consumer to consumer. Therefore, various test methods have been developed. Two major ones that are adopted by many international organizations are discussed below. The first is the BS EN ISO 105-C06:2010, Textiles-Tests for Color Fastness (Part C06: Color Fastness to Domestic and Commercial Laundering). The second is the AATCC Test Method 61-2003, Colorfastness to Laundering, Home and Commercial: Accelerated. There are various methods to assess the color fastness to washing as per ISO methods ranging from C01 to C06. A different soap percentage, temperature, time of washing, and level of washing are used in methods from C01 to C06 as shown in Table 7.2. However, C06 is the most frequently used method and is described in detail in Table 7.3.

In BS EN ISO 105-C06:2010, dyed fabric having dimensions of at least 10 cm × 4 cm are sewn together with standard multi-fiber strip. There are two types of multi-fiber strip containing six types of fibers; type DW has wool, while type TV is without wool. Loose dyed fibers can also be assessed by compressing the samples through padding and then sewing it between the multi-fiber strip. To test, the colored yarns can be converted to knitted fabric, or the above mentioned approach for loose fibers can also be repeated for yarns as well. The sample is washed in a washing machine at a speed of 40 revolutions per minute using one of the sets of conditions mentioned in Table 7.3. A washing solution is prepared by dissolving 4 g of detergent per liter of water. The reference detergent may be AATCC 1993, WOB (without optical brightener) or ECE (European Colorfastness Establishment) detergent with phosphates. Sixteen different subtests can be adopted as per requirement.

TABLE 7.2

Conditions of Color Fastness for C01–C06

Test	Liquor	Temperature (°C)	Time (min)	Reproduces Action of
C01	0.5% soap	40	30	Hand washing
C02	0.5% soap	50	45	Repeated hand washing
C03	0.5% soap 0.2% soda ash	60	30	Medium cellulosic wash Severe wool wash
C04	0.5% soap 0.2% soda ash	95	30	Severe cellulosic wash
C05	0.5% soap 0.2% soda ash	95	240	Very severe cellulosic wash
C06	4 g/L reference detergent + perborate	Various	Various	Domestic laundering

These test methods have a different liquor volume, washing temperature, chlorine level, perborate level, quantity of steel balls as well as pH.

Abrasive action on the dyed fabric is achieved by a low liquor level and by using a high number of steel balls. The gray scale rating for change of color and staining can be given by using the respective gray scales. A rating can be given on a scale of 1–5. The C06 method is equally effective for all type of dyes whether synthetic or natural (Mohsin et al., 2016b). The C06 test method also has the option of multiple tests, which are shown as M in Table 7.3. One multiple (M) test is equal to five washes. The AATCC Test Method 61 has five subtest options as shown in Table 7.4. A similar washing machine is used as in ISO C06; however, the sample size and washing conditions are different. The sample size is 50 mm × 100 mm for Test No. 1A and 50 mm × 150 mm for Test Nos 2A, 3A, 4A, and 5A.

7.1.3 Color Fastness to Light

This test is used to assess color fading when the sample is kept under a specific light source. The test samples are exposed to light for a certain time (24, 36, 48, 72 h, etc.) or to customer demand, under certain conditions of light source, temperature, and humidity, and compared with standard samples. A blue scale is used to determine the color change. This test is quite important for the dyestuff manufacturer, the dyeing units, and the retailer. Certain products like carpets, curtains, and upholstery require excellent light fastness due to their end use performance requirement.

The two most commonly used standard test methods are (1) BS EN ISO 105-B02:2014; Textiles, Tests for Color Fastness: Color Fastness to Artificial Light: Xenon Arc Fading Lamp Test and (2) AATCC Test Method 16. Various options are available in the test method, such as various light sources, various temperature levels, and various humidity levels, as shown in Table 7.5. Typically a xenon

TABLE 7.3

Test Conditions as per ISO C06

Test Number	Temperature (°C)	Liquor Volume (mL)	Available Chlorine (%)	Sodium Perborate (g/L)	Time (min)	Number of Steel Balls	Adjust pH to
A1S	40	150	None	None	30	10	Not adjusted
A1M	40	150	None	None	45	10	Not adjusted
A2S	40	150	None	1	30	10	Not adjusted
B1S	50	150	None	None	30	25	Not adjusted
B1M	50	150	None	None	45	50	Not adjusted
B2S	50	150	None	1	30	25	Not adjusted
C1S	60	50	None	None	30	25	10.5 ± 0.1
C1M	60	50	None	None	45	50	10.5 ± 0.1
C2S	60	50	None	1	30	25	10.5 ± 0.1
D1S	70	50	None	None	30	25	10.5 ± 0.1
D1M	70	50	None	None	45	100	10.5 ± 0.1
D2S	70	50	None	1	30	25	10.5 ± 0.1
D3S	70	50	0.015	None	30	25	10.5 ± 0.1
D3M	70	50	0.015	None	45	100	10.5 ± 0.1
E1S	95	50	None	None	30	25	10.5 ± 0.1
E2S	95	50	None	1	30	25	10.5 ± 0.1

TABLE 7.4

Test Conditions of AATCC Test Method 61

Test No.	Temperature		Total Liquor Volume (mL)	Detergent (% of Total Volume)	Available Chlorine (% of Total Volume)	No. Steel Balls	Time (min)
	°C (±2)	°F (±4)					
1A	40	105	200	0.37	None	10	45
2A	49	120	150	0.15	None	50	45
3A	71	160	50	0.15	None	100	45
4A	71	160	50	0.15	0.015	100	45
5A	49	120	150	0.15	0.027	50	45

TABLE 7.5
Machine Exposure Conditions as per AATCC Test Method 16

Component	Option 1	Option 2	Option 3	Option 4	Option 5
Light source	Enclosed carbon	Enclosed carbon	Xenon	Xenon	Xenon
	Continuous light	Alternate light/Dark	Continuous light	Alternate light/Dark	Continuous light
Black panel temperature,	63°C ± 3°C	63°C ± 3°C	63°C ± 3°C	—	—
light cycle	(145°C ± 6°C)	(145°C ± 6°C)	(145°C ± 6°C)	—	—
Black standard temperature,	—	—	—	70°C ± 1°C	60°C ± 3°C
light cycle	—	—	—	(158°F ± 2°F)	(140°F ± 8°F
Chamber air temperature Light cycle	43°C ± 2°C (110°F ± 4°F)	43°C ± 2°C (110°F ± 4°F)	43°C ± 2°C (110°F ± 4°F)	43°C ± 2°C (110°F ± 4°F)	43°C ± 2°C (110°F ± 4°F)
Dark cycle		43°C ± 2°C (110°F ± 4°F)		43°C ± 2°C (110°F ± 4°F)	
Relative humidity (%) Light cycle	30 ± 5	35 ± 5	30 ± 5	35 ± 5	30 ± 5
Dark cycle		90 ± 5		90 ± 5	
Light cycle, hours					
Light on	Continuous	3.8	Continuous	3.8	Continuous
Light off	—	1.0	—	1.0	—

(Continued)

TABLE 7.5 (*Continued*)

Machine Exposure Conditions as per AATCC Test Method 16

Component	Option 1	Option 2	Option 3	Option 4	Option 5
Filter type	Borosilicate	Borosilicate	—	—	—
Irradiance $W/m^2/nm$ (at 420 nm)	Not controlled	Not controlled	1.10 ± 0.03	1.10 ± 0.03	1.25 ± 0.2
Irradiance $W/m^2/$ (300–400 nm)	Not controlled	Not controlled	48 ± 1	48 ± 1	65 ± 1
Water requirements (input) Type		Demineralized, distilled, or reverse osmosis			
Solid ppm		Less than 17 ppm, preferably less than 8			
PH		7 ± 1			
Temperature		Ambient 16°C ± 5°C (61°F ± 9°F)			

arch-based light source exhibits similar spectral content to that of daylight and therefore is much preferred. In addition, an effective filter needs to be used between the lamp and specimen to control the intensive light. These variables can easily lead to significant changes in the light fastness rating. Therefore, it is important to mention which conditions have been used to expose the sample.

The tested samples as well as the standard blue wool reference are kept in a specific light source under the standard conditions. A certain portion of the samples is partially covered as per the standard and the remainder is exposed to the specific light source. A comparative subjective rating is given depending on the equal color change between exposed and nonexposed samples and standard specimens. A rating will be given on a scale of 1 (lowest) to 8 (highest). Half-a-scale rating such as 3–4 can also be given. The American and European scales use two different sets of reference standards. The European scale rating is given from 1 to 8, while for the American it is given from L2 to L9. It is important to note that these scales are not interchangeable and that the scale on which the rating is based should be mentioned.

7.1.4 Color Fastness to Perspiration

This test is used to assess the change in color of the fabric when exposed to perspiration. Tested samples are dipped in a solution that mainly consists of histidine. The sample is then placed in the perspirometer equipment, where it is treated with histidine solution; a standard sample is separately dried. A gray scale can be used to determine the change in color and staining. Two significant test methods in this area are the AATCC Test Method 15 and the BS EN ISO 105-E04. Two major types of perspiration are based on the solution pH. In an alkaline solution, a pH of 8 is maintained by using NaOH; other ingredients are 0.5 g of histidine monohydrochloride monohydrate, 5 g of sodium chloride, and 2.5 g of disodium hydrogen orthophosphate per liter. For acidic pH, a pH of 5.5 is maintained; other ingredients of the recipe are kept the same as above.

The fabric is thoroughly wetted with the desired acidic or basic pH solution, having a liquor ratio of 50:1, by dipping the sample in the solution for half an hour at ambient temperature. To remove the extra liquor, the sample is wiped through two glass rods and then placed in between the two plates of the perspirometer under the recommended pressure of 12.5 kPa. The sample is dried in an oven at 37°C for 4 h. The rating is given as 1–5 by using the gray scale.

7.1.5 Color Fastness to Sublimation

The color fastness of a dyed article is not only affected by washing, water, or rubbing but also by heat in many forms, such as by pressing, dry heat, or heat with moisture. Some dyes are sensitive to heat and hence can fade/bleed due to the effect of heat. This test is intended to assess resistance of color to the action of dry heat only and not by pressing. Mostly, there are two methods to evaluate color fastness to dry heat: ISO 105–P01, Color Fastness to Dry Heat

(Excluding Pressing) and AATCC Test Method 117-2004-Color Fastness to Heat: Dry (Excluding Pressing). The methods are applicable to textiles of all kinds and can be conducted at different temperatures depending upon the stability of fibers, which can be influenced by chemicals used during dyeing, printing, chemical processing, and by physical factors involved in color change and staining. Nondyed fabric attached to tested samples are exposed to heat. The rating is given using the gray scale.

A specimen of the sample under test is treated with dry heat for 30 s under a specified temperature and pressure. Three temperature options are 150°C ± 2°C, 180°C ± 2°C, and 210°C ± 2°C, and it is important to mention the temperature at which the sample is exposed as it can significantly alter the results. A pressure of 4 ± 1 kPa is applied to the sample during testing. The treated specimens are conditioned and evaluated using the gray scale for color change and staining on adjacent fabric.

7.2 Fabric Shrinkage

Typically, a decrease in fabric dimensions leads to fabric shrinkage. This is one of the critical issues for garment performance during customer use. There are few faults or drawbacks that can affect the quality of the garment, such as color fading or pilling, but fabric shrinkage can make the finished garment unusable. It is considered to be a leading problem. Mainly, it can arise during manufacturing of the garment or during washing by the end user.

One of the typically used test methods is the AATCC Test Method 135; Dimensional Changes of Fabrics after Home Laundering. The fabric to be tested should be 380 mm × 380 mm length and widthwise, and be marked as 250 mm apart. Each benchmark must be at least 50 mm from all edges. A washing load of 1.8 ± 0.1 kg is required and is made up of the tested sample as well as ballast. A reference detergent quantity of 66.0 ± 1 g is used. There are three machine washing cycles: normal, delicate, and permanent press. In addition, there are four washing temperatures: 27°C, 41°C, 49°C, and 60°C. A washed sample can be dried in four different ways: tumble, line, drip, and screen. It is necessary to mention that the washing cycle, temperature, and drying method that is used can significantly influence the results. The following formula is used to calculate the dimensional change

$$\text{Average }\%\text{DC} = 100\frac{(\text{B} - \text{A})}{\text{A}}$$

where

DC is the average dimensional change
A is the average original dimension
B is the average dimension after laundering

This method is equally effective for a range of fabrics made from various fibers like cotton, silk, and wool (Mohsin et al., 2014c, 2015). Another test method is BS EN ISO 5077 Textiles, Determination of Dimensional Change in Washing and Drying, which is also similar to AATCC Test Method 135. Most of the shrinkage tests follow the same instructions, apart from the washing conditions. Samples are typically conditioned before washing and should be conditioned again after drying and before taking the final measurements. The same formula as mentioned above is used to calculate the shrinkage.

7.3 Fire Retardancy Test

The flammability of textiles depends on various parameters like the type of fiber, weight, weave and construction, chemical finish, and source and extent of heat. Depending on the different end use requirements and government legislation, there are various testing methods to assess fire retardancy. The accuracy, reproducibility, subjective and objective type of testing for fire retardancy, and how closely they reflect the ratings as per real life scenarios are the important factors in fabric fire retardancy testing.

One of the most reliable and objective types of testing is the limiting oxygen index (LOI), assessed as per ASTM D-2863. A typical dimension of the sample is 80–150 mm in length and 10 mm in width. Each test specimen should be preconditioned for 88 h at 23°C ± 2°C and 50% ± 5% relative humidity prior to use. The flame fuel should be methane or natural gas of at least 97% purity, without premixed air. The fuel supply should be adjusted so that the flame projects 16 ± 4 mm vertically downwards from the outlet when the tube is vertical within the chimney and the flame is burning within the chimney's atmosphere. It is likely that, for materials where the oxygen index is known to within ±2% by volume, 15 test specimens will be sufficient. However, for materials of unknown oxygen index or that which exhibit erratic burning characteristics, between 15 and 30 test specimens are likely to be required. Mount a specimen vertically in the center of the chimney. Set the gas mixing and flow controls so that an oxygen/nitrogen mixture at 23°C ± 2°C, containing the desired concentration of oxygen, flows through the chimney at a rate of 40 ± 2 mm/s. Let the gas flow purge the chimney for at least 30 s prior to ignition of each specimen and maintain the flow without change during the ignition and combustion of each specimen. Apply the flame for up to 30 s, removing it every 5 s, just briefly, to observe the burning behavior and level of the specimen surface. The LOI value can be defined as the minimum level of oxygen required to continue the burning for 2 in. or 3 min, whichever comes first. Typically, textile fabrics having an LOI value greater than 20 will resist the fire to some extent. The greater the value of the LOI, the better will be the fire retardancy.

BS 5438 is used to assess the flame retardancy of the fabric when it is ignited from the face of the fabric or from the bottom (Mohsin et al., 2013d). A sample size length would be 200 mm with a width of 160 mm. The flame height is kept at 40 ± 2 mm. In this test method, the distance between the top of the flame burner tube and the edge of the fabric is kept at 20 mm. Typically, fabric is ignited for 10 s or for a predefined time. The damaged char length (mm) and width (mm) of the burnt sample is measured.

ASTM D 6413 is another method available to assess fire retardancy and is quite similar to BS 5438. A flame height of 38 mm is suggested in this method. The length of the sample is kept at 300 mm and the width at 76 mm. The tested sample is ignited for 12 s. Later the damaged char length and width is measured in mm. It is important to mention that in order to get the accurate and reproducible results from any of the above described tests, following the strict testing protocols as per standard is necessary.

7.4 Oil and Water Repellency Test

There are three main types of test methods available for assessing the water repellency of the specimen, which should be suitably preconditioned prior to testing under standardized conditions:

1. Class I spray tests for assessing rain impact
2. Class II hydrostatic pressure tests, which measure water penetration
3. Class III sorption of water due to immersion of specimen in water

The most widely used test methods are briefly discussed here.

7.4.1 Class I: Spray Tests to Simulate Exposure to Rain

In the AATCC Test Method 22, Water-Repellency: Spray Test, water is showered on the fabric specimen, which has been preconditioned for 4 h prior to testing, producing a wetted pattern. A rating will be given by comparison of the wetted pattern to standard chart pictures. This is a rapid, simple method, which is technically equivalent to ISO 4920 and BS EN 24920.

The AATCC Test Method 35, Water Resistance: Rain Test assesses fabric performance when it is sprayed with rain water as well as the pressure due to the rain's impact. This test is applicable to all type of fabrics whether treated with a water repellent chemical or not. The test specimen is conditioned at a relative humidity of 65% ± 2% and a temperature of 21°C ± 1°C for at least 4 h. The specimen is placed on a weighed blotter and water is showered on it when it is placed in the rain tester for 5 min. In this test method, rain impact can be varied by changing the height of the water from 60 to 240 cm. At the end, the blotter will be weighed again to assess the quantity of water that has leaked through the fabric. The fabric performance is assessed by various parameters by determining the maximum pressure where no penetration is observed, the effect of a change in pressure on fabric penetration, and the least pressure required for penetration of 5 g of water onto the tested specimen.

The AATCC Test Method 42, Water Resistance: Impact Penetration Test is also useful. The fabric resistance to impact by water is measured and used to predict the penetration of rain into the fabric. In this test, 500 mL of water is showered onto the sample at a height of 2 ft. The rest of the procedure is the same as that for Test Method 35.

Other standard test methods are ISO 9865, Textiles, Assessing the Repellency of Water by Bundesmann Shower Test and BS EN 29865; both determine the repellency of fabrics that are permeable to air. Water is filtered and de-ionized, which is then passed through jets of specific dimensions and sprayed onto the fabric surface. Four test specimens will be kept at a specific angle to the cups and are simultaneously exposed to a heavy rain shower of controlled intensity while the under-surface of each specimen is subjected to a rubbing action. Water that is passed through the fabric will be collected in the cup and later its volume will be measured. In addition, the amount of water that is retained by the test specimen will be measured by comparison of the weight of the fabric before and after the testing. Note that all rain simulation tests should, in theory, replicate rain conditions that occur in practice.

7.4.2 Class II: Hydrostatic Pressure Tests

For many high-performance fabrics that are rendered waterproof, a hydrostatic pressure test may be conducted in one of two ways:

1. By applying a gradual hydrostatic pressure on the fabric and assessing the minimum pressure necessary for penetration
2. By subjecting the fabric to a constant hydrostatic pressure for a lengthy time duration and assessing any penetration

Both the International Standard and the British Standard tests subject the specimen to gradual hydrostatic pressure and measure the pressure necessary for penetration. Two test methods are ISO 811: 1981, Water Textile Fabrics, Determination of Resistance to Penetration, Hydrostatic Pressure Test and BS EN 20811: 1992, Resistance of Fabric to Penetration by Water—Hydrostatic Head Test. In the AATCC Test Method 127, Water Resistance: Hydrostatic Pressure Test (related to ISO 811) the pressure of the water applied to the recessed base where the specimen is placed will be gradually increased as per the standard rate; the fabric surface will then be observed for any signs of penetration by water. The end point will come when penetration occurs for the third time (that is, three points of leakage) and is determined by the penetration pressure, which is measured in centimeters in a water gauge.

7.4.3 Class III: Sorption of Water by the Fabric Immersed in Water

The AATCC Test Method 70, Water-Repellency: Tumble Jar Dynamic Absorption Test assesses the absorption of water into the specimen under conditions similar to actual use.

Preconditioned and pre-weighed samples are kept in water for a specific time; extra water is eliminated by the wringer method and the sample is weighed again. The percentage weight increase of the specimen will reflect the sample's absorption.

The AATCC Test Method 118, Oil-Repellency: Hydrocarbon Resistance Test (technically equivalent to ISO 14419) can be used to assess the sample capacity for repellency of oil under specific conditions (Mohsin et al., 2013a, 2016a,c). Drops of the standard test liquids are assessed on the test fabric. The rating reflects the highest numbered test oil that is not able to wet the fabric.

The visual evaluation of the liquid drop on the fabric surface is graded as

A = Pass: clear well-rounded drop

B = Borderline pass: rounding drop with partial darkening

C = Fail: wicking apparent and/or complete wetting

D = Fail: complete wetting

The 3M test uses mixtures of Nujol Oil and *n*-heptane in various proportions numbered from 50 (100% Nujol) to 150 (100% *n*-heptane). It should be noted that the oil-repellency test is conducted under static conditions and depends completely upon the contact angle of the oil on the fibers.

7.5 Easy Care Performance Test

Easy care performance is a very generic term; many such performances fall into this category. However, typically there are three main performance tests: crease recovery, wrinkle recovery, and appearance of the fabric after laundering (durable press [DP] rating). These are discussed below.

7.5.1 Crease Recovery Test

BS EN ISO 3086 is the standard method to measure crease recovery and can be used to assess the crease resistance and wrinkle resistance of the specimen. Typically, it is used for comparison of the untreated fabric and cross-linker or easy care recipe treated fabric (Mohsin et al., 2013c). This method is equally effective for various chemical finished fabrics like high formaldehyde, low formaldehyde, zero formaldehyde, and softener treated fabrics (Mohsin et al., 2014b).

The fabric is conditioned for 24 h at a relative humidity of 65% and a temperature of 22°C prior to testing. Each specimen is 40 mm long and 15 mm wide. The sample is folded in half and a force of 10 N applied for 5 min. Then the fabric is removed from the load and clamped in the instrument for five more minutes to recover from the crease. The instrument dial is then rotated to keep the free edge of the specimen in line with the reference edge. Twenty specimens are taken from one sample: 10 from the warp and 10 from the weft. Five of the warp samples and five of the weft specimens are folded face to face, with the other five folded face to back. The total crease recovery angle is calculated by adding the mean value of the warp and weft. For the wet crease recovery angle measurement, each specimen is dipped in water for 1 min at room temperature and the excess of water is removed by gently pressing them between two tissue paper layers for 30 s. The wet crease recovery angle is then measured using the same method as that described above. This method is typically given preference over AATCC 66-1998 for the wrinkle recovery of the woven fabric, in which the equipment and the conditions are the same except that a 5 N force is applied in the AATCC method, whereas a 10 N force is applied using the British standard.

7.5.2 Wrinkle Recovery Test

The standard method of AATCC 128-1989 can be used for assessing the appearance of the fabric after wrinkling. The size of the sample is 15 cm × 28 cm. The fabric is wrapped between the top and bottom flange of the wrinkle tester and then the top flange is lowered. A weight of 3500 g is applied for 20 min, after which the weight is removed and the fabric is hung vertically in the long direction by clips on a cloth hanger. After 24 h, under standard atmospheric conditions, a comparative rating is given, using five three-dimensional plastic replica sheets, by two different observers. Two specimens are tested from each sample and the mean value is taken from the two ratings.

7.5.3 Appearance of the Fabric after Laundering (DP Rating)

AATCC 124 is used to check the appearance of the fabric after repeated home laundering. The Wascator FOM 71 MP laundering machine is used for washing. There are numerous wash cycles with variable temperatures and times, so it is quite important to mention which one is used. Sixty-six grams of the 1993 AATCC standard reference detergent is used in each wash cycle; 50/50 poly/cotton is used as a ballast to make the weight of one cycle equal to 1.8 kg, including the sample. There are various washing conditions and the one used needs to be mentioned. A comparative rating is provided by two observers using six different AATCC three-dimensional smooth appearance plastic replicas and identifying the closest resemblance. This rating is known as the DP rating.

7.6 Antimicrobial Test

Testing anti-microbial finishes on textiles is more difficult and more complicated than many other tests. There are few methods available for assessing the antimicrobial behavior of the textile. Agar-based zone of inhibition tests and bacteria counting tests are among the most popular. ISO/DIS 20645 and EN ISO 20645 test methods are related to agar diffusion, while ISO 11721 is a burial test—its first part is for assessing antimicrobial finish, while part two is related to long term antimicrobial resistance. The two main difficulties are the reproducibility of results and obtaining the results as per actual conditions. The current British Standard on preservative textile treatments (BS 2087) is purely analytical in content. It is based on field experience over many years and does not have any references to antimicrobial tests. While in some ways this is an unbeatable approach, it provides no scope for swift development. Laboratory testing is essential to provide reasonably rapid screening of new preservatives and to check the effectiveness of existing preservatives under new conditions or when used in combination with other finishes. In all antimicrobial testing, controls as well as treated materials must be tested.

In the agar plate method, a nutrient gel containing a micro-organism is poured onto a plate and, when set, a piece of the fabric under examination is put on the surface of the gel and the whole plate is then incubated under conditions ideal for microbial growth. Alternatively, the inoculum is sprayed onto the sample after it has been placed on the agar plate. For bacteria, this could be 18–24 h at 37°C; for fungi, 3–14 days at 28°C, or up to 4 weeks for materials such as PVC coated fabrics. At the end of the incubation period samples are assessed either visually or by performance loss. The visual assessment is normally a comparison between uninhibited growth of the micro-organism in the dish and the growth on or in contact with the sample. There may be a reduction in growth or a complete absence of growth on the sample. There may also be a zone of inhibition around the sample where the biocide has diffused into the gel and prevented the micro-organisms from developing. Large zones of inhibition are not desirable; they indicate that either the material has been over-loaded with biocide or that the biocide is diffusing rapidly and easily into the gel, an indication that durability will not be good. Because of possible variations, samples will always be tested at least in quadruplicate with controls to check the viability of the organism. Care must be taken to prevent contamination from outside the bacteria or fungi.

The agar plate method provides a reasonably economical and swift method for assessing materials. The test method is flexible and many variations are possible:

1. The test can use single bacteria or fungi or mixed cultures.
2. A wide range of textiles both coated and non-coated and cellulosic and non-cellulosic can be tested.
3. A mineral salt agar can be used so that the only carbon source is the material under test and development of the micro-organism will indicate that degradation is taking place.
4. Quite large numbers of samples can be handled so that it is possible to compare a range of treatments and their durability, such as leaching, washing, dry cleaning, and exposure to UV light.

Assessment by performance loss is done by measuring the loss of tensile strength or weight loss, though this type of assessment is not relevant to all materials. For instance, with a polyurethane-coated nylon fabric the nylon substrate may be completely unaffected by micro-organisms so that the tensile strength of the material is unaffected during tests and weight losses are minimal. However, fungal attack of the polyurethane can result in cracking and a loss of waterproofness. The agar plate method, when used with bacteria, relies on the biocide on the material under test migrating or diffusing into the agar gel. If the diffusion rate is very low, as with certain coated materials or with water-repellent finishes, misleading results may be obtained, particularly with bacteria with growth occurring before the biocide has had time to diffuse out. This problem can be partly overcome by holding a sample on the plate at 5°C for 24 h prior to incubation. The test organism does not develop during this period but the biocide on the material has more time to diffuse into the gel. AATCC 147 gives qualitative measurement of the textile antimicrobial performance. A sample having the dimensions of 25 mm × 50 mm is placed on plates containing sterile agar that have either gram positive or gram negative bacteria. For the remaining steps, the above procedure is adopted. The AATCC 100 method gives quantitative evaluation of textile antimicrobial performance. A sample having dimensions of 1.9 in. will be inoculated with bacteria containing a solution and is incubated for a time of 18–24 h at 37°C. The swatches are then extracted and the number of bacteria in the extract determined by serial dilutions placed on plates containing sterile agar and incubated for 48 h at 37°C. Calculations are made to find the percentage reduction of bacteria found on the antimicrobial treated textile compared to the number of bacteria found on an untreated textile of similar construction. The major challenge to such test methods are poor reproducibility and a vast difference between testing results as compared to real life condition results. Proper training of the laboratory staff along with the complete following of the testing procedure is critical in obtaining accurate and repeatable results.

References

American Association of Textile Chemists and Colorists (AATCC), 2016. AATCC technical manual. Research Triangle Park, NC.

British Standards Institution (BSI), 2016. BSI Group. Chiswick High Road, London, U.K.

International Organization for Standardization (ISO), 2016. ISO textiles. Vernier, Geneva, Switzerland.

Mohsin, M., Ahmad, S. W., Khatri, A., Zahid, B., 2013a. Performance enhancement of fire retardant finish with environment friendly bio cross-linker for cotton. *Journal of Cleaner Production* 51: 191–195.

Mohsin, M., Carr, C. M., Rigout, M., 2013b. Novel one bath application of oil and water repellent finish with environment friendly cross-linker for cotton. *Fibers and Polymers* 14: 724–728.

Mohsin, M., Carr, C. M., Rigout, M., 2013c. Development of zero formaldehyde easy care finishing system by using nano titanium dioxide with citric acid and its impact on physical properties. *Fibers and Polymers* 14: 1440–1444.

Mohsin, M., Farooq, A., Abbas, N., Noreen, U., Sarwar, N., Khan, A., 2016a. Environment friendly finishing for the development of oil and water repellent cotton fabric. *Journal of Natural Fibers* 13: 261–267.

Mohsin, M., Farooq, A., Ashraf, U., Ashraf, M. A., Abbas, N., Sarwar, N., 2016b. Performance enhancement of natural dyes extracted from acacia bark using eco-friendly cross-linker for cotton. *Journal of Natural Fibers* 13: 374–381.

Mohsin, M., Farooq, U., Iqbal, T., Akram, M., 2014a. Impact of high and zero formaldehyde crosslinkers on the performance of the dyed cotton fabric. *Chemical Industries and Chemical Engineering Quarterly* 20: 134–139.

Mohsin, M., Farooq, U., Ramzan, N., Rasheed, A., Ahmad, S., Ahsan, M., 2014b. Softeners impact on low or zero formaldehyde cross-linkers performance for cotton. *Industeria Textilia* 65: 134–139.

Mohsin, M., Farooq, U., Raza, Z. A., Ahsan, M., Afzal, A., Nazir A., 2014c. Performance enhancement of wool fabric with environment friendly bio cross-linker. *Journal of Cleaner Production* 68: 130–134.

Mohsin, M., Ramzan, N., Ahmad, S. W., Afzal, A., Quatab, H. G., Mehmood, A., 2015. Development of environment friendly bio cross-linker finishing of silk fabric. *Journal of Natural Fibers* 12: 276–282.

Mohsin, M., Rasheed, A., Farooq, A., Ashraf, M., Shah, A., 2013d. Environment friendly finishing of sulphur, vat, direct and reactive dyed cotton fabric. *Journal of Cleaner Production* 53: 341–347.

Mohsin, M., Sarwar, N., Ahmad, S., Rasheed, A., Ahmad, F., Afzal, A., Zafar, S., 2016c. Maleic acid crosslinking of C-6 fluorocarbon as oil and water repellent finish on cellulosic fabrics. *Journal of Cleaner Production* 112: 3525–3530.

8

Apparel and Home Textiles

Abher Rasheed and Ateeq ur Rehman

CONTENTS

8.1 Introduction

The objective of this chapter is to provide a basic understanding of the classification and inspection/testing related to the sewn product. The raw material received in a garment or home textile manufacturing unit is the finished fabric, sewing thread, fabric trims, and accessories. It is necessary to inspect or test the received raw materials to ensure that they meet a certain quality standard. To fail to do this could result in a final product of bad quality. Once the raw material has arrived, the production of an order is started. Inspection during production is a necessity so that the end product is made according to the customer specifications. A final inspection is carried out at the end of production. This chapter covers the inspection and testing of the raw material during and at the end of production.

8.2 Classification

Generally, sewn products can be classified on the basis of the raw material or the end use. The raw material for a clothing manufacturer is the finished fabric. The characteristics of finished fabrics depend upon several factors: fiber content, yarn properties, fabric structure, and finishing. The raw material for a product is chosen by keeping in mind the end use of the product. Fibers having different characteristics are chosen for different end uses. Cotton is used for better thermal comfort in summer clothing, while polyester may be used for a better fall of the fabric. Yarn properties play a very important role in the finished fabric properties. Fabrics made by using highly twisted yarns, for instance, may have a poorer fall than fabrics made with low twisted yarns. On the basis of the fabric structure, products can be divided into three categories: knitted, woven, and nonwoven. Knitted fabrics are known for their stretch, flexibility, and comfort. That's why knitted textiles are used in sports. Knitted fabrics, however, have poor dimensional stability. Woven fabrics have good dimensional stability and strength, while nonwoven fabrics are famous for being cheap. The products made of nonwoven fabrics are mostly disposable. Textile processing and finishing is the last stage where the raw material is treated and is an extremely important part of the textile supply chain. Greige fabrics can be dyed or printed according to the application to which it is to be used. Further, different types of finishes can also be applied during textile processing and finishing, for example, water repellent finish, fire resistant finish, or antimicrobial finish.

On the basis of function, textile products can be divided into two major categories: home textiles and garments. Home textiles are those products that we use in our homes, for example, towels, bed sheets, tablecloths, whereas garments/clothing are those products that are used for wearing, for example, trousers, skirts, shirts. Figure 8.1 shows the classification of home textiles.

Home textile products can be further divided into three categories: kitchen products, bedroom products, and bathroom products. Kitchen products consist of home textiles that are used in the kitchen, for example, tablecloth, aprons, and cleaning cloth. Different kitchen products are used for different functions. For example, a cleaning cloth having good absorbency will provide better cleaning efficiency. On the other hand, aprons and tablecloths should provide better protection against penetration so that clothes and tables may be protected from stains.

Most home textiles are used in the bedroom and may be termed bedroom products. Bed sheets, blankets/quilts, curtains, and carpets are some examples. Each bedroom product has its own importance and function. There are two types of bed sheets: the fitted sheet and the flat sheet. A fitted sheet is normally stitched with elastic at its corners so that it can be fitted on the mattress, while a flat sheet's sides are secured with lock stitch to prevent the

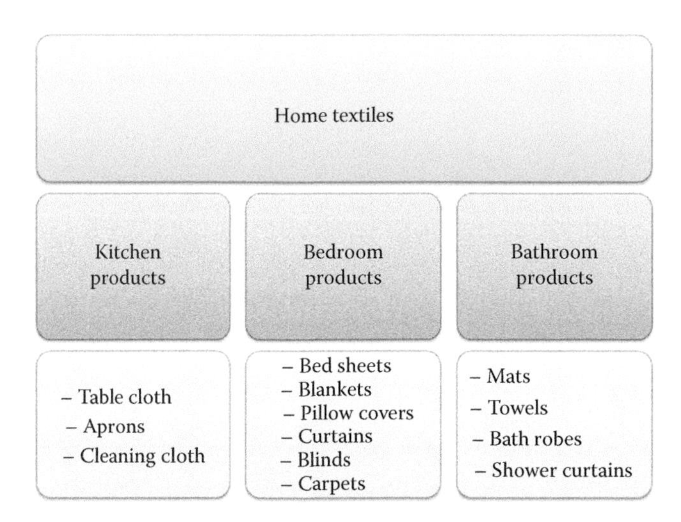

FIGURE 8.1
Home textile classification.

yarn from fraying during use. The mattress is covered with the fitted sheet first and then a flat sheet is placed on it. The function of both sheets is to avoid staining of the mattress. Blankets/quilts are used for thermal protection. To achieve thermal protection properties, quilts are filled with a sheet of polyester fibers. This sheet entraps a good amount of air, which is how it provides good thermal protection. Curtains come in many types. Their functions include the control of sunlight entering the room, privacy, and decoration. Carpets are used to cover the floor and may be used for multiple functions such as insulation and decoration.

Some examples of bathroom products are towels, shower curtains, and mats. Towels must have very good absorbency and a soft hand feel. That's why natural fibers are used to make the towels. Shower curtains should be waterproof or water resistant so that they can stop water being deposited in undesired areas of the bathroom. Mats are used for insulation and decoration.

Apparel products may be classified in several ways on the basis of wearing, fit, material, and function. Figure 8.2 depicts the classification of garments/clothing. Three types of apparel exist on the basis of their utilization: top, bottom, and undergarment. Top and bottom apparel are used to cover the body and to protect it from the weather. Tops are used on the torso while bottoms cover the lower body. Examples of tops are shirts, jackets, and coats, while trousers and skirts are examples of the bottom apparel. Undergarments, on the other hand, are worn under the top or bottom apparel. They are used to support different body parts or to transport moisture away from the body.

On the basis of fit, garments can be classified as slim fitted or loose fitted. Slim fitted garments are those that stick with the body while loose fitted garments are made with sufficient allowance that they do not reveal body shape.

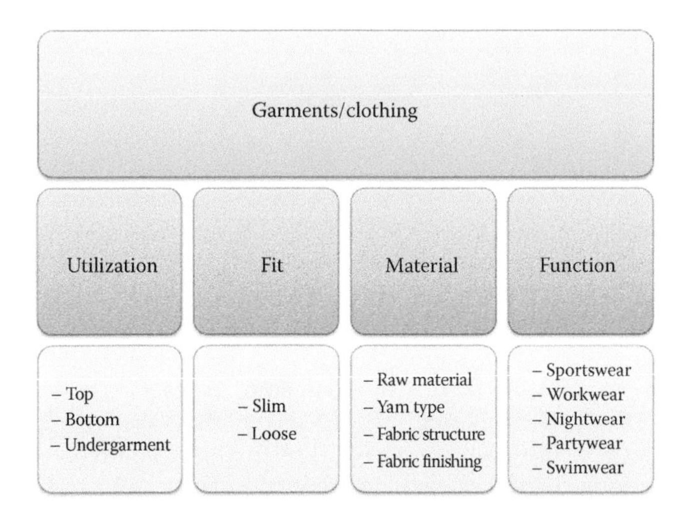

FIGURE 8.2
Classification of garments/clothing.

Garments can also be classified on the basis of the raw material with which they are made. Cotton and polyester are the most commonly used raw materials for clothing but other fibers are also used where required. For example, clothing made of nomex is used where there is a fire hazard. Different types of yarns can also be used to make different products. Most of the time, spun yarns are used to make conventional garments. However, filament yarns are used in technical products. Further, highly twisted yarns are used for summer clothing while yarns with a low twist may be used for winter clothing.

Fabric structure plays a significant part in defining end product characteristics. Fabric may be knitted, woven, or nonwoven. Knitted fabrics are known for their stretch, softness, and comfort. That's why they are abundantly used in sportswear where high stretch is required. In addition to that, knitted fabrics are also used to make undergarments in which softness and stretch are the basic requirements. Woven fabrics have better strength and dimensional stability than knitted fabrics. Further, a variety of weave designs exists, which provides several possibilities of surface designs. Trousers, formal shirts, coats, and workwear are a few examples of woven garments. The advantages of nonwoven fabrics are that they are cheap and their production speed is very high. Such fabrics were not used in clothing until recently. Nowadays, nonwovens are used to make disposable gowns.

Another possibility of classification of garments is by their function, such as formal wear, casual wear, nightwear, swimwear and workwear. These are purpose-built garments and are used to satisfy certain needs of the end user. Swimwear, for example, may be designed with good water resistance so that the swimmers may stay dry during swimming. Nightwear, on the other hand, needs to be soft and easy to wear to provide comfort during sleeping [1].

8.3 Fabric Testing

Fabric is the most important raw material for garments. It incurs around 70% of the garment cost and nearly 100% of the exposed area. All the other materials like trims, closures, and packing accessories are generally arranged on or around the fabric utilized in the garment. Fabric has to pass through maximum checks as the cost increases as it moves forward in the manufacturing process. A faulty fabric, if not screened out at the receiving phase, would involve further cutting, stitching, finishing, and packing costs. Therefore, it is ideal to check fabric at the source to save also on the transportation cost and time, in the case of faulty fabric. Another common practice, in local purchase, is to inspect the fabric right after it is received in the warehouse of an apparel manufacturing facility. A third party inspection may be involved where the prepared fabric has to be shipped to other countries.

The finished fabric may involve raw material irregularities that may become more apparent after dyeing and finishing. It also includes yarn defects like thick and thin places. In addition weaving or knitting imperfections occur, such as miss picks, broken ends, and miss stitches. Dyeing and printing process irregularities also impart adverse effects to finished fabric. Other testing requirements relate to shrinkage, skewness, strength, elasticity, pilling and abrasion, stiffness, weight of fabric, fabric quality, among others. Therefore, it is necessary for garment manufacturing that the provided fabric is free of all previous processing irregularities and that it complies with garment performance standards.

8.3.1 Visual Inspection and Grading

This is the most common test for deciding whether the fabric lot should be accepted or not and is based on visual inspection of the finished fabric. The fabric roll is loaded on one side of the inspection table that is equipped with the appropriate type and intensity of light. The fabric runs across the inspection table at a certain speed and an expert checks visually for predefined faults.

A certain number of points are allocated, depending on the dimension, severity, and nature of the faults, for a single meter of the fabric being inspected. More than one standard grading system exists. A four-point system, for example, is one in which the inspector allocates a maximum of four points to an individual meter inspected, according to predefined criteria. The total points are then counted over the roll length.

The procedure is repeated for all rolls taken in a selected sample from the whole population of the fabric lot. At the end, the total points and total length inspected are calculated. The results are then expressed in points per 100 square units (square yards or square meters) or per 100 linear units (meters or yards) of fabric. The decision to accept a fabric lot is made on the basis of the pre-settled mutual understanding of supplier and customer. For example, a result of 30 points per 100 linear meters will result in the rejection of the lot, if the supplier agreed to provide fabric at 20 points per 100 linear meters. The reference standard for inspection and grading of fabrics is ASTM-D5430 [2].

8.3.2 Fabric Color

A standard way for testing fabric color is to check the Pantone color reference number as described by the customer. The dyed fabric is matched to the relevant Pantone reference to decide on whether to accept the prepared fabric lot. Garment manufacturers usually check fabric color against pre-approved fabric swatches supplied by the customer. The fabric swatches are developed in the pre-production phase, according to the Pantone number given by the customer. Once the color swatches have been approved after multiple trials the supplier is asked to follow those approved. The supplier is expected to follow the same recipe as was used in dyeing the swatch fabrics. In reality, however, process variations occur, which put the relevant stakeholders in a decision-making situation. All variations, though minor, are sent to the customer for approval. The approved shades are given to the person checking fabric color. The inspector then decides on whether to accept on the basis of all the shades approved. If the provided roll matches with any of the approved shades, it will be considered valuable for further use, and vice versa. In addition the shade may vary within the roll. Therefore, it is more appropriate to check the fabric shade at first, last, and inner points of a single roll. Another common fault relating to color shade is the variation across fabric width. This fault typically arises when dealing with wider width fabrics like home textiles. Checking the shade from the left, center, and right of the fabric width gives appropriate information regarding variations.

8.3.3 Bursting Strength

A common consideration for strength parameters of textile fabrics is their bursting strength. This test is well known within the textile trades. The ball burst test is used to test the bursting strength of fabrics. A specimen of fabric taken from either a fabric roll or garment is placed between grooved plates of the testing equipment and fastened by means of screws. A polished and hardened steel ball attached to a pendulum actuating clamp of the machine is forced through the fabric, at right angles to the fabric plane, until the rupture occurs. The force reading gives the bursting strength of the fabric.

Two standard methods exist for this test: one with a constant rate of traverse, whose reference standard is ASTM-D3787, and another with a constant rate of extension (CRE), whose reference standard is ASTM-D6797 [3,4].

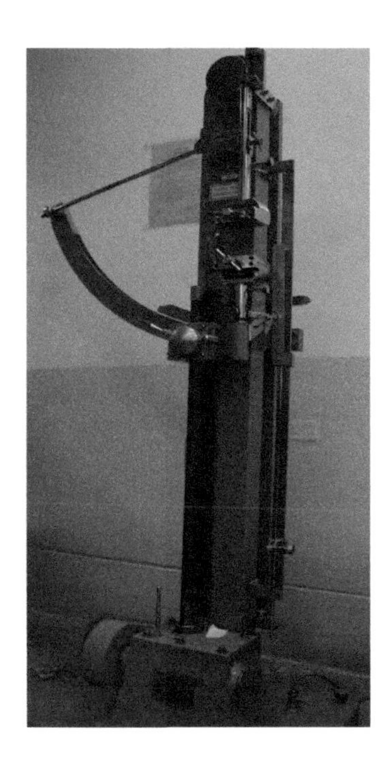

FIGURE 8.3
Tensile tester.

8.3.4 Tensile Strength and Elongation

Fabric in garments has to bear a number of forces throughout its performance life. A fabric, therefore, must satisfy the minimum level of strength to satisfy customer needs. The standard test involves checking more than one parameter of the fabric. For testing tensile properties, a test specimen of the fabric is clamped in the jaws of a tensile testing machine and a force is applied until the specimen breaks. The maximum amount of force exerted by the machine, recorded from the machine scale, gives the breaking force. A tensile tester is depicted in Figure 8.3.

Subtract the specimen's initial length from its length at breaking point. The difference expressed in percentage gives the elongation measurement. The method is applicable to most woven and nonwoven fabrics, knitted and stretch fabrics being an exception. The test should be performed as directed in standards ASTM-D5035 (strip method) and ASTM-D5034 (grab method) [5,6].

8.3.5 Tearing Strength

There are three methods to test the tearing strength of a fabric. The first method uses a falling pendulum (Elmendorf) type tester, which is shown in Figure 8.4. The tester includes a stationary clamp, a clamp carried on a

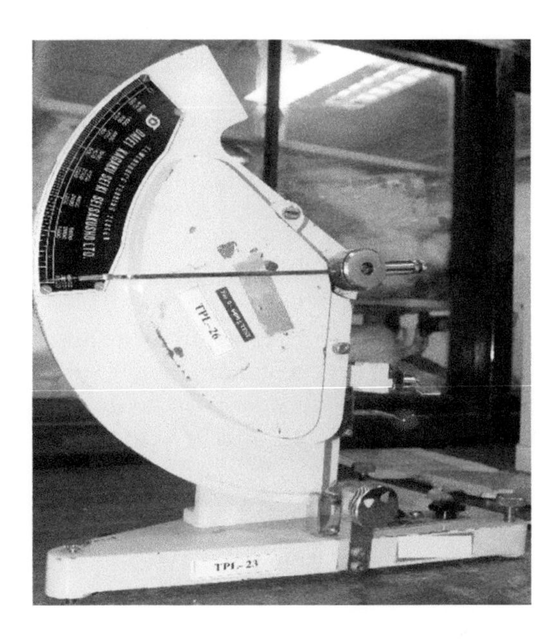

FIGURE 8.4
Tear tester.

pendulum that is free to swing on a bearing, and a mechanism to level, hold, then instantly release the pendulum to measure the force. The test specimen with a precut central slit is held between two clamps and the specimen is torn through a fixed distance. The resistance to tearing is then calculated using Equation 8.1. The reference standard is ASTM-D1424 [7].

$$F_t = \frac{R_s \times C_s}{100} \tag{8.1}$$

where
- F_t is the tearing force in cN (centi-Newton)
- R_s is the scale reading
- C_s is the full scale capacity in cN or lbf

In the second method, the tensile testing machine with a CRE is used. It is named the tongue procedure. A rectangular specimen with a cut in the short side is prepared. One tongue of the specimen is gripped in the upper jaw while the other is gripped in the lower jaw of the machine. The jaws are continuously moved apart with the application of force to propagate the tear. At the same time the instantaneous force is recorded and graphed. The average of the highest peaks in the graph gives the value of the tearing strength. The standard reference method is ASTM-D2261 [8].

The third type of test is known as the trapezoid procedure. An outline of a trapezoid is marked on a rectangular fabric specimen. A slit is made on the smallest base of the trapezoid to start the tear. The nonparallel sides of the marked trapezoid are gripped in the parallel jaws of the machine. Force is applied to increase separation continuously, between the jaws, to propagate the tear. At the same time the instantaneous force is recorded and graphed. The average of the highest peaks in the graph gives the value of the tearing strength. The standard reference method is ASTM-D5587 [7].

8.3.6 Yarn Slippage

One of the major issues in sewn fabrics is the yarn slippage. The fabrics in garments and other sewn products should exhibit sufficient resistance to yarn slippage along the seams. The test exists to determine the resistance to slippage of warp yarn over filling yarns, or filling yarns over warp yarns, using a standard seam.

For testing the slippage of warp yarns over weft yarns a specimen is cut from the fabric with its larger side along the fabric weft. The specimen is then folded and a standard seam, parallel to the warp yarn, is made at a particular distance from the folded edge of the specimen. The specimen is then clamped between the two jaws of the tensile tester in a standard way such that the seam ideally stays at the center of the jaws. The load–elongation curve of the fabric is superimposed over a load–elongation curve of the same fabric with a standard seam sewn parallel to the yarns being tested. The resistance to yarn slippage is reported as the load at which a slippage of a specified size is seen. The test should be performed as directed in standard ASTM-D434 [9].

8.3.7 Fabric (Weight) Areal Density

Fabric mass per unit area is a very important consideration in the selection of fabric for a particular end use. Garments made of fabric with higher areal densities are not used in the summer season, and vice versa. To determine the mass per unit area of a certain fabric a standard specimen is prepared first. This is weighed and the results are expressed in mass per unit area. The standard test method is ASTM-D3776 [10].

A common approach is to use a grams per square meter (GSM) cutter, as shown in Figure 8.5. The cutter prepares a circular specimen of diameter 113 mm. The weight of this specimen in grams, multiplied by a factor of a hundred, gives the weight (in grams) of the fabric in a square meter.

8.3.8 Bowing and Skewness

Bowing is the displacement of filling yarns (woven) or courses (knitted) from an imaginary line perpendicular to the fabric selvage. When the same displacement is in angular form, it is regarded as skewness. Bowing and skewness disturbs the grain line of the garment patterns and causes discomfort and improper functioning of the final garment. In addition, it also diminishes

FIGURE 8.5
GSM cutter.

the aesthetics of the garment. These measurements are equally important for accepting a fabric lot.

To measure the bowing a steel tape may be used. The straight edge of the steel tape is placed across the fabric width to measure the distance between the two points where a selected weft yarn or course of fabric meets the two edges or selvages, denoted by "BL" in Figure 8.6. The greatest distance between the straight line and the marked filling or course line is measured parallel to the fabric selvage, as indicated by "D" in the diagram.

The fabric bow is then calculated by

$$\text{Bow}(\%) = 100\left(\frac{D}{BL}\right) \tag{8.2}$$

where

D is the bow depth

BL is the base length or the straight distance between the points of the marked filling yarn/course

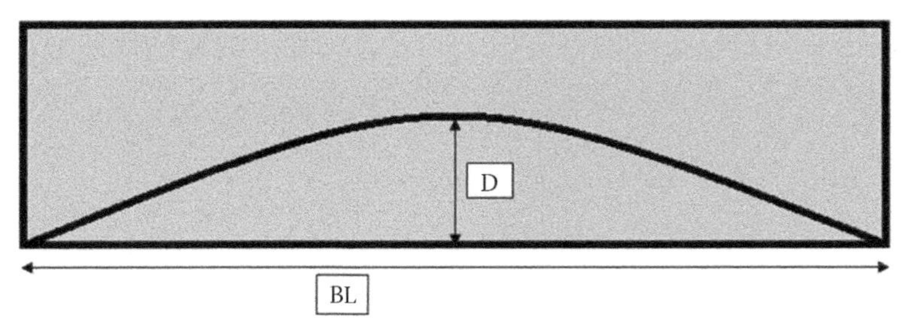

FIGURE 8.6
Measuring bow in fabric.

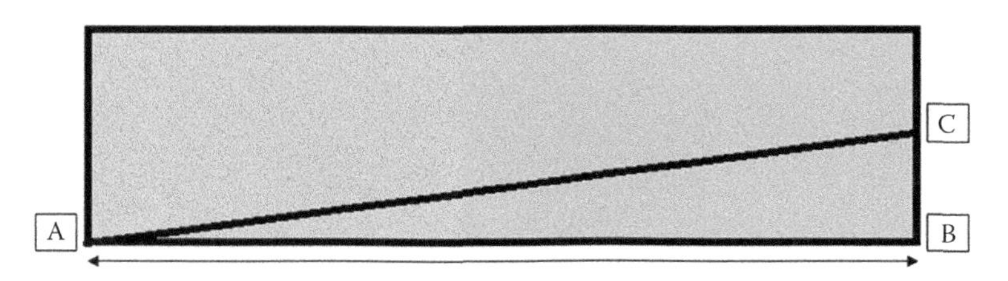

FIGURE 8.7
Measuring skewness in fabric.

To measure skewness the straight line distortion of the marked filling or course, from a line perpendicular to the selvage, is measured. Figure 8.7 shows the measurement of skewness in fabric.

Equation 8.3 is used to calculate the skewness.

$$\text{Skew}(\%) = 100\left(\frac{AB}{BC}\right) \tag{8.3}$$

where
AB is the fabric width, perpendicular to the selvage
BC is the skew depth

The reference standard test method is ASTM-D3882 [11].

8.3.9 Air Permeability

This is the ability of a fabric to allow air passage perpendicular to the fabric plane. This is an important testing requirement regarding the acceptance/rejection of commercial shipments. It is also important for garment manufacturing as the fabric with too little air permeability will not allow sweat to evaporate and hence create body irritation and odor along with the accumulation of perspiration. Therefore, fabric may be required to pass a certain level of air permeability before it is accepted to make a certain garment. The rate of air flow passing perpendicularly through a known fabric area is adjusted to obtain a prescribed air pressure differential between the two fabric surfaces. From this rate of air flow, the air permeability of the fabric is determined. The scale provides a reading in $cm^3/s/cm^2$ or $ft^3/min/ft^2$. An air permeability tester is illustrated in Figure 8.8. The reference standard test method is ASTM-D737 [12].

8.3.10 Dimensional Changes after Home Laundering

Fabric shrinkage is one of the prime considerations for the selection of fabrics in apparel manufacturing. If the appropriate shrinkage is not considered

FIGURE 8.8
Permeability tester.

the garment may run short on the wearer's body after washing. For that purpose, the shrinkage along both warp and weft is calculated and adjusted in garment specifications in the cutting room. The stitched garment may appear enlarged in some dimensions, but it adapts to the original customer requirements after washing. Similarly, it is important to measure fabric elongation.

To calculate fabric shrinkage, a square of a particular size is cut from the individual fabric roll. A further square is marked inside at a certain distance from the edges, as shown in Figure 8.9. This is done because the fabric may experience fraying in washing and run short on dimensions, which will not represent the true shrinkage percentage. The marked sample is then washed and dried as per standard conditions. The difference in measurements is used to calculate shrinkage by

$$\text{Percentage dimensional change}\,(\%) = \frac{\text{Change in dimension}}{\text{Original dimension}} \times 100 \qquad (8.4)$$

The reference standard is AATCC TM 135 [13].

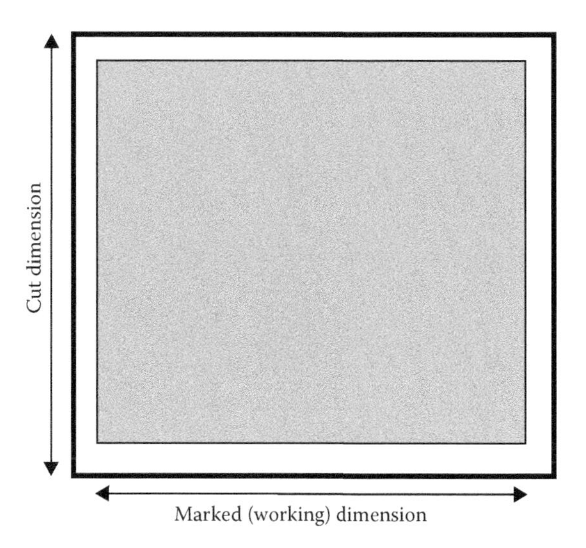

FIGURE 8.9
Sample preparations for dimensional change testing.

8.4 Sewing Thread Testing

Sewing thread bears less than 5% of the total sewn product cost but contributes to its major functional and aesthetic properties. It can be said confidently that thread can alone make or break a sewn product on the basis of its characteristics. These can be spun, core spun, or filament threads. In spun threads, yarns made of staple fiber or filaments are twisted and plied to form yarns. Two and three-ply threads are most commonly in use for industrial sewing. The core spun threads offers high seam strength. Filament threads can be monofilament, multifilament, or textured. Cotton, polyester, and nylon are the most widely used raw materials. In general, synthetic fibers offer better performance in comparison to their respective natural threads due to their high tenacity, resistance to abrasion, and chemicals; they are also less affected by moisture, mildew, insects, and microorganisms.

The sewing thread in garments and other sewn products has to exhibit a few or even all of the following characteristics during the process of manufacturing and throughout its life-span:

- Appropriate strength
- Color fastness to laundering, dry cleaning, and water migration
- Resistance to abrasion
- No/low shrinkage

- Resistance to chemicals
- Zero/minimum metamerism

The following variables can be checked to satisfy the required customer or consumer needs.

8.4.1 Thread Twist

Thread twist is measured by the number of turns in unit length of thread. Usually it is measured in twists per inch (TPI), however other units may also be used. Twist keeps the fibers and ply yarns together to increase thread strength; however, an optimum level of twist exists. Too high a twist can result in looping and may prohibit stitch formation. Too low a twist will result in loose fibers and yarns that may fray and break. Following variables are checked to assure the right selection of sewing thread.

8.4.2 Twist Direction

There are two types of twist direction: S-twist and Z-twist. A direction of twist or thread spiral that is parallel to the center part of the letter "S" is known as an S-twist. A direction of the twist or thread spiral that is parallel to the center part of the letter "Z" is known as a Z-twist. Figure 8.10 shows an S and a Z-twist.

FIGURE 8.10
S and Z-twist.

Twist direction can be checked by an empirical method. For that you must hold the particular length of thread between thumb and first finger of both hands. By rotating the thread in a right and/or left direction, it can easily be seen as to which direction it is being untwisted. An S-twist will untwist in a right direction, and vice versa. For a single needle lock stitch and other machines it is recommended to use thread with a Z-twist. An S-twist will be untwisted in the sewing process as the machine imparts a right twist to the thread during stitch formation. For a two-needle lock stitch, however, both bobbins unwind in the opposite direction; therefore, both S and Z-twisted threads should be used. Alone, an S-twisted thread is preferred for flat lock and cover stitch machines. The more sophisticated way is to use a digital twist tester.

8.4.3 Twists Per Inch

Manual and electronic twist testers are available on the market. A specified length of thread is provided between two jaws of the tester and the thread is given a rotation opposite to the twist direction until individual plies are parallel and free from twist. The total number of revolutions count divided by the total length of thread in inches gives the TPI. Manual and digital twist testers are depicted in Figures 8.11 and 8.12, respectively. The reference standard methods for twist testing are ASTM-D1423 for plied threads and ASTM-D1422 for singles. The twist tester provides accurate twist direction and TPI. The TPI is less commonly measured than twist direction [14,15].

8.4.4 Twist Balance

The tendency of sewing thread to twist in loop form is also an important parameter in thread performance. This may result in excessive kinking and snarling during actual sewing. To test the twist balance, unwind the thread

FIGURE 8.11
Manual twist tester.

FIGURE 8.12
Digital twist tester.

from the package in the same way as it is supposed to be on the sewing machine. Approximately one yard of length is sufficient to perform this testing. The two ends are then kept apart by nearly 4 in. and the thread is allowed to adopt the form of a loop. The whole loop is then suspended in a draft free environment to check the twist on itself. A twist tester may also be used to determine twist balance. The reference standard is ASTM-D204 [16].

8.4.5 Strength

The sewing thread experiences friction while passing through various parts of the stitching machine, such as thread guides, thread take-up levers, tension disks, and the needle eye. It also has to face a high temperature at the needle point, in high speed stitching. In addition, it experiences forces during the garment wet process. The sewing thread should be strong enough to stay firm through all of these situations. The tensile properties of thread are direct measures to make the appropriate selection of thread. A certain elongation is also necessary in sewing thread for particular seam types. A low strength of thread may save minor initial costs but adds up to high rework owing to thread breakage during sewing and broken stitches in the consequent washing and finishing process. A too strong thread, however, incurs excessive cost, hence is a waste of resources. Therefore, one must know the required match between seam performance and tensile properties of sewing thread.

The testing machine takes a certain length of yarn between two jaws and applies force until the thread breaks up. This gives the following results corresponding to a loaded thread package:

- Tensile strength
- Tenacity
- Elongation

The reference testing standards are ASTM-D204 and ASTM-D2256 [16,17].

8.4.6 Thread Uniformity

A uniform thread diameter is an ideal requirement for a smooth sewing operation and a clean seam finish. In actuality, however, a thread may include a number of imperfections like thick and thin places and knots, in severe cases. Such irregularities may cause excessive thread breakage in the sewing process.

A physical examination, at a certain accepted quality level (AQL), of thread arriving as thread cones is useful to highlight such irregularities. A sample size of cones is selected from a particular population batch and are brought into appropriate light conditions. The number of knots is observed by rotating the cones at 360° in the hands, by a quality control person. After that the thread is unwound and the number of imperfections are noted over the length of yarn. The same is repeated for other cones in the sample size and a written report is made. The lot is accepted if the quality faults are within acceptable limits. Otherwise the lot is rejected and sent back to the supplier for replacement.

8.4.7 Thread Size

It is important to use thread of the appropriate *diameter* for a better sewing performance and aesthetics. A larger cross-section of thread may cause excessive displacement of fabric yarns and result in seam puckering. Moreover, it is also an important parameter where thread cover factors and contrasting colors are to be considered.

For testing the thread diameter a thickness gauge is used. Thread segments are placed on the stage of the thickness gauge to determine the diameter of the provided thread sample. Another way (the optical method) is to place the thread segments on the rotatable microscope stage where their diameters are measured using a calibrated eyepiece. This detailed experiment should be performed as directed in standard ASTM-D204 [16].

Thread diameter is generally specified in terms of a "ticket number," which is a commercial numbering system. These are manufacturers' identification for a particular thread and are based on the fixed weight system. The higher the ticket number, the finer the thread will be, and vice versa. A tex number can be converted to a ticket number by

$$\text{Ticket number} = \frac{1000}{\text{Tex number}} \times 3 \tag{8.5}$$

For example, a 40 tex will be equivalent to a 75 ticket number. Similarly, reverse calculations can be done to calculate a tex number for a required ticket number.

The standard testing involves the determination of a yarn number (linear density expressed in tex number) and then conversion to a ticket number.

To determine the yarn number, a measured length of preconditioned thread is wound on a reel as skeins and the weight is taken. The resultant yarn number is expressed in tex. The reference standard methods are ASTM-D204 and ASTM-D1907 [16,18].

The standard practice to determine the thread ticket number involves the calculation of the weighted average of a tex number and then finding the respective ticket number using standard ASTM-D3823 [19].

8.4.8 Needle Size

Inappropriate needle size causes thread breakage and poor seam performance. Thread has to undergo countless penetrations across the material along with the needle during the sewing process. An improper needle eye size, shape of blade, needle point, and scarf size may contribute to improper stitch formation, broken thread, and/or skip stitching. To check for the appropriate needle size, the thread is held at a 45° angle, after passing through the needle eye. The needle is then allowed to slide down from the top end. The appropriate needle should slide down due to its own weight without significant hindrance.

8.4.9 Shrinkage

Shrinkage in sewing thread can adversely affect seam appearance through seam puckering. An excessive tension on the thread in the sewing machine may cause shrinkage. In more depth, the differential shrinkage of the fabric and the stitching thread may deteriorate seam performance and aesthetics.

The standard test method for testing single strand of thread involves tying a thread strand in a loop and measuring it under a prescribed tension before and after exposure to boiling water or dry heat. The change in length gives the shrinkage percentage. This detailed experiment should be performed as directed in reference standard ASTM-D204 [16].

8.4.10 Colorfastness

The garment along with the sewing thread has to undergo wet processing that may result in color bleeding. Similarly, dry cleaning may also cause alteration of shade and staining. In addition, thread staining may be the result of treating the garment under home care conditions like keeping it in a wet state for a specified time. This holds true for natural threads, synthetic threads, and also combinations of both.

To test colorfastness to laundering, the sewing thread, in contact with a multifiber test cloth, is washed with home washing and drying machines with or without bleach under conditions intended to reproduce the effect of home laundering on sewing thread. The alteration in shade of the sewing thread and the degree of staining of the multifiber test cloth are graded by reference to the AATCC Gray Scale for Color Change or to the AATCC Chromatic Transference Scale, as appropriate. The reference standard is ASTM-D204 [16].

Colorfastness to dry cleaning for a sewing thread is performed in contact with multifiber cloth. The alteration in shade of the sewing thread and the degree of staining of the multifiber test cloth are graded as above.

Colorfastness to water migration is also checked. A sewing thread in contact with multifiber cloth is washed in a home washing machine and retained in a wet condition for a specified period of time. The degree of staining of the multifiber test cloth is graded by reference to the AATCC Chromatic Transference Scale. The reference standard is ASTM-D204 [16].

8.5 Seam Testing

A seam is a series of stitches that may be functional or ornamental. Generally a seam is made to join two or more pieces of fabric in garment construction. The aesthetics and performance of the garment are strongly dependent on the seam's appearance and performance characteristics. The considerations related to seam appearance include colorfastness and the area covered under the seam, as was discussed earlier in the thread testing section. Performance characteristics are more related to strength, elongation, and elasticity.

8.5.1 Stitch and Seam Type

Seam engineering, the determination of the appropriate stitch type, seam configuration, and thread type for a particular assembly requires a thorough understanding of many variables. Seams may be required to exhibit a certain level of strength, elasticity, durability, security, service life, and appearance. A single type of stitch and seam may not offer strength and flexibility at the same time. One may need to consider certain tradeoffs while making a particular selection of stitches and seams. Moreover, the place of stitching in a garment also requires certain characteristics.

A standard classification along with a method of determination for each stitch and seam type can be followed from standard ASTM-D6193 [20].

8.5.2 Seam Efficiency

Standard seam efficiency testing can be performed by using a tensile testing machine. First of all, fabric without a seam is placed between the clamps of a tensile testing machine and force is applied until failure occurs.

The amount of force required to break the fabric sample is noted. Then, a sewn sample of woven fabric is prepared as explained in ASTM-D1683. The sewn sample size is 4 in. × 8 in. Seam allowance should be 1/2 in. To test seam strength, the fabric specimen containing the seam is placed between the two jaws of a tensile testing machine such that it is directed toward the rear of the machine, using vertical alignment guides. The seam should also be equidistant from the upper and lower clamps of the testing machine. The force is applied until the thread breaks and seam failure occurs. The amount of force required to break the seam is noted. The seam efficiency is then calculated using Equation 8.6. It is worth noting that this test is not applicable on bound seams [21]

$$\text{Seam efficiency}\,(\%) = \frac{\text{Seam strength}\,(\text{N})}{\text{Fabric breaking force}\,(\text{N})} \times 100 \qquad (8.6)$$

8.5.3 Seam Inspection

Apart from seam efficiency, seams are inspected for various parameters. Seam slippage, as described in fabric inspection, is a major fault, though one not caused by the seam itself, yet it destroys the aesthetics and functioning. A miss stitch is a major seam fault owing to which the seam fails to grip one or more plies of the fabric in the seaming process. A seam may also be inspected for its appearance. Sometimes the lower thread appears on top and deteriorates the seam's appearance. Such a problem is usually caused by improper tension setting of the machine and can be rectified with a little care. Ideally, the lower and bottom threads should meet at the center point of the fabric plies. Seams are also inspected for too many protruding fibers on the sewing thread, which indicates rough contact with the surfaces of the guides or the needle eye. Friction can be overcome by lubrication and appropriate thread selection. Finally, seams are also checked for the appropriate stitch and seam types with respect to the location of usage or customer directions.

8.5.4 Bursting Strength and Elongation

Bursting strength is the force required to rupture the seam by distending it with a force applied at a right angle to the plane of the sewn fabric. The seam sample, of a certain size, can be taken from an already stitched garment. A ball burst device can be used to measure the bursting strength. The apparatus consists of a ring clamp of a certain diameter and an opposite steel ball that penetrates into the sample in the ring clamp. This testing is significant to measure the rupture force and extensibility of seams in knit or woven stretch fabrics. The standard test method for the experiment is ASTM-D3940 [22].

8.6 Testing for Trims and Accessories

8.6.1 Zipper Testing

Zippers are means of closure in a garment. They provide the ease of putting a garment on and off with little effort. They also offer flexibility to allow garment opening to avoid excessive heat inside. Zippers also provide closure with zero visibility, which is essential for some garments like the front zippers in trousers and jeans. In addition, these can be used on pockets and even for aesthetic purposes (non-functional) in garments. Apart from garments they are widely used in tents, handbags, and luggage bags. Depending upon the area and item they are used on, zippers may be required to fulfill certain aesthetic and functional necessities.

The *strength of a zipper* is measured for various purposes, generally in terms of the strength of chains and elements, the holding strength of the stops, the holding strength of separable units, resistance to compression, resistance to twist, resistance to the pulling-off of the slider pull, and so on. Further details of these tests can be seen in ASTM-D2061 [23]. Several tests are performed to determine the strength of the chains and other zipper parts, which are explained in the following.

8.6.1.1 Cross-Wise Strength

This is the ability of the zipper to withstand lateral stress. It is measured with a tensile testing machine equipped with clamps having special jaws. A 1 in. sample of zipper chain is fixed between the jaws and loaded until it is destroyed. This test is important to measure the zipper's resistance against failures like tape rupture, unmeshing, or element separation when the zipper is exposed to side stresses during usage. Figure 8.13 illustrates the zipper cross-strength tester.

8.6.1.2 Element Pull-Off Strength

This is the gripping strength of an element around the bead. It is measured by pulling off a single element from the bead at right angles to the stringer by using a specially designed fixture to the tensile testing machine (Figure 8.14). It is used to measure the resistance of the element to being pulled or fractured by side stress during usage of the zipper.

8.6.1.3 Element Slippage Strength

This is the resistance of the element to longitudinal movement along the bead of the tape (Figure 8.15). It is determined with a tensile tester equipped with a special attachment.

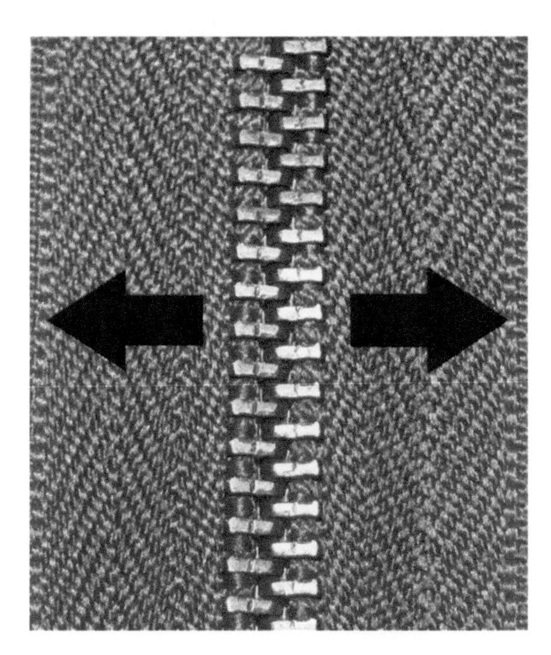

FIGURE 8.13
Zipper cross-strength testing.

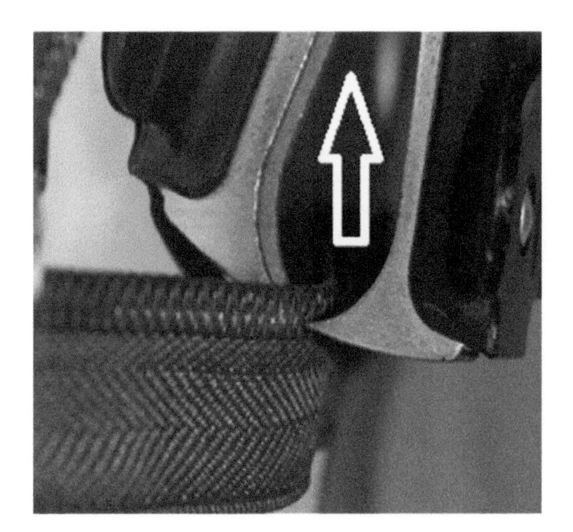

FIGURE 8.14
Zipper pull-off testing.

FIGURE 8.15
Element slippage test.

The ability of stops to perform their intended functions is determined by five different methods, which are explained as follows.

8.6.1.4 Top Stop Holding

This test may be used to determine the top stop attachment strength, which measures the ability of the top stop to prevent travel of the slider beyond the end of the chain. A tensile tester may be used to determine the top stop strength with a special attachment on the upper jaw that holds the slider puller as shown in Figure 8.16. The lower jaw is set 3 in. apart and force is applied until failure occurs. The maximum force and failure type is noted.

8.6.1.5 Bottom Stop Holding, Slider

This test determines the bottom stopper strength, which measures the ability of the bottom stop attachment to resist stress caused by longitudinal force to it by the slider. To check bottom stop holding strength the slider is brought to the lowest position at the bottom stop holder. The puller is attached to a specially designed fixture at the upper jaw of the tensile tester. The two stringers are then placed in the lower jaw with equal lengths between the

FIGURE 8.16
Top stop holding test.

jaws (3 in. apart). The angle includes the stringers and should be such that no elements meet either at the flanges or the diamond. An increasing load is applied until failure occurs. The amount of force and nature of the failure is recorded.

8.6.1.6 Bottom Stop Holding, Cross-Wise

This test determines the bottom stop attachment strength, which measures the ability of the bottom stop to resist side stresses. The slider is removed from the zipper and elements adjacent to the bottom stop are removed to a half inch. The side tapes of the zipper are then placed in the jaws of the tensile tester such that the bottom stopper is at the center, vertical, and horizontal. An increasing load is applied until failure occurs. The force and nature of the failure is recorded. Figure 8.17 depicts the bottom stop holding, cross-wise test mechanism.

8.6.1.7 Bottom Stop Holding, Stringer Separation

Position the bottom of the slider against the bottom stopper. The stringers are placed in opposite jaws of the tensile tester such that the slider

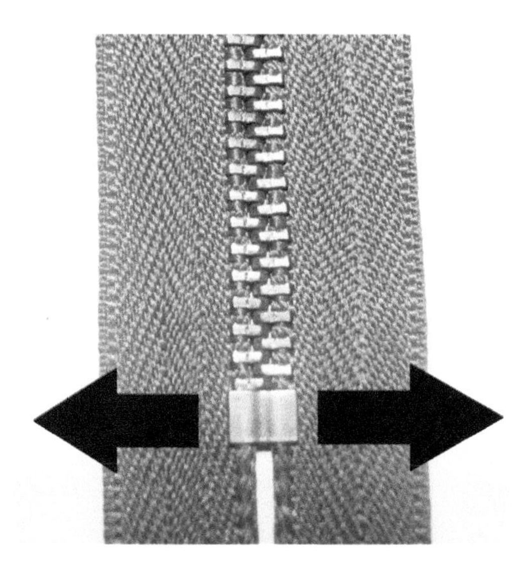

FIGURE 8.17
Bottom stop holding, cross-wise test.

stays exactly at the center with the jaws 3 in. apart. An increasing load is applied until failure occurs and the nature of the failure and force applied are recorded. The bottom stop holding, stringer separation mechanism is shown in Figure 8.18.

8.6.1.8 Bridge Top Stop Holding, Stringer Separation

The bridge top stop attachment strength measures the ability of the bridge top stop to remain in place, holding the stringers together and limiting slider travel when the stop is stressed through stringers. The zipper is placed with its stringers open and clamped in opposing jaws of the tensile tester. Position the stop along the axis of clamps and midway between them. An increasing load is applied until failure occurs. The holding strength of the separable units is measured through three test methods that simulate the various stresses faced by the zipper over long term usage.

8.6.1.9 Fixed Retainer Pull-Off Test

This test is used to measure the resistance of the fixed retainer to displacement on the tape bead when a longitudinal force is applied. A fixture is attached to the testing machine. After separating the stringers the fixed retainer is placed on the upper edge of the slot of the fixture such that the upper edge of the lower clamp is 3 in. apart from the lower edge of the fixed retainer. An increasing load is applied until failure occurs. The force applied and nature of the failure is noted.

FIGURE 8.18
Bottom stop holding, stringer separation.

8.6.1.10 Separable Pin, Pull-Off Test

This measures the ability of the separable pin to resist displacement on the tape bead, when a longitudinal force is applied. To perform the experiment, after separating the stringers, a few teeth adjacent to the separable pin are removed. The fixture of the separable pin is fixed in the upper clamp of the machine while the separable pin rests on the fixture notch. The clamp holds the stringer on the other end such that the distance between the top of the lower clamp and the seated end of the pin is 3 in. An increasing force is applied until failure occurs. The force applied and nature of the failure is recorded.

8.6.1.11 Separating Unit, Cross-Wise Test

The tapes of the zipper are secured in the clamps of the testing machine such that the edges of the jaws are parallel and 3 mm apart. The separating unit is positioned to align the separating pins to the sides of the front jaws. An increasing load is applied until failure occurs. The nature of the failure and force applied are recorded.

The resistance of the slider to the compression test is used to determine its crushing resistance. For example, pressing the end item may cause the inappropriate functioning of the zipper or even failure. A compression testing machine is used for this purpose. The lower platen of the machine is cushioned with a neoprene pad. The specimen is laid on the pad and a load is applied at a rate of traverse of approximately 13 mm/min until the load reaches the applicable specification. The effects are measured on the ability of the zipper to operate.

Slider deflection and recovery measures the ability of the slider to retain control of the chain by the resistance of the slider planes against an applied opening of a spreading force. In one procedure the force is applied to the mouth of the slider while in other it is applied through the slider pull and the backplane of the slider.

8.6.1.12 Slider Deflection and Recovery (Mouth)

A slider deflection mouth fixture is used on the tensile testing machine. The appropriate diamond spacer and nibs are selected first. The diamond spacer is inserted into the slider transversely and an initial force of 4.4 N is applied and the deflection indicator dial is set to zero. A specified load, corresponding to the size of the zipper being tested, is applied at a constant rate of traverse of approximately 13 mm/min. The deflection of the dial indicator is recorded. The load is then reduced to 4.4 N again and the measurement on the dial indicator is taken again to note any reading above zero. This is recorded as the permanent set.

8.6.1.13 Slider Deflection and Recovery (Pull)

The slider deflection fixture is fastened to the lower clamp of the tensile testing machine and the slider is placed on the fixture in such a way that the fixture enters the mouth first and the diamond enters the notch of the top plate. The puller is attached to the upper clamp of the testing machine in such a way that it is at right angles to the slider body. If the design allows the lengthwise travel of the slider then it must be at a position midway along its whole traveling length. The dial gauge is adjusted so that it is on the top surface of the top plane and as close as possible to the mouth end. Apply an initial load of 4.4 N and set the dial to zero. Now, apply the appropriate load for the corresponding zipper size, at a constant rate of traverse of approximately 13 mm/min. The deflection is read and recorded from the dial indicator. The load is brought to 4.4 N again to read the measurement on the dial indicator. Any reading above zero is recorded as the permanent set.

The resistance of slider and puller against twist is the test that determines the slider–puller assembly strength to resist deformation or rupture against a torsional stress faced during its usage in the end product. The pull test fixture assembly is used to perform the relevant test. The slider is attached to the pull twist test fixture. The puller is positioned in the clamp such that the front surface of the clamp is aligned to the midpoint of the pull. A testing drum is turned clockwise to a certain point and the dial is set at zero. A torque wrench is used to apply an appropriate torsional force at 9° per second. The torque is removed and the testing drum is rotated by hand so that the clearance is taken up. The amount of permanent twist is recorded from the dial. The same is repeated for the second sample, except that the direction of drum rotation is anticlockwise rather than clockwise. The failure or deformations are noted along with any permanent twist. The standard testing method is ASTM D-2062 [24].

In addition to the above, zippers can also be tested for their operability, such as the sticking of the zipper at the stops, zipper closing and opening, and separator functioning. Apart from this, zippers may need to be tested for their colorfastness to washing, dry-cleaning, and light, the durability of the finish of the zipper to laundering, and also the colorfastness of the zipper tapes to crocking.

8.6.2 Testing for Buttons

Buttons are the most common means of closure for garments, especially for men. There are various types of buttons depending on the materials used in their manufacturing, such as metallic and plastic buttons. These may also be differentiated with respect to their types, such as sew through buttons, flange buttons, snap buttons, and shank buttons. Each of these involves its own performance requirements and hence different testing techniques. Some of the most common testing methods related to buttons are discussed here.

The impact resistance test is used on the sew-through flange button. It determines the ability of a button to resist breaking under impact. An impact resistance tester is used. Vernier calipers are used to determine the button diameter in mm, which is divided by a constant 0.635 mm to obtain the relevant ligne size. The specimen is then placed in the centering device for the ligne size of the button so that it lies at the center of the vertical tube. A mass is allowed to fall from a predetermined height onto the button. The specimen is then analyzed with a 5× magnifying glass for cracking, chipping, and breakage.

The tension strength test is used to determine the resistance of the bridge of a sew-through flange button to strain that may cause it to fall off the garment to which it is attached. A tensile tester with a working principle of CRE is used for this test. The button is sewn to the fabric and positioned in the clamp of the testing fixture. The clamped fixture is positioned on the testing machine and two ends of the fabric are placed into the open jaws of the lower clamp with proper alignment. It is then pulled until failure occurs. The force used and the kind of damage is noted. This detailed testing method should follow ASTM D-6644 [25].

8.6.3 Polybag Inspection

Polybags are used to protect prepared merchandise from external effects like dust and stains. They also provide compact packaging, easy handling, and the transportation of goods. These are used to preserve the aesthetic properties of prepared merchandise and range from the packaging of individual items to bulk packaging bags. For the garment industry, polybags are made up of polypropylene or polyethylene. Home textile items like bed sheets, quilts, and quilt covers are usually packed in bags made from polyvinyl chloride (PVC). The important variables are polybag thickness and sealing or zipper performance characteristics. The sealers are chosen for one time or multiple time usage where it is required to open and seal the bags more than once.

The polybag thickness is generally measured with a screw gauge and is termed the "micron value." The supplier of bags is supposed to use the predecided sheet thickness which is inspected by textile and apparel manufacturers upon delivery arrival to their respective premises. Any deviation in the micron value may result in a price penalty, rejection of the lot, or even termination of the business if the problem is ongoing.

Other issues related to polybags may involve zipper color fastness, which has already been described in this chapter. In addition, if polybags are printed they should also be inspected for color shade, text, and colorfastness to the relevant performance conditions.

8.6.4 Stiffener and Carton Inspection

The use of corrugated stiffeners and cartons is very common in the packaging of textile and clothing products. Stiffeners are inspected to check for deformation against impact from various angles and at different points. The ideal stiffener should retain its desired shape against possible impacts faced in material handling and transportation. This quality is very much necessary to preserve the aesthetic properties of packed goods like shirts and beds sheets.

Cartons are subject to much environmental variability during material handling and transportation. These include changing temperature and humidity conditions which are a menace to carton strengths. A crash test is the most common in which the carton is subjected to a force between two flat surfaces until damage occurs. One more important consideration is to test for edge crushing which determines the resistance to damage when an impact acts on a corner of the carton.

8.6.5 Woven Labels

Woven labels are most commonly used inside garments. The main types include care labels, brand labels, size labels, operator labels, and labels containing style description. Labels are inspected to check for size and text. The color of the label and text are also important as well as font size and style. Any abnormality is reported for quick fixing or for replacement of the whole lot.

8.6.6 Printed Materials

A number of printed materials are used in apparel and textiles and in the manufacturing and packaging processes. Insert cards, hang tags, price stickers, stock keeping unit (SKU) stickers, labels, and printed bags are the most common. These are all inspected to check if there is any error in the text (spelling and wording) and formatting (font size, style, etc.). In addition, the color of printing and the shade of the color may also be among points of concern from the customer's point of view.

8.6.7 Paper and Card Stuff

Most of the insert cards and hang tags used in the garment industry are paper-based products. These are generally negotiated in terms of paper weight (GSM). Surface finish and textured materials are also used. Sometimes embossing or engraving is also done to attain a certain aesthetic and hand feel.

8.7 Clothing Inspection

Garment manufacturing involves minimum automation in its manufacturing processes as compared to yarn manufacturing, fabric manufacturing, and textile processing. A single garment may consist of more than a hundred operations and each of them involves manual operator handling which creates variation in output. A number of factors contribute to the quality of stitching, including operator skill, machine setting, thread quality, reference standards being followed, and technical support to the operators. Operators are hired by passing them through certain examinations and further training on the machines, yet it is common to observe variations in sewn products with respect to the finished measurements, stitching aesthetics, and performance. For example, improper handling at the overlock machine may result in cutting edges being more or less than the requirement and consequently a variation in the finished measurement of the garment. In addition, a blunt needle or improper thread tension setting, if ignored by the operator, also deteriorates the quality of the product. It is therefore necessary to pay strong attention to quality management. To ensure quality conformity to customer requirements, inspection is done at various levels.

8.7.1 Cut Part Inspection

Cutting measurements are more than the finished garment measurements; this is to accommodate stitching and shrinkage. If seam allowances are not sufficient, yarn slippage may occur, causing a seam failure. Although cutting automation exists, only a few large manufacturers can afford it, owing to its higher initial cost. Most of the small and medium size apparel manufacturers still execute cutting process using straight knife cutter. As human hands are involved in cutting the chances of error increase. To ensure the proper size of the cut parts it is necessary to check for appropriate sizes. Although a problem at this level is acute, it is easier to deal with here rather than in the finished garments. Special instructions may also be issued to deal with certain problems in cut parts, to be followed while stitching. In other cases pieces may be recut to replace the faulty pieces.

Apart from this, inspection of cut panels also highlights yarn and fabric faults in particular cut panels which were overlooked or left intentionally to be addressed at the next stage. This is done with an expectation that the fault may fall within a buffer zone and be removed without too much effort. Such faulty panels are replaced with new panels cut from fresh fabric. The result is a fault free cutting feed to the production floor.

8.7.2 Inspection at the Sewing Floor

The main stitching faults associated with the sewing floor are

- Seam puckering owing to improper thread tension or machine adjustment
- Skip stitches occurring due to machine parts not being synchronized
- Broken stitches caused by heat, improper needle size, a blunt needle, poor quality thread, or improper thread tension
- Measurement shortage caused by excessive trimming of fabric at machines
- Non-standard stitch density (stitches per unit length), generally caused by operators rushing the pieces under the needle
- Variable stitch density due to machine disorder
- Unsecured seam end points when back tack function is not operated
- Shade variation due to not following the panel numbering
- Wavy seams caused by inadequate operator skills
- Needle holes caused by a blunt needle and also by keeping the machine running when thread breaks occur or when the thread cone/bobbin needs to be replaced/refilled
- Unbalanced stitch caused by dislocation of upper and lower thread lock position owing to machine setting
- Thread fusing due to excessive heat at needle point.

One or more of the above mentioned faults may occur in a single piece of garment or even in a single operation. These faults are checked throughout the stitching process (inline) and also when they are offline after completion (end line). Both of them are discussed as follows.

8.7.3 Inline Inspection

There have been different viewpoints on whether inline inspection is worth doing or not. Yet it is common to observe, in apparel manufacturing, inline quality control activities with different names and criteria. Inline quality inspection is to check the quality of products right at the workstation where it is being manufactured. A certain sample of semi-furnished products is selected after regular intervals, from each work station, to check whether the required parameters are being met or not. The essence of inline quality checking is to screen out and rectify observed problems at a particular workstation. In the case of any abnormality, the relevant supervisor is informed to investigate and root out the problem cause to ensure that it does not happen again. The quality control person usually checks the seam aesthetics (shape, size, and joints), stitch density, and alignment of stitches.

8.7.4 End Line Inspection

Inspection is carried out at the end of each section of a sewing line and/or at the end of a whole production line when the garment is fully prepared. Generally, the small part section, front section, and back section have their respective check points. Each end line checker is responsible for major and minor faults in view of customer definitions of various faults. In a nutshell, measurements and stitching faults (broken stitches, puckered seams, seam appearance, skip stitches, and improper shape) are checked at this point. A number of faulty pieces are sent back to the respective machine operator for mending while the remainder are rejected for being unrectifiable.

8.7.5 Inspection in Washing

The finishing of the garment is carried out after end line inspection. Garment washing (if involved) is done before finishing but after stitching. In some cases, garment dyeing is also done after stitching the whole garment to impart certain aesthetics. Washing has its own quality conformance levels with respect to color and shade matching. In the case of garment dyeing certain Pantone numbers may need to be followed and matched, failure of which may lead to shipment rejection. Regarding washing there are certain aesthetic and hand feel criteria that must be achieved in view of the approved standards made by the customer. Any abnormality which is uncontrollable should be approved for acceptance by the customer before the shipment is made ready for customer inspection, otherwise there may ensue payment penalties, cancelation of shipment, and/or future business loss.

8.7.6 Inspection at Finishing

Industrial washing causes garments to pass through heavy throws while in the washing, squeezing, and drying process which may lead to certain faults. A slightly broken stitch may get wider, for example. Therefore, garments are checked to see if any rework is needed after the washing process. Buttons and other metallic items are attached after washing usually. The garment is then pressed and declared ready for final check. At this stage garments are particularly checked for their size specifications as washing and steam pressing causes fabric shrinkage. The appropriate size labels are also checked on each garment and then packing is carried out.

8.7.7 Shipment Audits

These are usually conducted by third party auditors. They are generally native and occasionally from other countries and are hired by the customer to help in ensuring his or her required parameters are being met by the manufacturers. Once the inspection is ready the auditor is called upon to cross-check the items. At the decided date and time the inspection team arrives and selects a sample from the offered shipment lot. The sample is generally selected at the mutually accepted quality level. The inspection is done to check for any faults regarding yarn, fabric, stitching, washing, finishing, and packing, including all kinds of materials. A list of detailed faults is created while checking and an analysis report is made either to accept or reject the prepared consignment. If the consignment is found to be up to the mark it then proceeds to shipment through logistics. If not, it stays at the manufacturers end for removal of the highlighted problems before being offered for inspection again.

References

1. A. Rasheed, Clothing, in, Y. Nawab (ed.), *Textile Engineering: An Introduction*, Berlin, Germany: De Gruyter, pp. 111–132, 2016.
2. ASTM-D5430-93, Standard test methods for visually inspecting and grading fabrics, ASTM, West Conshohocken, PA, 2000.
3. ASTM-D3787, Standard test method for bursting strength of fabrics—Constant-rate-of-traverse (CRT), ASTM, West Conshohocken, PA, 2001.
4. ASTM-D6797, Standard test method for bursting strength of fabrics—Constant-rate-of-extension (CRE) ball burst test, ASTM, West Conshohocken, PA, 2002.

5. ASTM-D5035-11, Standard test methods for breaking force and elongation of textile fabrics (strip method), ASTM, West Conshohocken, PA, 2015.
6. ASTM-D5034-95, Standard test methods for breaking strength and elongation of textile fabrics (grab test), ASTM, West Conshohocken, PA, 2001.
7. ASTM-D1424, Standard test method for tearing strength of fabrics by falling-pendulum type (Elmendorf) apparatus, ASTM, West Conshohocken, PA, 1996.
8. ASTM-D2261-96, Standard test method for tearing strength of fabrics by the tongue (single rip) procedure (constant-rate-of-extension tensile testing machine), ASTM, West Conshohocken, PA, 2002.
9. ASTM-D434, Standard test method for resistance to slippage of yarns in woven fabrics using a standard seam, ASTM, West Conshohocken, PA, 1995.
10. ASTM-D3776-96, Standard test methods for mass per unit area (weight) of fabric, ASTM, West Conshohocken, PA, 2002.
11. ASTM-D3882, Standard test method for bow and skew in woven and knitted fabrics, ASTM, West Conshohocken, PA, 1999.
12. ASTM-D737, Standard test method for air permeability of textile fabrics, ASTM, West Conshohocken, PA, 1996.
13. AATCC TM 135, Dimensional changes of fabrics after home laundering, AATCC, 2004.
14. ASTM-D1423/1423M, Standard test method for twist in yarns by direct-counting, ASTM, West Conshohocken, PA, 2016.
15. ASTM-D1422/1422M, Standard test method for twist in single spun yarns by the untwist-retwist method, ASTM, West Conshohocken, PA, 2013.
16. ASTM-D204, Standard test method for sewing threads, ASTM, West Conshohocken, PA, 2002.
17. ASTM-D2256, Standard test method for tensile properties of yarns by the single-strand method, ASTM, West Conshohocken, PA, 2002.
18. ASTM-D1907, Standard test method for linear density of yarn (yarn number) by the skein method, ASTM, West Conshohocken, PA, 2001.
19. ASTM-D3823, Standard practice for determining ticket numbers for sewing threads, ASTM, West Conshohocken, PA, 2001.
20. ASTM-D6193, Standard practice for stitches and seams, ASTM, West Conshohocken, PA, 1997.
21. ASTM-D1683, Standard test method for failure in sewn seams of woven apparel fabrics, ASTM, West Conshohocken, PA, 2004.
22. ASTM-D3940, Test method for bursting strength (load) and elongation of sewn semas of knit or woven stretch textile fabrics, ASTM, West Conshohocken, PA, 1983.
23. ASTM-D2061, Standard test methods for strength tests for zippers, ASTM, West Conshohocken, PA, 2003.
24. ASTM-D2062, Standard test methods for operatability of zippers, ASTM, West Conshohocken, PA, 2003.
25. ASTM-D6644-01, Standard test method for tension strength of sew-through flange buttons, ASTM, West Conshohocken, PA, 2002.

Section II

Testing of Technical Textiles

9

Composite Materials Testing

Khubab Shaker and Yasir Nawab

CONTENTS

The composite may be defined as a macro-scale combined assembly of two or more components, having properties that none of the constituents possesses individually [1]. Composites are not isotropic materials, but are orthotropic, in general. They are fabricated to meet the design and functional properties of the structure. The constituent categories of composite material are termed the reinforcement and the matrix. Usually, the reinforcement is a collection of fibers, which are responsible for the mechanical strength of the material [2]. The fibrous reinforcements are used in a number of forms, for example, fabric, unidirectional, and chopped fibers. The orientation and composition of reinforcing fibers defines the performance properties of composite material. The fiber orientation is also an indicator of composite strength. The composite shows its best properties parallel to the fiber axis, because the fiber can bear the maximum load along its length. Any change from the fiber axis reduces the load bearing capability of the composite drastically. The other constituent, namely, the matrix, serves as a binder and helps to hold the reinforcing fibers together [3]. In addition, it also transfers the applied load to the reinforcement, which carries the applied stress. The composite properties also depend on the mechanical properties of the matrix.

9.1 Physical Testing

The physical testing of a composite material investigates the effect of physical phenomena on the performance properties of the material. It determines one or more properties of a given material according to a predefined practice. The physical testing involves its morphology, density/specific gravity, resin/fiber/volatile/void content, water absorption, moisture content, thermal properties, and so on.

9.1.1 Surface Morphology

Morphology is the term used for the study of the structure or shape of a part. In the case of composite materials, surface morphology comprises high resolution 3D imaging by use of sophisticated microscopes. The exposed surface of the sample or product is viewed for details that cannot be seen with the naked eye. This technique produces very minute details of the specimen surface even at the nanometric scale. The scanning electron microscope is the most powerful tool for examining the surface morphology of composite materials.

9.1.2 Analytical Testing

9.1.2.1 Density of Sandwich Core Materials

Material density is one of the fundamental physical properties that may be used to characterize the sandwich core materials, along with other properties. Most of the structural properties of sandwich core materials (e.g., stiffness, strength) are related to its density. ASTM C271/C271M-16 is the standard test used to determine the density of sandwich core materials [4]. This test method consists of environmentally conditioning a sandwich core specimen, weighing the specimen, measuring the length, width, and thickness of the specimen, and calculating the density.

The test specimen must have a square or rectangular cross-section. The minimum specimen size is 300 mm × 300 mm (length by width). The caliper is used to measure the thickness of the specimen. An air circulating oven or vacuum drying chamber is used that is capable of maintaining the temperature to ±3°C and is used for the conditioning of the specimens to one of the following conditions

- Standard atmospheric conditions of 23°C ± 3°C and 50% ± 5% relative humidity
- In oven-dried equilibrium at a temperature of 105°C ± 3°C
- In oven-dried equilibrium at a temperature of 40°C ± 3°C

After conditioning, the specimens are cooled at room temperature and measurements are taken. The density is calculated using

$$d_{SI} = \frac{W \times 10^6}{lwt}$$

where

d_{SI} is the material density in kg/m^3
W is the final mass in grams
l is the final length in mm
w is the final width in mm
t is the final thickness in mm

All these parameters are recorded after conditioning.

9.1.2.2 Constituent Content of Composite Prepreg

Prepreg is a fibrous reinforcement impregnated with a polymeric matrix system, generally in the form of a sheet, tape, or tow. It is used to fabricate the composite material without the infusion/application of more resin, by just laying it in the mold. The fiber, matrix solids, and matrix content of the composite material prepreg are determined using the ASTM D3529M-10 standard test method [5]. The determination of the content of matrix solids depends on the determination of the content of volatiles.

The prepreg specimen of a specific area is weighed and the matrix is removed by the appropriate procedure. In one method, the specimen is subjected to a suitable solvent that affects the matrix (but not the reinforcement) and removes it by dissolving it. The remaining reinforcement is then dried and weighed. The most commonly used matrix solvents are effective for many thermosetting matrices, including acetone, methyl pyrrolidinone, methyl ethyl ketone, dimethylformamide, dichloromethane (methylene chloride), and methyl isobutyl ketone.

The specimen is weighed initially and then placed in a separate container using a minimum of 100 mL of solvent for each specimen. Specimens remain in the solvent for at least 3 min. The specimen is then placed in a fritted glass crucible and dried in a circulating oven at a temperature and time sufficient to remove the solvent (a minimum of 100°C for 5 min). The crucible is then cooled with the specimen remains in a desiccator to ambient temperature for a minimum of 5 min. The fiber remains are then examined to evaluate completeness. The fibers should be easily separable, with no evidence of binding action between them.

In the other method, the specimen is placed in a muffle furnace for a specific time and at a specific temperature (at which the reinforcement remains unchanged) until the matrix can be entirely removed as ash residue. The specimen exposure temperature is 500°C in an air environment for up to 6 h. The remaining reinforcement is weighed. The minimum specimen size is 80 mm × 80 mm, with a minimum mass of 1 g.

The matrix content is calculated as

$$\text{Matrix content, MC} = \frac{M_i - \left(M_f - M_c\right)}{M_i} \times 100$$

where

M_i is initial mass (grams) of specimen

M_c is the mass (grams) of the container

M_f is the combined mass (gram) of container and reinforcement

The content of matrix solids (dry resin content) is calculated as

$$\text{Matrix solid content, MS} = \frac{M_i\left(1 - V_c/100\right) - \left(M_f - M_c\right)}{M_i\left(1 - V_c/100\right)} \times 100$$

where V_c is the content of average volatiles (percentage of weight). The content of average volatiles of the specimen is calculated according to ASTM D3530 [6].

Fiber content is calculated as

$$\text{Fiber content, FC} = \frac{M_f - M_c}{M_i} \times 100$$

This information is further used to determine the areal weight of the fiber using

$$\text{Fiber areal weight, FAW} = \frac{M_f - M_c}{A}$$

where A is the nominal area of the specimen in m^2.

9.1.2.3 Content of Volatiles of Composite Material Prepreg

When the content of matrix solids is requested, adjacent samples are tested for the content of volatiles according to ASTM D3530-97 (Reapproved 2015) [6]. The average mass loss due to volatiles is subtracted from the average resin content result, and this result is expressed as a percentage of the initial specimen mass as the content of matrix solids, which is applicable mainly to thermosetting matrices.

The samples are placed on the rack and then into a preheated oven set at the nominal cure or consolidation temperature. The specimen is placed in such a way that the maximum surface area is exposed to the circulating heat. After 15 min, the specimens are removed from the rack and placed in a desiccator. After cooling to ambient temperature, the specimens are weighed within 1 min of removal from desiccator. The volatile contents are calculated as

$$\text{Volatile content}, V_c = \frac{M_i - M_f}{M_i} \times 100$$

9.1.2.4 Ignition Loss of Cured Reinforced Resins

The test is carried out according to ASTM D2584-11 [7]. The specimen of 2.5 × 2.5 cm (approximately 5 g) is placed in a crucible. The crucible is heated at 500°C–600°C for 10 min and the specimen is allowed to burn completely, leaving behind remains of carbon and ash. This remaining carbon is then reduced to ash by heat in a muffle furnace at 565°C ± 28°C; it is then cooled in a desiccator, and weighed to obtain the residue weight.

The percentage of ignition loss of the specimen is calculated in terms of weight using

$$\text{Ignition loss \%} = \frac{W_1 - W_2}{W_1} \times 100$$

where

W_1 is the specimen weight

W_2 is residue weight (both in grams)

9.1.2.5 Void Content of Reinforced Plastics

ASTM D2734-09 is referred to determine the void content of the reinforced plastics [8]. This method is based on the density of reinforcement, resin, and the composite material, which are separately measured. Firstly, the resin content of the composite material is determined and a theoretical density of composite is calculated. This theoretical density of composite material is compared to its measured density. The difference between both values gives the void content of the composite material. A well fabricated composite material must have a void content of 1% or less, while a poorly prepared composite has a much higher void content.

In this test method, it is assumed that the density of resin is the same in the composite material as it is in a larger mass. Density is determined for the bubble-free resin pieces, cured under similar conditions of temperature and time as that of the composite. Density values provided by the manufacturer are also acceptable. Theoretically, the density of a composite material is calculated as

$$\text{Theoritical density of composite, } T = \frac{100}{(R/D)+(r/d)}$$

where

R is the percentage of resin in the composite
D is the density of resin
r is the percentage of reinforcement in the composite
d is the density of reinforcement

Now, the void content in a composite material is determined as

$$\text{Void content, } V = \frac{100(T_d - M_d)}{T_d}$$

where

T_d is the theoretical density of the composite
M_d is the measured density of the composite

9.1.2.6 Constituent Content of Composite Materials

There are two test methods reported for the determination of the contents of the composite material constituent in ASTM D3171-15 [9]. According to the first method, the matrix is physically removed by digestion, ignition, or carbonization, in such a way as to leave the reinforcement unaffected. The reinforcement is then weighed and finally the matrix or reinforcement content (by volume or by weight) is determined along with the percentage of void volume. The volume percentage is calculated only if the densities of both the composite and the reinforcement are known. An additional calculation for void volume may be made if the density of the matrix is known or determined.

The other test method applies to laminated composite materials with reinforcement of known areal density. It determines the matrix or reinforcement content (by volume or by weight) and cured ply thickness, based on the measured thickness of the laminate. This method is not applicable for the determination of void contents. The length, width, and thickness of the flat composite material are measured using a caliper. Now, the density of this composite specimen is calculated in g/cm^3 as

$$\text{Composite density, } \rho_c = \frac{M_i}{A \times h \times 1000}$$

where M_i is mass (grams) of a specimen having area A (m^2) and thickness h (mm).

Using this information, the reinforcement content is calculated as

$$\text{Weight of reinforcement, } W_r = \frac{A_r \times N \times 0.1}{\rho_c \times h}$$

where

A_r is the areal weight (g/m^2) of one reinforcement sheet

N is the number of plies in the test specimen

The content of the reinforcement volume is calculated according to

$$\text{Volume of reinforcement, } V_r = \frac{A_r \times N \times 0.1}{\rho_r \times h}$$

The matrix content in terms of percentage weight and volume is calculated using

$$\text{Weight of matrix, } W_m = 100 - \left(\frac{A_r \times N \times 0.1}{\rho_c \times h}\right)$$

$$\text{Volume of matrix, } V_m = \frac{W_m \times \rho_c}{\rho_m}$$

The cured ply thickness (mm) is calculated from the measured laminate thickness, h

$$h_p = \frac{h}{N_p}$$

where

h_p is the cured ply thickness (mm)

N_p is the number of plies in the laminate

The test methods described here are used primarily for a two-part composite material system, and cannot be applied to a composite material having a third constituent, that is, filler particles. Special provisions are required to extend these test methods for filled composite material systems (having more than two constituents).

9.1.3 Thermal Properties

9.1.3.1 Transition Temperatures, and Enthalpies of Crystallization and Fusion

The transition temperatures and enthalpies of crystallization and the fusion of polymers are determined according to ASTM D3418-15 using differential scanning calorimetry [10]. The test method involves heating or cooling the test material at a precise rate, under a controlled flow rate of purge gas. The test material is monitored continuously using suitable sensing devices. The absorption or release of energy by the test specimen giving an endothermic or exothermic peak corresponds to a transition. The areas under a fusion endotherm or a crystallization exotherm of the test material are compared with the respective areas of a well characterized standard material.

Thermal analysis is a quick tool to measure the transitions in material produced by chemical or morphological changes. These changes occur when the polymer is heated/cooled through a definite temperature range. Changes in material properties during these transitions, like heat flow, temperature, and specific heat capacity, are determined. Differential scanning calorimetry is a useful tool to identify specific polymers, polymeric alloys, and polymeric additives that exhibit thermal transitions. This technique is helpful for measuring the chemical reactions (oxidation, curing of thermoset resin, thermal decomposition, etc.) that affect/cause certain transitions. Further details can be found in the Chapter 2 dedicated to polymeric materials testing.

9.1.3.2 Linear Thermal Expansion of Solid Materials

ASTM E831-14 is the standard test method to determine linear thermal expansion using a thermo-mechanical analyzer [11]. To conduct this test, the specimen is subjected to heating at a constant rate. The change in specimen length as a function of temperature is then recorded electronically, and the coefficient of linear thermal expansion is calculated from this data. The specimen length is measured in the direction of expansion to ±25 μm at 20°C–25°C, and recorded as its initial length.

The specimen is then placed in a specimen holder under the probe in such a way that a temperature sensor is also in contact with the specimen. The furnace is moved to enclose the specimen holder, and an appropriate contact load is applied to the sensing probe to make sure that it is in contact with the specimen. The specimen is then heated to the desired temperature level, at a constant rate of 5°C/min, and the change in specimen length with temperature are recorded. Meanwhile, the change in the length of the specimen holder is also recorded without the specimen, and the measured ΔL for the specimen is corrected, especially for low expansion specimens.

The mean coefficient of linear thermal expansion, in μm/(m °C), for a desired temperature range is calculated as

$$\text{Coefficient of thermal expansion}, \alpha_m = \frac{\Delta L_{sp} \times k}{L \times \Delta T}$$

where

k is the calibration coefficient
L is the initial specimen length (m) at room temperature
ΔL_{sp} is the change in specimen length (μm)
ΔT is the temperature difference (°C) over which this change is measured

9.1.3.3 Compositional Analysis by Thermogravimetry

ASTM E1131-08 (2014) describes a general method to determine the amount of medium and highly volatile matter, combustible material, and ash content of compounds [12]. It is done by the incorporation of the thermogravimetry technique. This test method is beneficial in performing a compositional analysis of the composite material.

The specimen of 10–30 mg is carefully placed in the specimen holder, and its initial mass is recorded. Heating is initiated within the desired temperature range, and the change in mass of the specimen is recorded continuously over the temperature interval. The mass loss is expressed in either milligrams or as a percentage of mass of the original specimen mass. The whole process is carried out in an inert environment (nitrogen). After establishment of mass loss in the range 600°C–950°C, the environment is switched from inert to reactive (air or oxygen). The analysis is complete upon the establishment of a mass loss plateau following the introduction of the reactive gas.

The highly volatile matter is represented by a mass loss measured between the starting temperature and temperature X. The content of highly volatile matter may be determined by

$$V = \frac{W - R}{W} \times 100$$

where

V is the content of highly volatile matter (%)
W is the original specimen mass (mg)
R is the mass measured at temperature X (mg)

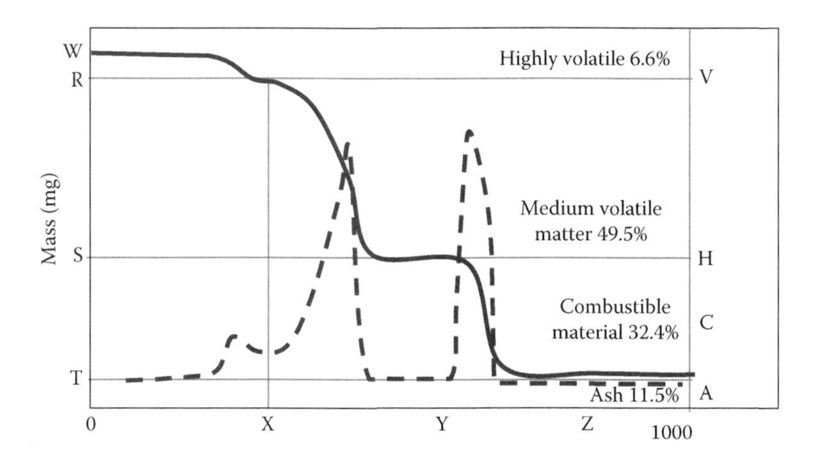

FIGURE 9.1
Sample thermogravimetric curve.

The medium volatile matter is represented by the mass loss measured from temperature X to temperature Y. The content of medium volatile matter can be determined using

$$O = \frac{R - S}{W} \times 100$$

where
 O is the content of medium volatile matter (%)
 S is the mass measured at temperature Y (mg)
 W is the original specimen mass (mg)

The content of combustible material is represented by the mass loss measured from temperature Y to temperature Z. This region corresponds to the mass loss as a result of the oxidation of carbon to carbon dioxide (Figure 9.1). The content of the combustible material may be calculated by

$$C = \frac{S - T}{W} \times 100$$

where
 C is the content of combustible material (%)
 T is the mass measured at temperature Z (mg)
 W is the original specimen mass (mg)

9.1.4 Moisture Absorption

ASTM C272/C272M-12 is the standard test method used for water absorption of core materials for sandwich materials [13]. The test specimens have a square or rectangular cross-section, having dimensions 75 × 75 × 13 mm (length × width × thickness). The specimen is then oven dried either at 50°C ± 3°C for 24 h or at 105°C ± 3°C for 2 h, depending on the water absorption behavior of the material. The specimen is then conditioned using one of the following environments:

- *Twenty-four-hour immersion*: The specimens are immersed in a container horizontally under a 25 mm head of water for 24 +1/−0 h at a temperature of 23°C ± 3°C. Afterwards, specimens are removed and all surface water is wiped off with a dry cloth until no visible water is present, and their weight is recorded.
- *Elevated temperature humidity*: The standard conditioning environment is 70°C ± 3°C and 85% ± 5% relative humidity for 30 days. The specimen is dried with dry cloth and the weight is recorded.
- *Maximum percentage mass gain*: The specimen is immersed in a container horizontally under a 25 mm head of water for 48 +1/−0 h at a temperature of 23°C ± 3°C. The weight is recorded after drying. The specimens are then placed back into the water and the process is repeated until the mass gain after the last 48-h interval is less than 2% of the entire mass gain of all the previous intervals.

The percentage increase in mass is calculated as

$$\text{Increase in mass, \%} = \frac{W - D}{D} \times 100$$

where
W is the specimen mass (gram) after immersion and blotting
D is the pre-immersion mass (grams) of the specimen

ASTM D5229/D5229M-14 is the other test method used to determine the moisture absorption properties of the polymeric composite material and its moisture equilibrium conditioning [14]. A material is said to be in a state of moisture equilibrium if the change in moisture content of the material is less than 0.020% over each of two consecutive reference time period spans. It is desirable to plot and examine the mass change versus the $\sqrt{\text{time}}$ curve prior to deciding that the specimen has reached moisture equilibrium.

This method requires a length to thickness proportion of 100:1 to minimize the impact of the edges. However, distinctive length to thickness proportions are utilized, for instance relating to 50:1 (for 1 mm thick), 25:1 (for 2 mm thick), and 12.5:1 (for 4 mm thick) proportions. Existing composite examples have a 5 mm thickness and a 100 mm length, so the length to thickness proportion is 20:1 (5 mm thickness). The specimen should have a mass of at least 5 g and its thickness should not vary by more than ±5% over the surface of the specimen.

Subsequent to cutting and cleaning, the surfaces of the specimens need to be cleaned with methanol to uproot any soil and oil build-up from machining and are afterward dried in a stove at 60°C. The masses of the composite examples are measured intermittently whilst drying, which proceeds until there is no further change in the weight. Once dried, the initial weights of the examples are recorded and the specimens are quickly put into a standard environmental system. A temperature of 20°C ± 2°C and a relative stickiness of 65% is used for the environment. The adjustment in mass is measured by utilizing a mechanical logical parity with a precision of 0.1 mg. The specimen's weight is taken periodically after different intervals of time until they reach equilibrium or a certain behavior was established. A graph is then drawn between the weight gains versus the square root of time. The percentage moisture regain is plotted against $\sqrt{\text{time}}$.

The rate of water absorption in the composites is computed by the weight difference between the specimens immersed in water and the dry samples using

$$\text{Mass change \%, } \Delta M = \left| \frac{W_t - W_i}{W_i} \right| \times 100$$

where

W_i is the initial specimen mass (gram)
W_t is the current specimen mass (gram)

The initial moisture content of the test specimen is

$$\text{Initial moisture content \%} = \frac{W_{ar} - W_{od}}{W_{od}} \times 100$$

where

W_{ar} is the as-received specimen mass (gram)
W_{od} is the oven-dry specimen mass (gram)

The diffusivity of the specimen is calculated as

$$D_z = \pi \left(\frac{h}{4M_m} \right)^2 \left(\frac{M_2 - M_1}{\sqrt{t_2} - \sqrt{t_1}} \right)^2$$

where

h is the average specimen thickness (mm)
M_m is the percentage of effective moisture content at equilibrium
$M_2 - M_1$ is the initial slope of the moisture absorption curve
$\sqrt{t_2} - \sqrt{t_1}$ is the linear portion of the curve ($\sqrt{s^{-1}}$)

9.2 Mechanical Characterization

9.2.1 Tensile Testing

The standard test method ASTM D3039/3039M is used to determine the in-plane tensile properties of polymer matrix composite materials [15]. A rectangular thin strip of composite material having a uniform rectangular cross-sectional area is used as a test specimen. This strip is clamped in the jaws of a tensile testing machine and the load is applied in a preset way. The force and elongation values are determined at regular intervals of time and a curve is plotted. The maximum tensile load, which the specimen carries before mechanical failure, is attributed to the tensile strength of the composite material. Meanwhile, the stress–strain curve plotted during the test helps us to obtain additional information like the ultimate tensile strain, transition strain, tensile modulus, and the Poisson's ratio of the composite material. Five specimens are tested for each sample and their average value is reported, along with the standard deviation.

The specimen dimensions (thickness, width, and length) are selected in a way to promote failure in the gauge section, as shown in Figure 9.2. The test specimen must be a statistical representative of the bulk material. This may be assured by careful selection of the specimen width, such that its cross-section contains a sufficient number of yarns/fibers. The materials such as 2D laminates, fabric-based composites, or randomly reinforced materials are successfully tested without tabs. However, tabs are strongly recommended for the testing of unidirectional materials so as to cause failure in the fiber direction. The recommended specimen geometry configurations for different materials are given in Table 9.1.

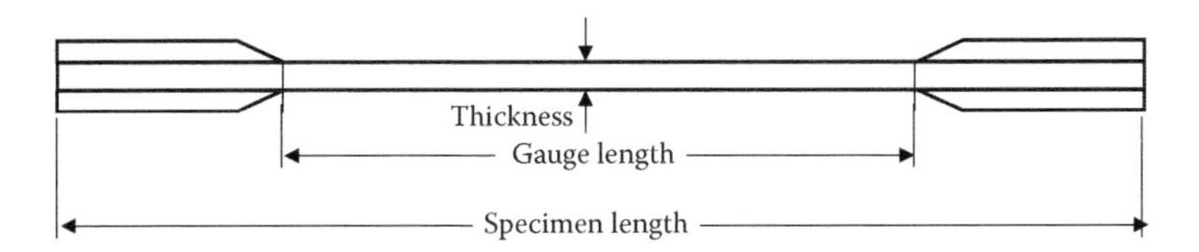

FIGURE 9.2
Cross-sectional view of test specimen.

TABLE 9.1
Recommended Specimen Geometry Configurations

Fiber Orientation	Width (mm)	Specimen Length (mm)	Thickness (mm)
0° unidirectional	15	250	1
90° unidirectional	25	175	2
Balanced and symmetric	25	250	2.5
Random discontinuous	25	250	2.5

The specimen area (as $A = w \times h$, in mm²) after machining and conditioning is also determined prior to the test at three places in the gauge section. The strain rate is selected so as to produce failure within 1–10 min of the test; and in the case of constant head speed tests, the head displacement rate is set to 2 mm/min.

The ultimate tensile strength, F_{ut} (MPa), of the specimen is calculated by using the maximum force, P_{max} (N), at failure and the initial area, A (mm²), of the specimen. The stress at any point is equal to the force at that point divided by the area. Similarly, strain at any point may also be calculated:

$$F_{ut} = \frac{P_{max}}{A}$$

$$\sigma_i = \frac{P_i}{A}$$

$$\varepsilon_i = \frac{\delta_i}{L_g}$$

The tensile modulus of elasticity (E) is defined as the ratio of change in stress to a corresponding change in strain. The following relation is used:

$$\text{Tensile modulus of elasticity, } E = \frac{\Delta\sigma}{\Delta\varepsilon}$$

where

$\Delta\sigma$ is the change in stress (MPa) between two selected points
$\Delta\varepsilon$ is the change in strain between two selected points

ASTM D5766/D5766M-11 determines the open-hole tensile strength of multidirectional polymer matrix composite laminates reinforced by high-modulus fibers [16]. The uniaxial tensile test is performed on the laminate in accordance with ASTM D3039, with a centrally located hole. The only acceptable failure mode for ultimate open-hole tensile strength is one that passes through the hole in the test specimen. The test results are highly affected by the ratio of specimen width to hole diameter (w/D). This ratio is maintained at 6. Results are also affected by the ratio of hole diameter to thickness (D/h). The preferred ratio is in the range from 1.5 to 3. The configuration of the specimen is shown in Figure 9.3. All the calculations are performed in a similar way to that of ASTM D 3039. While performing the calculations, the area of the specimen is the gross cross-sectional area, ignoring the hole.

9.2.2 Compression Testing

There exists a number of test methods to determine the in-plane compressive properties of polymer matrix composite materials. These standards differ in

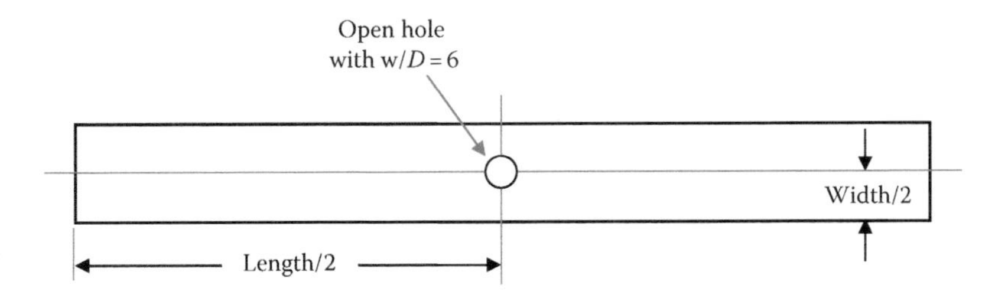

FIGURE 9.3
Cross-sectional view of the open-hole test specimen.

the type of force transfer, that is, by loading or shearing. For example, the compressive force is transmitted into the specimen by end-loading, through shear at the wedge grip interface or by combined shear and end-loading in test methods ASTM D695 [17], ASTM D3410M [18], and ASTM D6641M, respectively [19].

According to the standard test method ASTM D695, the specimen for compression strength testing is either in the form of a right cylinder or a prism, with its length twice its principal width or diameter. For a cylindrical specimen, the preferred specimen has a diameter of 12.7 mm and a height of 25.4 mm, while in the case of a prism it is 12.7 × 12.7 × 25.4 mm. When the standard specimens (right cylinders or prisms) cannot be obtained due to the thinness of the material (less than 6.4 mm), alternative specimens are used. For 3.2–6.4 mm thick materials, the specimen will consist of a prism having a cross-section of 12.7 mm by thickness of material and a length of 12.7 mm. For materials less than 3.2 mm thick, a compressometer or similar device is used. The supporting jigs are used to support the specimen during testing. The standard speed of testing should be 1.3 ± 0.3 mm.

In ASTM D3410, a flat rectangular rip of composite material having a uniform cross-section is subjected to a shear force that acts along the wedge grips (a specially designed fixture). The ultimate compressive strength of the material is the maximum force carried before failure.

9.2.3 Flexural Testing

The flexural properties are investigated by a three-point bending test (ASTM D7264) [20] or a four-point bending test (ASTM D6272) [21]. In flexural testing, a rectangular bar rests on two supports and is loaded at one or two points (by means of loading noses) at an equal distance from the adjacent support point, depending on the test type. The distance between the loading noses (the load span) is either one-third or one-half of the support span as shown in Figure 9.4.

The specimen dimensions are subject to a span-to-thickness ratio of the composite material. In the three-point bending test, the standard specimen width is 13 mm, the standard thickness is 4 mm, and the recommended

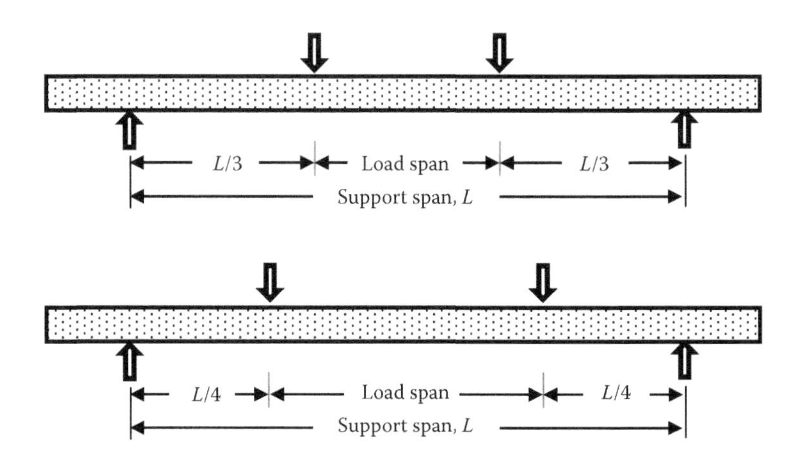

FIGURE 9.4
One-third of support span, and one-half of support span.

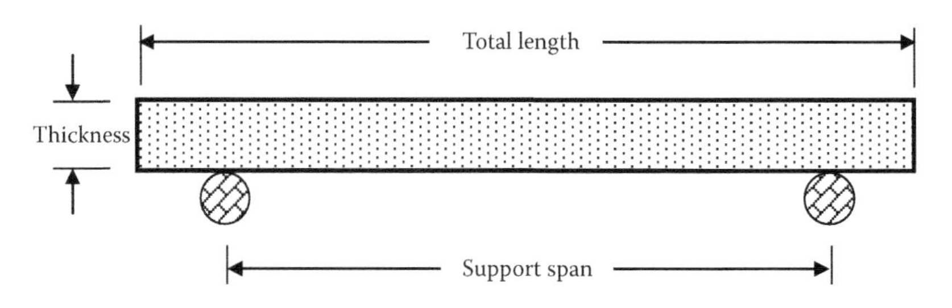

FIGURE 9.5
Sample specifications for flexural testing.

standard span-to-thickness ratio is 32:1. The specimen length calculated using the span-to-thickness ratio is termed the support span and the overall specimen length is about 20% more than the support span (Figure 9.5). The standard allows one to use any other specimen thickness while using the standard span-to-thickness ratio (32:1). In addition, there are certain optional span-to thickness ratios of 16:1, 20:1, 40:1, and 60:1 with the condition that it must be noted in the report. The test speed is adjusted at a crosshead rate of 1.0 mm/min.

The flexural strength test results are reported in terms of maximum applied force and deflection. In addition, the load–deflection curve is also obtained, as a representative of the composite behavior. The relations are reported to calculate the flexural strength, maximum strain, and flexural modulus of elasticity using this information.

The flexural strength, σ, is calculated using

$$\text{Flexural strength}, \sigma = \frac{3PL}{2bh^2}$$

where
  L is the support span (mm)
  b is the specimen width (mm)
  h is the specimen thickness (mm)
  P is the maximum applied force (N)

The flexural stress at any point can also be calculated using the above relation, with the value of applied force at that point.

The maximum strain, ε, is the strain at the outer surface of the specimen and is calculated using

$$\text{Maximum strain}, \varepsilon = \frac{6\delta h}{L^2}$$

where
  L is the support span (mm)
  h is the specimen thickness (mm)
  δ is the mid-span deflection (mm)

The flexural modulus is the ratio of change in stress to the corresponding change in strain. It is recommended to use a strain range of 0.002 (from 0.001 to 0.003) for the calculation of the flexural modulus. The following relation is used to calculate the flexural modulus

$$\text{Flexural modulus of elasticity, } E_f = \frac{\Delta\sigma}{\Delta\varepsilon}$$

where
  $\Delta\sigma$ is the change in stress (MPa) between two selected points
  $\Delta\varepsilon$ is the change in strain between two selected points (nominally 0.002)

9.2.4 Impact Testing

The impact resistance of fiber reinforced polymer matrix composites is determined by ASTM D7136/D7136M-15 [22]. In impact testing, a concentrated impact is made on a flat composite plate using a drop-weight device having a hemispherical impactor. The potential energy of the drop-weight is defined by the mass and drop height of the impactor, and is specified prior to the test. The damage resistance is quantified in terms of the resulting size and type of damage in the specimen.

The response of a laminated plate to an out-of-plane drop-weight impact is dependent upon many factors, such as laminate thickness, ply thickness, stacking sequence, the environment, geometry, impactor mass, striker tip geometry, impact velocity, impact energy, and boundary conditions.

Prior to testing, specimen width and length is measured at two different points, while thickness of specimen is measured at four places near the impact location. The drop height to produce the specified impact energy is calculated using

$$H = \frac{E}{m_d g}$$

where
H is the drop height (m) of the impactor
m_d is the mass of the impactor (kg)
g is the acceleration due to gravity, 9.81 m/s^2

The impactor is positioned at the calculated height and dropped toward the specimen. The force versus time data is recorded during contact, either continuously or at regular intervals.

The impact velocity of the impactor is performed using

$$v_i = \frac{W_{12}}{(t_2 - t_1)} + g\left(t_i - \frac{(t_1 + t_2)}{2}\right)$$

where
v_i is the impact velocity (m/s)
W_{12} is the distance (m) between the leading edges of the first and second flag prongs
t_1 is the time taken for the first flag prong to pass the detector
t_2 is the time taken for the second flag prong to pass the detector
t_i is the time of initial contact obtained from a force versus time curve

The impact energy can be calculated as

$$E_i = \frac{m v_i^2}{2}$$

where
E_i is the measured impact energy (J)
m is the mass of the impactor (kg)

The other two methods to determine the impact properties are EN-ISO 179 (Charpy impact test) and EN-ISO 180 (Izod impact test) [23]. In the case of Charpy impact testing, the test specimen is supported near its ends as a horizontal beam, and is impacted upon by a single blow of the striker. The line of impact lies midway between the supports. The test specimen is

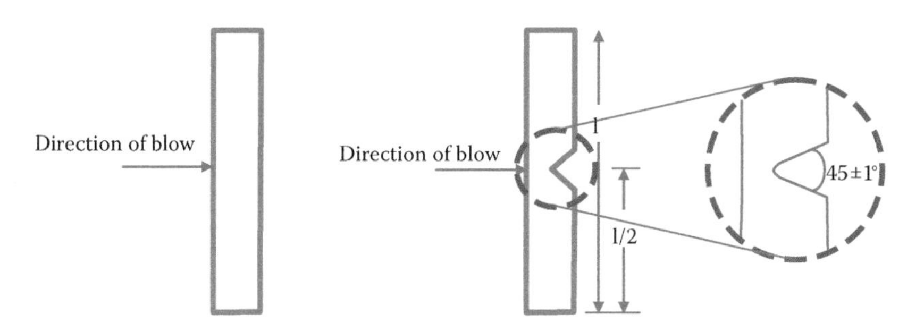

FIGURE 9.6
Specimen specifications for Charpy impact testing.

either notched or flat, as shown in Figure 9.6. The radius of the notch base may be either 0.10, 0.25, or 1.0 mm.

Now, the Charpy impact strength of the test specimen is determined using the following equation for an unnotched specimen

$$a_{cU} = \frac{E_c}{h \cdot b} \times 10^3$$

For a notched specimen, the equation becomes

$$a_{cU} = \frac{E_c}{h \cdot b_N} \times 10^3$$

where
E_c is the corrected energy (J) absorbed by breaking the specimen
h is the thickness (mm)
b is the width (mm)
b_N is the remaining width (mm) of the test specimen

9.2.5 Hardness Testing

The hardness of a material is defined as its resistance to localized plastic deformation (scratch or a small dent). To approximate this resistance, certain quantitative techniques have been developed where a small indenter is forced into the surface of a material and the depth or size of the indentation is measured. This indentation corresponds to a hardness number of the material. The softer the material, the larger and deeper is the indentation (and the lower the hardness number).

Hardness is a measure of the resistance of a material to penetration by a sharp object (Figure 9.7). It falls into three categories, namely, macro-hardness, micro-hardness, and nano-hardness. Macro-hardness expresses

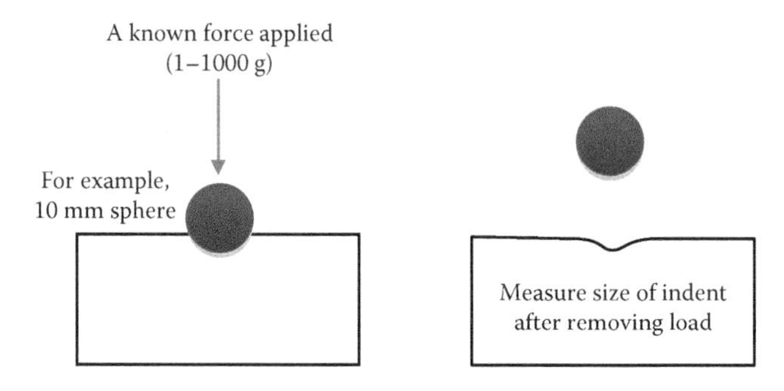

FIGURE 9.7
Schematic of hardness test.

the overall bulk hardness of materials and is measured using loads greater than 2 N. Micro-hardness is measured typically using loads less than 2 N, and nano-hardness is measured on a 1–10 nm length scale using very small loads (~100 μN).

There exists a number of test methods to determine the hardness of a material. For example, the Brinell hardness (ASTM E10), Rockwell hardness (ASTM E18), Vickers hardness, and Knoop hardness (ASTM E384). The ASTM E140 presents hardness conversion tables for the relationship among different hardness types [24].

9.2.6 Shear Testing

ASTM D3518/D3518M-13 is used for the in-plane shear response of polymer matrix composite materials by a tensile test using a ±45° laminate in x direction [25]. The test method is the same as that of ASTM D3039, except for the stacking sequence of the laminate. The maximum in-plane shear stress is calculated as

$$\tau_{12}^{m} = \frac{p^{m}}{2A}$$

$$\tau_{12i} = \frac{P_i}{2A}$$

where

τ_{12}^{m} is the maximum in-plane shear stress (MPa)
P^{m} is the maximum force (N)
τ_{12i} is the shear stress (MPa) at the ith data point
P_i is the force (N) at the ith data point
A is the cross-sectional area (mm^2)

The engineering shear strain at each required data point is determined using

$$\gamma_{12i} = \varepsilon_{\chi i} - \varepsilon_{yi}$$

where
γ_{12i} is the engineering shear strain at the *i*th data point
$\varepsilon_{\chi i}$ is the longitudinal normal strain at the *i*th data point
ε_{yi} is the lateral normal strain at the *i*th data point

The shear modulus of elasticity is calculated as

$$G_{12} = \frac{\Delta\tau_{12}}{\Delta\gamma_{12}}$$

where
G_{12} is the shear modulus of elasticity (GPa)
$\Delta\tau_{12}$ is the difference in the applied engineering shear stress between the two shear strain points (MPa)
$\Delta\gamma_{12}$ is the difference between the two engineering shear strain points

ASTM D3846-08 (Reapproved 2015) determines the in-plane shear strength of polymer matrix reinforced composites [26], by applying a compressive load to a notched specimen. Failure of the specimen occurs in the shear between two centrally located notches machined halfway through the specimen's thickness and spaced a fixed distance apart on opposing faces. The maximum load carried by the specimen is recorded. The failed (sheared) area length is determined. The ratio of maximum shear load to the product of the specimen's width and length of the failed area is the in-plane shear strength of the specimen.

9.2.7 Peel Testing

The known methods of measurement include the force required to bend the separated layers, in addition to that required to separate them. The bond strength or ply adhesion of similar laminates made from flexible materials is determined by ASTM F 904-98 (2008) [27]. The separation of plies of the test specimen is initiated mechanically or chemically. For example, adhesive tape applied to both sides and pulled apart, or separation initiated by making a heat seal and pulling it apart. In terms of chemical means, the best solvent is determined by trial and error method that may initiate the separation by immersion of the strip end.

The separated ends of the test specimen are clamped in the jaws of a tensile testing machine, with an original jaw distance of 25.4 mm. The unseparated portion of the test specimen is either left loose to move around freely, supported at 90° to the direction of draw by hand, or by some mechanical means. The test rate is set to 280% ± 10% mm/min, and the force required to separate 3 in. of test specimen are noted. The average force is expressed in N/m, g/25.4 mm, or pounds force/inch.

9.3 Non-Destructive Testing

Non-destructive testing does not affect the structural integrity of the sample, for example, weighing or measuring. The major objectives of non-destructive testing are initial inspection of test samples to confirm structural integrity, monitoring sample tests in progress, analyzing the reasons for failure, and so on. Some of the commonly used non-destructive testing techniques include ultrasonication, radiography, shearography, thermography and visual inspection.

9.3.1 Ultra-Sonication

This is the most commonly used technique for inspection of composites. The commonly used frequency range for this test is 5 MHz or less. Short pulses of ultrasound (a few microseconds) are passed into the composite material and detected after having interrogated the structure. It is necessary to avoid frequencies at which resonance occurs between ply interfaces. This may be a contact or non-contact testing. Contact testing involves the scanning of a test area by a probe manually, and therefore requires skilled operators. Non-contact testing has the advantage of speed and is commercially used, for example, the laser ultrasonic inspection system. Another more practical solution is immersion ultrasonic testing for very small samples. This technique is very helpful in the determination of delamination, voids, foreign inclusions, or cracks.

9.3.2 Radiography

This is the development of beams of ionizing radiation for the non-destructive testing of structures. Two types of radiation are used for composite material, namely, x-rays and neutrons. The parts of the specimen that have different radiation absorption properties can be discriminated in an image formed by the beam transmitted through the specimen onto film. The variations in the intensity of unabsorbed radiation appears as shades of gray in the developed film. The radiograph is evaluated on the basis of comparison of a known and an unknown specimen.

9.3.3 Shearography

A near-surface flaw decreases the local strength of composite material and the surface deforms under loading. This technique is based on the optical interference of sheared and direct images. The specimen has a laser beam projected on it and the scattered light is projected onto an image plane by a shearing lens. The interference of sheared and direct images gives a map of the distribution of local surface strain. A stress distribution is applied to the surface and another pattern is recorded. The superimposition or difference of the stressed and unstressed states produces a fringe pattern. In the pattern, each fringe represents a line of strain.

9.3.4 Thermography

Thermography involves flaw detection by monitoring the flow of heat over a surface when introduced to an external temperature gradient. The presence of flaws disrupts the normal pattern of heat flow. This method is more sensitive to flaws near the surface. Modern thermographic systems use infrared cameras to detect the radiated heat. These methods are classified broadly into two, that is, active and passive methods. The active methods produce and maintain thermal gradients by the application of cyclic stress. In passive methods, thermal gradients are produced from a transient change.

9.3.5 Visual Inspection

Visual inspection allows the examination of surface flaws. In addition, optical aids like microscopy may be used to improve the probability of detection. Normal eyesight can easily indicate the impact damage or delamination. It is an inexpensive, simple, and rapid method of finding the flaws in a composite material.

References

1. F. C. Campbell, *Manufacturing Process for Advanced Composites*. New York: Elsevier, 2004.
2. D. Gay, S. V. Hoa, and S. W. Tsai, *Composite Materials*. Boca Raton, FL: CRC Press, 2003.
3. D. D. L. Chung, *Composite Materials*, 2nd edn. London, U.K.: Springer, 2010.
4. ASTM C271-16, Standard test method for density of sandwich core materials. ASTM International, West Conshohocken, PA, 2016.
5. ASTM D3529M-10, Standard test methods for constituent content of composite prepreg. ASTM International, West Conshohocken, PA, 2010.
6. ASTM D3530-97, Standard test method for volatiles content of composite material prepreg. ASTM International, West Conshohocken, PA, 2015.
7. ASTM D2584-11, Standard test method for ignition loss of cured reinforced resins. ASTM International, West Conshohocken, PA, 2011.
8. ASTM D2734-09, Standard test methods for void content of reinforced plastics. ASTM International, West Conshohocken, PA, 2009.
9. ASTM D3171-15, Standard test methods for constituent content of composite materials. ASTM International, West Conshohocken, PA, 2015.

10. ASTM D3418-15, Standard test method for transition temperatures and enthalpies of fusion and crystallization of polymers by differential scanning. ASTM International, West Conshohocken, PA, 2015.
11. ASTM E831-14, Standard test method for linear thermal expansion of solid materials by thermomechanical analysis. ASTM International, West Conshohocken, PA, 2014.
12. ASTM E1131-08, Standard test method for compositional analysis by thermogravimetry. ASTM International, West Conshohocken, PA, 2014.
13. ASTM C272-12, Standard test method for water absorption of core materials for sandwich. ASTM International, West Conshohocken, PA, 2012.
14. ASTM D5229-14, Standard test method for moisture absorption properties and equilibrium conditioning of polymer matrix composite materials. ASTM International, West Conshohocken, PA, 2014.
15. ASTM D3039-14, Standard test method for tensile properties of polymer matrix composite materials. ASTM International, West Conshohocken, PA, 2014.
16. ASTM D5766-11, Standard test method for open-hole tensile strength of polymer matrix composite. ASTM International, West Conshohocken, PA, 2011.
17. ASTM D695-15, Standard test method for compressive properties of rigid plastics. ASTM International, West Conshohocken, PA, 2015.
18. ASTM D3410-03, Standard test method for compressive properties of polymer matrix composite materials with unsupported gage section by shear. ASTM International, West Conshohocken, PA, 2008.
19. ASTM D6641-14, Standard test method for compressive properties of polymer matrix composite materials using a combined loading compression (CLC). ASTM International, West Conshohocken, PA, 2014.
20. ASTM D7264-15, Standard test method for flexural properties of polymer matrix composite materials. ASTM International, West Conshohocken, PA, 2015.
21. ASTM D6272-10, Standard test method for flexural properties of unreinforced and reinforced plastics and electrical insulating materials by four-point bending. ASTM International, West Conshohocken, PA, 2010.
22. ASTM D7136-15, Standard test method for measuring the damage resistance of a fiber-reinforced polymer matrix composite to a drop-weight impact event. ASTM International, West Conshohocken, PA, 2015.
23. ISO 179-1, Plastics—Determination of charpy impact properties. ASTM International, West Conshohocken, PA, 2001.
24. ASTM E140-12b, Standard hardness conversion tables for metals relationship among brinell hardness, vickers hardness, rockwell hardness, superficial hardness, knoop hardness, scleroscope hardness, and leeb hardness. ASTM International, West Conshohocken, PA, 2013.
25. ASTM D3518-13, Standard test method for in-plane shear response of polymer matrix composite materials by tensile test of a ±45° laminate. ASTM International, West Conshohocken, PA, 2013.
26. ASTM D3846-08, Standard test method for in-plane shear strength of reinforced plastics, 2015.
27. ASTM F904-98, Standard test method for comparison of bond strength or ply adhesion of similar laminates made from flexible materials. ASTM International, West Conshohocken, PA, 2008.

10

Nonwovens

Alvira Ayoub Arbab and Awais Khatri

CONTENTS

10.1 Introduction

Nonwoven fabrics represent the third group of fabric forms after woven and knitted. According to the American Society for Testing Materials (ASTM D 1117-80) "a nonwoven is a textile structure produced by the bonding or interlocking of fibers, or both, accomplished by mechanical, chemical, thermal or solvent means and combinations thereof." A nonwoven fabric is a porous felt or sheet of oriented or random fibers merged by chemical adhesion, mechanical needle punching, and thermal fusion. One of the key benefits of nonwoven construction is that it is generally fabricated in a single continuous process directly from the fibrous materials to the finished product.

Moreover the high production rate and low labor cost prompt the invention of novel nonwoven products. Nonwoven fabrication methods can produce a wide range of unique characteristic fabrics. Suitable fibrous materials, web formation, bonding agents, and finishing treatments of nonwovens can alter their wide range of properties. The exceptional features of nonwoven fabrics make them ideal for numerous day-to-day applications. Nonwovens can be engineered into medical textiles, geotextiles, and insulation and filtration media supplies.

10.2 Market Overview and Application

Up to the 1990s, most of the world's nonwoven industry was based in those areas where the production and processes of nonwoven fabrication was matured and conceived. The major production share was in the United States, Japan, and European countries. Small scale businesses, sometimes part of a textile company's enterprises, operated a limited range of production based on dry laid web, chemical calendaring, thermal bonding, and needle punching. However, the large scale production companies DuPont, Asahi, Freudenberg, Kimberly-Clark, Ahlstrom, Polymer Group Inc., and British Board of Agreement (BBA) are responsible for high-end production and major research and development at the commercial level. A noteworthy patents estate has been developed for the production of high quality nonwoven products by Kimberly-Clark.

At the present time, the nonwoven industry has a major market share: the United States, China, Japan, and Europe hold 90% of total production and sales of nonwoven goods. Remarkably, many local traders began to expand globally with their own industries. The latest estimates according to official Association of Non woven Fabric Industry (INDA) and European Disposable and Non woven Association (EDANA) figures put the total production of nonwoven ranges at over 3.3 million tonnes. Western Europe holds a 33% major share of the production of nonwovens, while North America holds 31%, the Asia-Pacific region holds 25%, and 11% progress in Nonwoven. Europe, the leading nonwoven producer, has recently overtaken the United States in the installation of multiple new machinery and invention. The overall production of nonwovens in Europe grew by 2.5% in volume to reach 2,378,700 tonnes in 2016.

The impact of the extensive application of nonwovens in medical supplies, geotextiles, filtration, and home furnishing has drawn major attention. From household wipes to technical application, nonwovens offer solutions in every corner. As well as its success in Western markets, Western Europe has maintained production and sales of nonwoven household wipes throughout the world. Sales of nonwoven cleaning wipes have leveled off at $3 billion in retail value during recent years. Personal care wipes, baby wipes, and cosmetic wipes have topped $8 billion in recent years.

Nonwoven filtration media have shown continuous growth driven by customer demand for clean air and drinking water as well as increased fuel efficiency in vehicles and infrastructural improvements. According to the data released by INDA the nonwoven filtration market grew from $3.5 billion in 2014 to $4.6 billion in 2019. Needle punched fabrics are on the path of growth and global producers in the market are optimistic about the future. According to the recent estimation by EDANA, needle punched fabrics were among the top products in Europe in the year 2014. The data also show that the needle punching process grew by 9.1% from 2013 to 2014.

The production of spun laced, wet laid, and laminate nonwoven films, primarily used in surgical gowns, drapes, and sterile tray wraps, were products gaining in demand in the United States. These high-end products tend to prevent potential blood-and alcohol-borne pathogens from coming in contact with surgical staff and reduce small particle contamination of the surgical site by staff and the environment. According to the data published by INDA, the fast production rate means that the medical nonwovens market is forecast to grow by 4% per year from 271,000 tonnes in 2012 to 332,000 tonnes in 2017. There is also considerable growth forecast in the area of nonwoven film laminates. Furthermore, current trade and industry developments in China promote the production of nonwovens for filters and disposable medical textiles. With respect to customer requirement, China maintains standards that encourage the individual to purchase convenient items, promoting the increased production of disposable products. According to the latest 2016 report by INDA, banana-fiber-based hygiene products, facial exfoliating cushions, and stretchable adhesives for disposable wear are top of the novel nonwoven products.

Japan is one of the leading producers of hygienic baby diapers. The production volume of these increased significantly between 2012 and 2015. Total production increased by 2.8% to reach 14.7 billion units. The growth of baby diapers shows a growing export business of 246,000 tonnes. The largest percentage of Japanese diapers is exported to China and represents 165,000 tonnes. Japanese diapers tend to be of higher quality than Chinese ones of export quality. Consequently, many Chinese buyers choose Japanese-made diapers even though they are more costly. On the other hand, India has shown a high production rate of baby diapers, disposable wipes, and hygiene products in recent years. The growing rate of disposable hygiene products in India is expected to exceed China by 2035. The growing market demand for sanitary and hygiene products in India offers a very promising rise in the future of the nonwovens industry. According to data released by INDA, production lines of needle punched nonwovens, installed in May 2014, represent 50 crore INR of total investment with a target to exceed 100 crore INR over the next 3 years.

Nonwoven automotive products in North America have been substantially increased due to the amount of nonwovens going into vehicles.

However, North America has lagged behind Europe where green technology fuel and green energy recyclability have been regulated by government. Automotive nonwovens are 15%–30% lighter than conventional materials and reduce the weight of a vehicle by 2 kg, resulting in 30% less impact on the environment.

10.3 Nonwoven Technology

10.3.1 Raw Materials

Fundamentally all organic and nonorganic fibers and filaments are used in nonwoven fabric manufacturing. The selection of fibrous materials depends on the required properties of the fabric, the cost factors, and the demands of further manufacturing. Cotton is one of the most widely used materials for nonwoven baby diapers, baby wipes, and hygiene products. The recent explosion in the diversification of sanitary nonwovens has spurred interest in cotton as a valuable, natural, and sustainable component fiber. Cotton nonwovens were exclusively made from combing waste, to produce soft and absorbent fabrics with good bulk and bonding properties.

The huge application of nonwovens to cleaning purposes, wound dressings, and filtration media has regenerated the demand for manmade fibrous materials so that they now hold the major share. Spun laced viscose rayon is extensively used in hygiene and medical applications, due to its being highly moisture absorbent, soft, and biodegradable. Dull, high bulk, soft, and absorbent viscose rayon are aimed at medical and hygiene purposes. Their low cost, ease of processing with all types of web formation, and web bonding methods make them an ideal choice for fabrication. For filtration of hot, saturated, and volatile materials, coarse denier (more than 40 denier) viscose rayon filaments are used and recommended for gravity filtration. Viscose filtration products are biodegradable, cost effective, can hold a high liquid capacity, and are resistant to coolant continuants. Acetate multilayered lofted structures with a continuous fibrous matt are highly recommended for acoustic damping, absorbent cores, thermal insulation, and wound care applications. In synthetic fibers, polyester is most widely used for the fabrication of tea bags and sanitary and filtration purposes. Due to the low cost and high suitability for many product applications, polyester filaments and staple cut fibers are generally used for polyester tapes in electrical applications. Polypropylene also represents a major segment in durable products such as floor coverings, geotextiles, hygiene disposal, and needle punched geotextiles. Polyamide holds a 20%–25% part in nonwovens for interlinings. Figure 10.1 shows sections of the applications of nonwoven fabrics for different end uses.

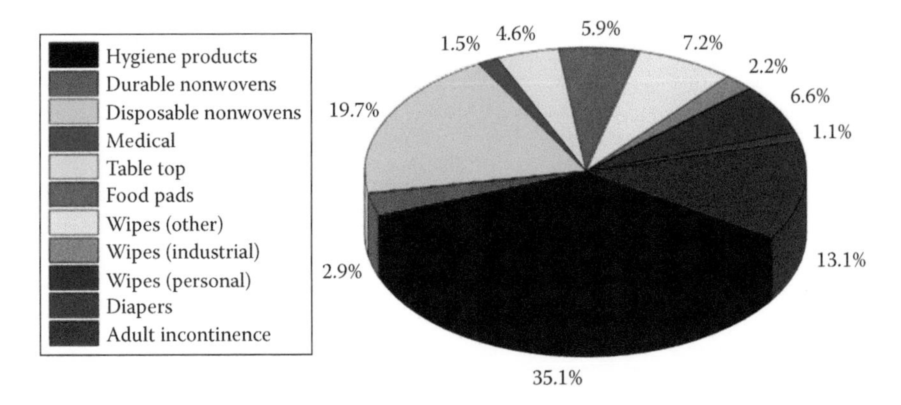

FIGURE 10.1
Sections of the applications of nonwoven fabrics for different end uses.

10.3.2 Web Formation Methods

Nonwoven web formation methods are classified according to the form of raw materials chosen for the specific application. Staple fiber and filaments are used to fabricate nonwoven webs.

Web formation techniques for staple fibers are

- Dry-laid web formation
- Wet-laid web formation
- Polymer-laid web formation

Dry-laid web formation is one of the old techniques and is very similar to the felting process. For the production of dry-laid web, carding machines and web lappers are used to layer the fibrous batt. The fibrous web layers are subsequently felted using heat, moisture, and agitation. The dry-laid web formation technique, such as fiber preparation, blending, carding, and garneting are innovations of the textile industry. These processes prepare staple fibers, blend them, and layer the fiber batt in a dry state. In dry-laid web formation, the fibers are collected into a web form by parallel lapping, cross-lapping, or aerodynamic (air laid) lap forming and then bonded by means of mechanical needles, hydro-entanglement, chemical adhesives, and thermal bonding methods. The simple way of fabricating a nonwoven web is by the parallel lapping of carded webs. In this method, several carded webs are layered over each other to form the fibrous batt structure. In this procedure, the mass per unit area of the batt can be increased by laying several webs over each other. A parallel-laid web formation system is extensively used in the production of fleeces for relatively light weight adhesive bonded fabrics used for cleaning cloths and hygienic products. The production setup for dry-laid webs is raised slightly above floor level to permit the working of the conveyor

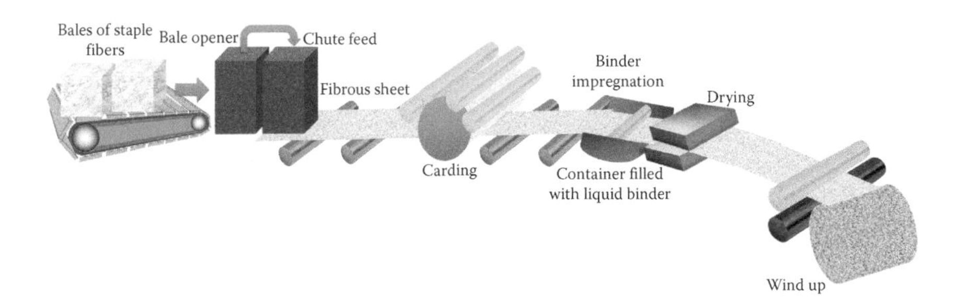

FIGURE 10.2
A schematic illustration of dry-laid web formation.

belt to pass underneath. The consecutive webs fall onto the conveyor lattice forming a batt with a condensed web. In the cross-laying method the carding machines are placed at right angles to each other to the main conveyor. The carding web moves transversely across the main conveyor belt, which moves to form a zigzag design sheet formation. These kinds of webs have high stress dispersion abilities as compared to the parallel style web forming system. Moreover, only a few zigzag layers can produce lighter weight fabrics with high stress recovery. An air-laid web forming system utilizes the air current dispersing method to lay and blend man-made fibrous webs. Very low weight fabrics with dimensional stability can be fabricated by this method. Usually a very low amount of binder is required to bind air-laid webs. Figure 10.2 shows a schematic illustration of dry-laid web formation.

The wet-laid web forming system is designed to fabricate short fibers dispersed in liquid, which are subsequently layered. Figure 10.3 shows a schematic illustration of wet-laid web formation. The wet-laid method is specifically suitable for the large scale production of disposable products, such as tea bags, aprons, gloves, napkins, and surgical gauze.

Polymer-laid or spun laced nonwoven webs are fabricated directly by polymer extrusion. In the basic polymer-laid method, sheets of synthetic filaments are polymer extruded and delivered onto a moving conveyor belt to form a randomly oriented continuous filament web. Figure 10.4 shows a schematic illustration of spun-laid web formation. Polymer-laid webs are used in a variety of applications including surgical packs and gowns, wipes and sponges, chemical barriers, and protective clothing. The spun laced method was first successfully commercialized by Dupont in the United States.

10.3.3 Web Bonding Methods

In order to incorporate strength with flexibility, nonwoven webs are bonded by means of mechanical, chemical adhesive, and thermal bonding systems. The high degree of Van der Waals forces and hydrogen bonding is a prerequisite for the bonding of a cellulosic fibrous web. The degree of bonding is

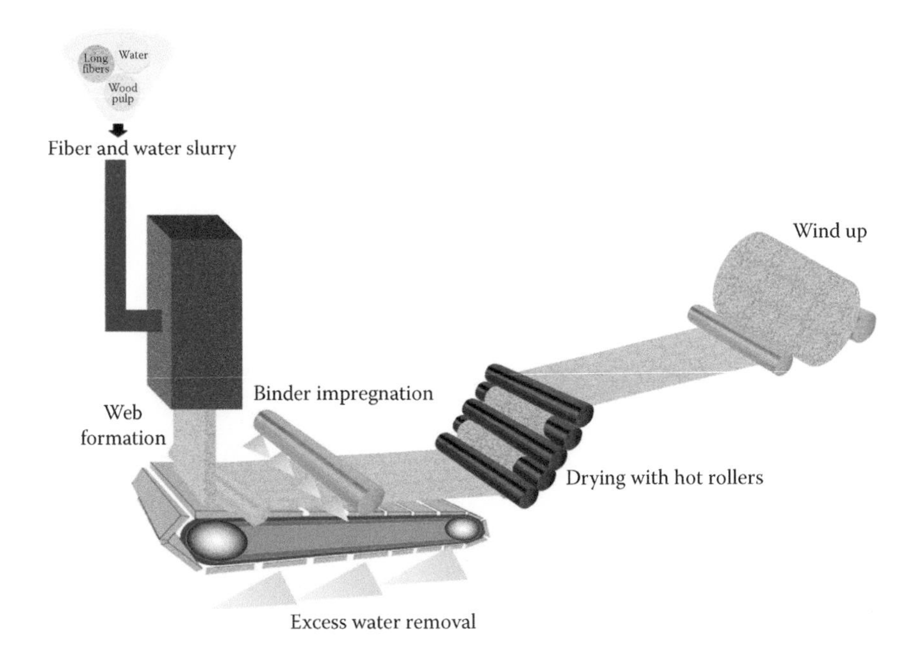

FIGURE 10.3
A schematic illustration of wet-laid web formation.

a basic input that determines the mechanical, dimensional, and absorbency properties. The bonding agent further influences the fabric's softness, flexibility, porosity, thickness, and loft properties. Web bonding process can be carried out simultaneously with web formation or separately. However, in certain nonwoven products two or more kinds of bonding methods are utilized to bond the fibrous batt.

In mechanical bonding a fibrous batt is bonded by means of needle punching, hydroentangling, and stich bonding. Needle punching is the oldest fabrication technique for nonwoven fabrics. This method is used to punch or bind dry-laid or spun laced webs. The triangular cross-sectioned barbed design needles are pushed through the cross-laid webs, capture fibers from the web, and then cross through the web. The consolidating effect of these plucked fibers provide strength in needle punched web. Needle punched webs have adequate flexibility in all directions with high opacity. However, different types of fibrous materials are utilized to fabricate the composite structure.

High strength needle punched fabrics are mostly used for geotextiles. The hydroentanglement bonding method uses the mechanical entanglements of staple fibers by means of forced jets of water droplets. The jets of water produce the coherence in the web and by means of hydraulic pressure fibers are entangled into the web structure. However, further adhesive bonding is required to bond the fibrous web fully. Stich bonding is used to bond

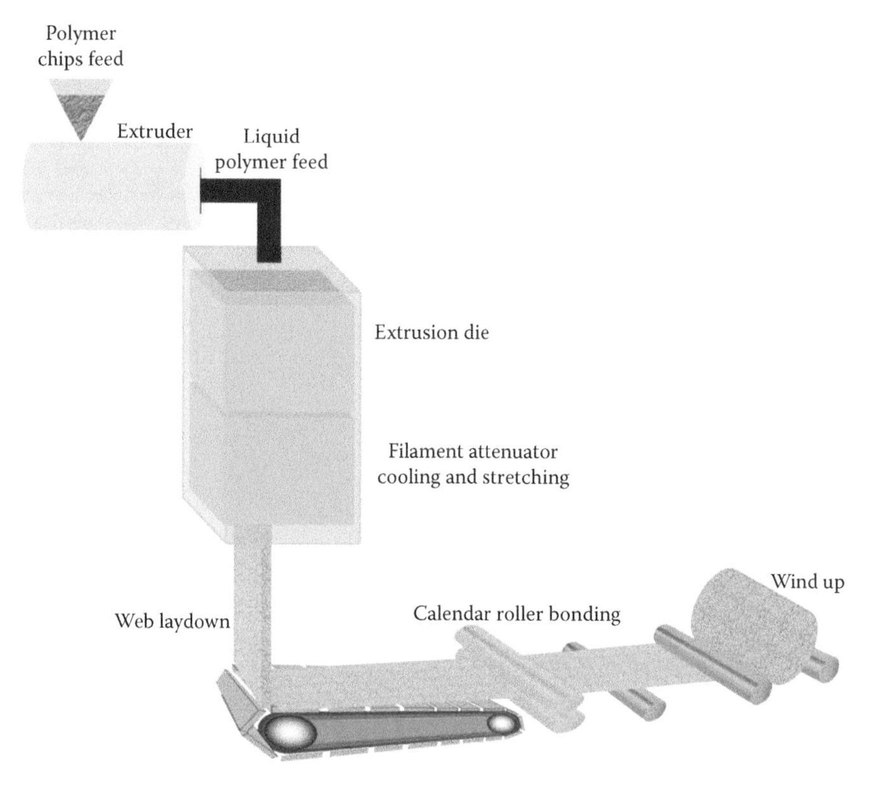

FIGURE 10.4
A schematic illustration of spun-laid web formation.

cross-laid webs by means of a warp stich. The machinery used in stich bonding is the modified form of warp knitting machine, which binds the fibrous layers by knitting columns of stitches.

Chemical bonding involves the application of chemical adhesives by means of complete saturation, spraying, printing, power, and foaming methods. There should be an adequate compatibility between adhesives and fibers. Usually the same polymeric materials are used to bind fibers having the same chemical formula. The chemical bond is the effect of physical and chemical forces that act as a boundary layer between the two polymers.

Thermal bonding utilizes thermoplastic and filaments to melt and fuse them together by means of heat and pressure. Bi-component filaments are used to bond webs where one component acts as a thermoplastic to facilitate thermal bonding while the other acts to enhance the quality of the required end product.

10.3.4 Finishing Methods

A variety of finishing treatments are used to incorporate softness, drab ability, absorbency, water repellency, flame retardancy, and UV absorption.

Mechanical finishing treatments include wrenching and craping, splitting, perforating, grinding, and velouring. Chemical finishes use various chemicals for the end use. Antistatic properties, water repellency, and flame retardancy are obtained by using organic and inorganic chemicals.

10.4 Composite Nonwovens

Composite nonwovens are the most important form of textiles and have a significant role due to their facile fabrication techniques. A composite nonwoven is a superficial method of coalescing system of divergent webs to obtain precise and unique functional properties. This system provides unique characteristics, such as mechanical stability, heat resistivity, thermal conductivity, antimicrobial activity, and interlaminar shearing force stability, which is impossible with only one kind of web system. The simplest way to form a nonwoven composite structure is to bond two or more webs in a laminating line. The webs are bonded to each other by means of hot melted glue or latex adhesives. In some cases, electrostatic forces or Van der Waals forces or electrostatic attraction is sufficient to hold different fibrous layers to obtain the required nonwoven composite structure. Moreover, different webs are taken off separately unwound strands before being finally wound onto a single wind up stand. Needle punching, hydroentanglement, and stich bonding are the ways to fabricate a nonwoven composite structure. Usually such a structure is hydroentangled with a scrim material. The scrim is usually a woven textile material, sandwiched between two nonwoven webs. Figure 10.5 shows the morphology of a nonwoven composite layered structure formed

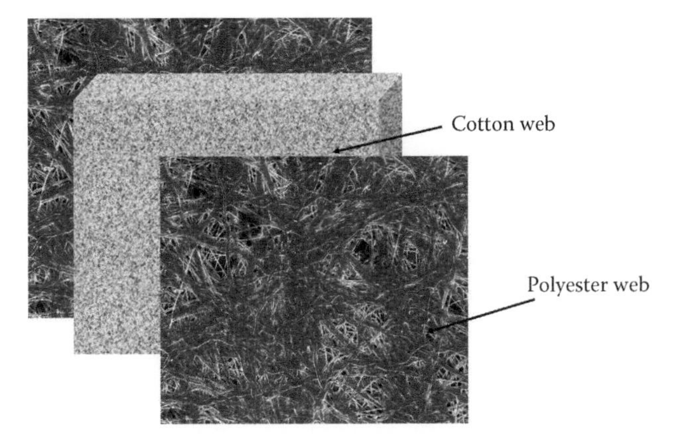

FIGURE 10.5
The morphology of a nonwoven composite layered structure formed by polyester and cotton webs.

by polyester and cotton webs. This method provides extra strength to a nonwoven composite structure. Polypropylene, polyester, polyethylene, and nylon are the most widely used to combine with organic and inorganic fibers for maximizing the functionality of a composite structure. The application of these forms of nonwovens is expected to have high demand, exclusively in technical textiles, that is, automotive textiles, households, building and construction reinforced materials, and surgical and filtration applications. For the facilitation of nonwoven composites, cost saving in processing and raw materials must be carried out so as to be able to fabricate advanced composite structures for versatile application in the current market. In this way, novel nonwoven composites will gain widespread recognition in the future market.

10.5 Characterization of Nonwovens

Nonwoven fabrics are bonded together by means of chemical, thermal, or frictional methods. The characterization of nonwovens is essential for analyzing the types, size, shape, rigidity, and density of bond structure. The bond structure is characterized in two ways: rigid solid bonds and flexible elastic joints. For this purpose, the choice of bond structure depends on the norm of manufacturing processes and end use application. In needle punched or hydroentangled nonwoven fabrics, the bond is formed by interconnectivity of fibrous layers. In needle punched fabrics, the bonds are flexible and essential fibers are capable of interchanging with the bonding limits. The extent of the bonding points is affected by the fabrication process, that is, the depth of needle barb penetration relative to fiber thickness, size of needle, total number of needles that stab into the web, and punch density. In this case, bond points depend on the fluid jets, specific energy, and number of injectors. However, in chemically bonded fabrics, the fibrous layers are bonded by means of surface adhesion or cohesion of polymers, in which the small proportion of fiber matrixes are bonded and restricted to move only with the bonding points. Bonding points depend on the method of binder application, that is, print bonding, spray bonding, powder bonding, or saturation bonding. Thermoplastic filament spun laced nonwoven fabrics are fabricated via the melting of polymers. In this case, bonds are formed at fiber crossover points; as a result fibers associated with melting polymer surfaces cannot move independently. The bond point area, land area, pressure, and the size of adhesive units determine the nature of bonding points. Stich bonded nonwoven fabrics are stabilized by a warp knit yarns through the fibrous layers. In this way, the bonding units are bendable and joined by yarns and fibers. The nature of solid bond points in nonwoven fabrics can be physically characterized by the tensile properties, tearing and tensile strength, elasticity, and durability. The measure of bonding may be directly examined by means of microscopic examination of the fiber's cross-sectional area and diameter.

10.6 Testing of Nonwovens

Numerous testing systems and procedures have been established for the characterization of nonwoven fabrics:

- The standard testing methods developed by the International Standards Organization (ISO), British Standard European Norm (EN/BS), American Society for Testing and Materials (ASTM), and ANSI systems
- The standard testing methods developed by the INDA, EDANA, and AATCC industrial associations
- The nonstandardized testing methods developed for research and development

The standards testing system for nonwovens are defined as appropriate testing methods developed for reproducibility and considered to offer reliable results with precise control for use in the exchange of nonwoven goods. Industrial test methods are typically designed for routine characterization that deals with the assessment, benchmarking, and quality control of the semi-finished end product.

10.6.1 Dimensional Properties

The dimensional properties of nonwoven fabrics are often characterized in relation to fiber orientation distribution, pore size distribution, and fabric uniformity.

10.6.1.1 Fiber Segment Orientation Distribution

The fiber segment orientation of nonwoven fabrics can be determined by the web fabrication and bonding systems. The fibers in nonwoven webs are oriented in two or three dimensions. Fiber segment orientation in a nonwoven can occur potentially in any three-dimensional way, therefore the characterization of fiber alignment is complex and costly. However, the oriented dimension and orientation angle can be determined. In this way, three-dimensional needle punched webs can be categorized as an arrangement of two-dimensional layers joined by fibers oriented perpendicularly to the nonwoven fabric. Moreover, the arrangement of fiber orientation in three-dimensional needle punched fabrics can be described by determining the orientation of fibers in two dimensions in the structure of fabric lanes.

10.6.1.2 Pore Size Distribution

Pore size distribution has become an essential technique in determining the concrete performance of different porous nonwoven sheets used for filtration, separation, reinforcements, and medical textiles. The porous morphology and structure of a nonwoven web is very complex for determining pore size distribution of two or three types of fibers. The pore size distribution of a nonwoven web can be calculated using density methods, gas expansion and adsorption, optical methods, electrical resistance, image analysis, optometry, and porosimetry. The opening porous surface and opening pore size is examined by the passage of sphere-shaped dense glass beads of different sizes (50–500 mm) from the largest pore size of the fabric within quantified circumstances. Further, pore size distribution can be calculated by means of dry sieving, wet sieving, and hydrodynamic sieving methods.

10.6.1.3 Fabric Uniformity

Fabric uniformity is the consistency of web thickness over a specified area. It can also be elucidated as the variation of the weight or density of the nonwoven fabric calculated directly by taking the specimen from different sections of the fabric. The difference in uniformity depends on the specimen size and weight. Therefore, to investigate the uniformity of nonwoven fabric, variation in optical density, the gray level intensity of the fabric, and electromagnetic ray absorption systems are considered.

10.6.2 Mechanical Properties

Mechanical properties such as tensile strength, bending, rigidity, compression, and stiffness demonstrate the functionality of nonwoven fabrics. The mechanical properties of nonwovens are tested by machine direction, cross-direction, and bias direction. Numerous testing systems are recognized for the analysis of mechanical properties of nonwovens. Generally, strip and grab tests are used for analysis. In grab test methods, the central width of the specimen is clamped by jaws at a fixed distance. The specimen extends outside the jaws' width. According to the standard test method, the width size of the nonwoven sample is 100 mm, whereas the clamping width in the central section of the fabric is 25 mm. The fabric is stretched at a rate of 100 mm/min (ISO standard) or 300 mm/min (ASTM standard). The discrete distance of two clamps is 200 mm (ISO standard) or 75 mm (ASTM). Nonwoven fabric samples typically contribute a maximum force before rupture occurs in the web layers. In the strip test, the fabric specimen is held between two clamps. In this case, the specimen width is 50 mm (ISO standard) or 50 mm (ASTM standard), though the stretch rate and clamp distance will be kept the same as in the grab test.

10.6.3 Wetting and Liquid Absorption

The wetting of nonwoven fabrics refers to the ability of fabric to absorb the liquid, which is analyzed by the equilibrium of surface energies in the interference of air, liquid, and solid constituents. Nonwoven fabrics are porous and have the ability to absorb a great amount of liquid. There are two key ways of transporting moisture in nonwoven fabrics. One of the systems demonstrates the absorption of moisture driven by means of capillary pressure in a porous fabric, although the moisture absorption method is performed via a negative capillary pressure gradient. In the second type of moisture absorption, the liquid absorption is the pressurized flow in which the liquid is governed by the external forces of the pressure gradient. In this case, one edge of the fabric specimen is dipped in a liquid. The liquid moves into the fabric structure and is known as wicking, the rate of which can be studied by a linear rate of absorption of liquid in a strip specimen of nonwoven fabrics. In an upward wicking test, the sample is conditioned at 20°C and 65% relative humidity for 24 h. The fabric sample strip is held vertically with its lower part dipped in distilled water or a dye solution. After a fixed time, the absorbed liquid height in the fabric specimen is measured.

Bibliography

Agarwal, B. D., L. J. Broutman, and K. Chandrashekhara. *Analysis and Performance of Fiber Composites*. John Wiley & Sons, Wiley Publishers, Hoboken, NJ, 2006.

Albrecht, W., H. Fuchs, and W. Kittelmann. *Nonwoven Fabrics: Raw Materials, Manufacture, Applications, Characteristics, Testing Processes*. John Wiley & Sons, Wiley Publishers, Hoboken, NJ, 2006.

Berman, M. H. S., D. D. Doshi, and T. F. Gilmore. Nonwoven continuously-bonded trilaminate. Google Patents, 1989, U.S. Patent No. 9,572,721.

Bhat, G. S. and S. R. Malkan. Extruded continuous filament nonwovens: Advances in scientific aspects. *Journal of Applied Polymer Science* 83(3) (2002): 572–585.

Butin, R., J. Harding, and J. Keller. Non-woven mats by melt blowing. Google Patents, 1974, U.S. Patent No. 9,572,721.

Chatterjee, P. K. and B. S. Gupta. Chapter I—Porous structure and liquid flow models. *Textile Science and Technology* 13 (2002): 1–55.

Chhabra, R. Nonwoven uniformity-measurements using image analysis. *International Nonwoven Journal* 12(1) (2003): 43–50.

Everhart, C. H., D. O. Fischer, F. R. Radwanski, and H. Skoog. High pulp content nonwoven composite fabric. Google Patents, 1994, U.S. Patent No. 9,572,721.

Hartmann, L. Process of producing non-woven fabric fleece. Google Patents, 1970, U.S. Patent No. 9,572,721.

Hearle, J. W. S. and M. A. I. Sultan. 19—A study of needled fabrics. Part I: Experimental methods and properties. *Journal of the Textile Institute* 58(6) (1967): 251–265.

Hsieh, Y. -L. and B. Yu. Liquid wetting, transport, and retention properties of fibrous assemblies. Part I: Water wetting properties of woven fabrics and their constituent single fibers. *Textile Research Journal* 62(11) (1992): 677–685.

Hughes, G. H. Stitched composite nonwoven fabric having a self-bonded fibrous supporting layer and outer fibrous layers. Google Patents, 1972, U.S. Patent No. 9,572,721.

Hutten, I. M. *Handbook of Nonwoven Filter Media*. Elsevier, Oxford, U.K., 2007.

Kanafchian, M. and A. K. Haghi. Understanding nonwovens: Concepts and applications. In, Berlin, A. A., R. Joswik, and V. N. Ivanovich (eds.), *Engineering Textiles: Research Methodologies, Concepts, and Modern Applications*. Apple Academic Press, pp. 1–57, 2015.

Kissa, E. Wetting and wicking. *Textile Research Journal* 66(10) (1996): 660–668.

Krčma, R. *Manual of Nonwovens*. Textile Trade Press, University of Minnesota, Minneapolis, MN, 1971.

Lee, Y. M., J.-W. Kim, N.-S. Choi, J. A. Lee, W.-H. Seol, and J.-K. Park. Novel porous separator based on Pvdf and Pe non-woven matrix for rechargeable lithium batteries. *Journal of Power Sources* 139(1) (2005): 235–241.

Lünenschloss, J. and W. Albrecht. *Non-Woven Bonded Fabrics*. Ellis Horwood, John Wiley & Sons, Wiley Publishers, Hoboken, NJ, 1985.

Mao, N. and S. J. Russell. A framework for determining the bonding intensity in hydroentangled nonwoven fabrics. *Composites Science and Technology* 66(1) (2006): 80–91.

Mohammadi, M., P. Banks-Lee, and P. Ghadimi. Air permeability of multilayer needle punched nonwoven fabrics: Theoretical method. *Journal of Industrial Textiles* 32(1) (2002): 45–57.

Pourdeyhimi, B., R. Dent, and H. Davis. Measuring fiber orientation in nonwovens. Part III: Fourier transform. *Textile Research Journal* 67(2) (1997): 143–151.

Purdy, A. T. Developments in non-woven fabrics. *Textile Progress* 12(4) (1983): 1–86.

Russell, S. J. *Handbook of Nonwovens*. Woodhead Publishing, Cambridge, U.K., 2006.

Schwartz, P. *Structure and Mechanics of Textile Fibre Assemblies*. Elsevier, Woodhead Publishing, Cambridge, U.K., 2008.

Smith, P. A. *Handbook of Technical Textiles*, Vol. 12, p. 130, 2000.

Suskind, S. P., S. L. K. Martucci, and J. Israel. High strength hydroentangled nonwoven fabric. Google Patents, 1989, U.S. Patent No. 9,572,721.

11

Medical Textiles

Nauman Ali and Sheraz Ahmad

CONTENTS

11.1 Introduction

Medical science and textile technology merged and introduced a new field called "medical textiles." Medical textiles have a huge market due to their extensive demand and use. As they support healthcare, a growing niche of medical textile products accounts for 20% of technical textiles. These are used for hygiene and in hospitals and the healthcare sector. In addition, they are used in hotels and other locations where sanitation is required. This sector is growing most rapidly within the technical textile market [1]. This growth is due first to the fact that living standards are rising and people are more conscious about their health and hygiene. Secondly, infection rates in hospitals are increasing, so there is a need to develop more medical products that may serve the purpose of protecting people from viruses, for example, wipes, wadding, gauze, wound dressings, scrubs, masks, laboratory coats, and many other products.

Medical textile products can be woven, knitted, braided, and nonwoven. They are also used in fibrous or yarn form as their medical properties can be embedded or sprayed on them, for example, textiles with elastomeric yarn are used in maternity belts, neck wraps, and surgical bras worn outside the body.

11.2 Essential Properties

Sterility is a constituent element of medical textile products, so the polymers used must be able to withstand the harsh physical and chemical conditions that are generally found in a sterilization process.

Depending on application, the most important requirements for textiles used in the medical area are fluid resistance or absorbency, strength, lightness, elasticity, and softness. Medical textile products are biodegradable and biocompatible in nature; they are designed to provide comfort and protection. Moreover, textile products that are used in the medical field need to be nonallergenic, nontoxic, noncarcinogenic, nonirritant, and also durable. These requirements are met by using appropriate polymers processed into fibers and textiles that possess the right structures to meet the needs of various end users.

11.3 Classification of Medical Textile Products

The major products of the medical textile industry are classified into four divisions, namely, implantable materials, nonimplantable materials, extracorporeal devices, and healthcare and hygiene products.

11.3.1 Implantable Materials

Implants include the replacement of damaged blood vessels and segments of the large arteries. It is feasible to manufacture a vascular prosthesis only 2–3 mm in diameter. Implantable materials are used to repair affected parts of the human body. They can be used as wound sutures or used in replacement surgery, such as vascular grafts and artificial ligaments. The material must be biocompatible if it is to be accepted by the body. The International Organization for Standardization (ISO) developed many standards for assessing the biocompatibility of medical devices. In particular, ISO 10993-1:2009 [2] clearly outlines the primary aim to protect humans from possible biological risks that may result from the use of medical appliances.

Examples of implantable materials include: sutures, artificial tendons, artificial ligaments, artificial skin, artificial lumina, eye contact lenses, orthopedic implants, artificial joints, artificial bones, cardiovascular implants, vascular grafts, and heart valves.

11.3.2 Nonimplantable Materials

Nonimplantable materials are used on the body; most of the time they have direct contact with the human skin. They include wound dressings, absorbent pads, simple and elastic bandages, plasters, gauze, pressure garments, wadding, and orthopedic belts.

Surgical dressings are one of the main types of nonimplantable medical textile products that are used to cover up, guard, and hold the injured body part. These dressing also help to absorb fluid coming out of a wound.

11.3.3 Extracorporeal Devices

Extracorporeal devices are used to maintain the function of essential bodily organs and include artificial kidneys (dialyzers), artificial livers, and mechanical lungs. An artificial kidney is a small instrument that is almost the size of a two-cell flashlight, made with thin hollow polyester or cellulose fibers; it is used to remove waste products from the patient's blood. An artificial liver consists of hollow cellulose fibers to separate and dispose of and supply fresh patient plasma. A mechanical lung is made with hollow polypropylene fibers or a hollow silicone membrane; it is used to remove carbon dioxide from the patient's blood and supply it with fresh oxygen.

11.3.4 Healthcare and Hygiene

The healthcare and hygiene sector is very significant for medical textile products. Large numbers of healthcare and hygiene products are used in the operating theater and in hospital wards. These products are either washable or disposable. Products used in an operating theater include surgical gowns, caps, masks, patient drapes, and cover cloths. Surgical gowns and masks act as a blockade to prevent the release of pollutant particles into the air as they are a possible source of infection to patients. Disposable surgical gowns that are nonwoven are also used in hospitals for the same purpose. These gowns, surgical masks, and caps are also used to prevent contamination. They should have a high level of air permeability and high filter capacity, as well as being lightweight and nonallergenic. Cover cloths and surgical drapes are used to cover the patient and the areas around him or her. They must be completely resistant to bacteria and permeable to bodily perspiration and wounds. Medical textile products used in hospital wards include bedding, clothing, mattress covers, incontinence products, clothes, and wipes.

11.4 International Testing Standards for MEDITECH Products

As medical textile products are used for health care and hygiene, they have very strict testing standards used to check the performance of the product at different levels.

11.4.1 Face Masks

A face mask is a major product that is used to protect patients and surgical staff from the transfer of microbes, bodily liquids, and particulate material. It is also used in daily life to protect people from harmful atmospheres.

In the United States, the American Society for Testing and Materials (ASTM) established qualifications for face masks and are referenced by the Food and Drug Administration, which is the required standard in the country. ASTM F2100-07 [3] specifies the performance necessities for face masks with five basic criteria:

1. Bacterial filtration efficiency
2. Particulate filtration efficiency
3. Fluid resistance
4. Delta P (pressure differential)
5. Flammability

According to the standard performance specifications in the United States, face masks are classified into three levels:

1. Low barrier
2. Moderate barrier
3. High barrier

Basic test methods that are used to evaluate face masks include air permeability, bacterial filtration percentage (ASTM F 2101) [4], and splash resistance (ASTM F 1862) [5].

11.4.2 Surgical Gowns

Different tests are carried out to assess the performance of surgical gowns

1. *Spray impact penetration test* (AATCC 42) [6]. A synthetic blood is sprayed on the stiff surface of the test sample backed by a weighted blotter. The blotter is then reweighed to assess the water penetration and the sample is categorized as shown in Figure 11.1.

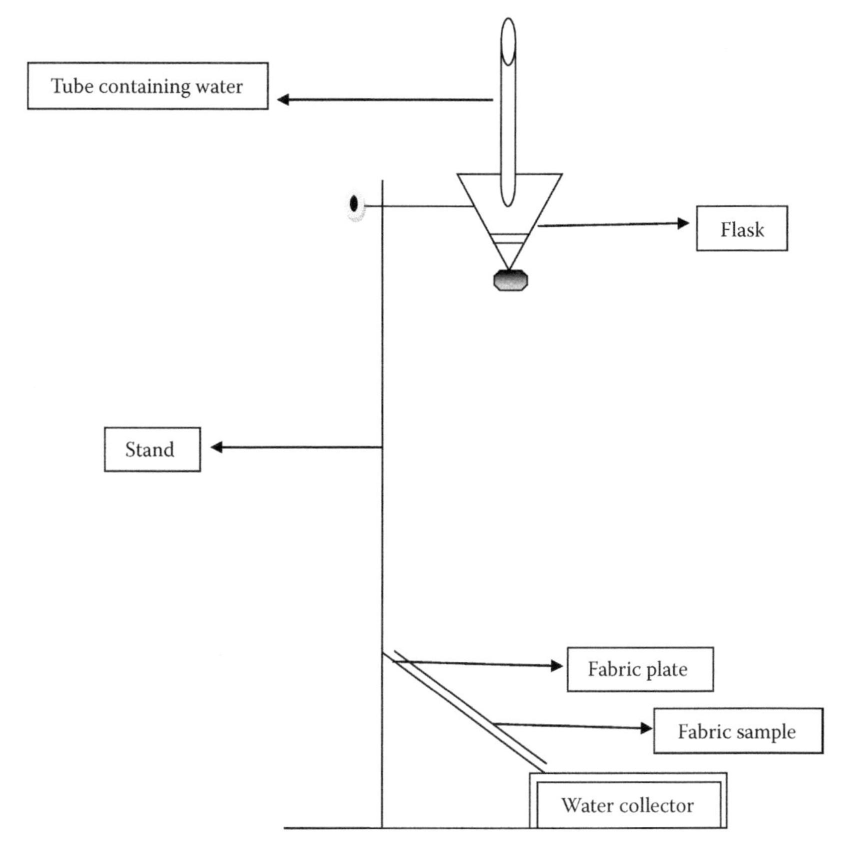

FIGURE 11.1
Impact penetration test.

2. *Hydrostatic head test*. One side of the test sample is directed toward hydrostatic pressure that is increasing at a constant rate until there is leakage at three points on its facade.
3. *Resistance to synthetic blood* (ASTM F 1670) [7]. The sample is directed toward synthetic blood for some specific time and pressure. A manual examination is carried out to assess when penetration occurs. Any proof of synthetic blood penetration represents failure. The result of this test is reported as fail or pass.
4. *Viral penetration resistance* (ASTM F 1670-1) [7]. This test is used to determine the resistance of the material used in a protective wearable against the penetration of blood-borne pathogens. The detection of viral penetration determines the pass/fail of the protective wearable.

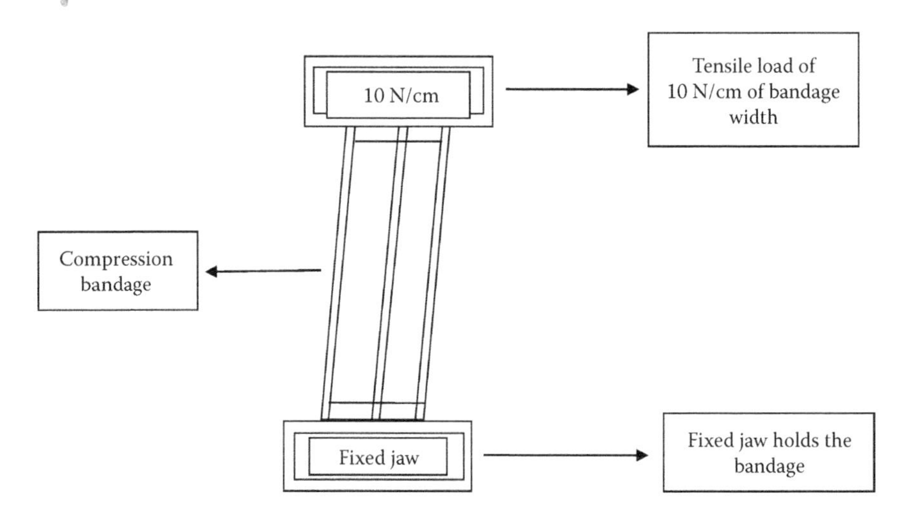

FIGURE 11.2
Testing for the elongation properties of a bandage.

11.4.3 Compression Bandages

The elongation of compression bandages is determined in a tension test, with a tensile load of 10 N/cm across the bandage's width and, as used in praxis, with the normal load of 3 N/cm across the bandage's width. Figure 11.2 shows testing for elongation properties of a compression bandage.

11.4.4 Surgical Sutures

Tensile strength, firmness, good knotting, and easy handling are the most important properties of surgical sutures. To assess these properties, a test is performed as per the standard testing method ASTM D3217-07 [8], which is used to measure the breaking tenacity of manufactured textile fibers taken from filament yarns, staple, or tow fibers, either crimped or uncrimped, and tested in either a double loop or as a strand formed into a single overhand knot. When testing surgical sutures, the samples are first cut to a length of 6 cm. For each suture type, knotted and knotless samples should be prepared separately. The samples are then put into tubes containing seven different medical liquids. They are kept at temperatures of 21°C and 4°C, since medical operations are often carried out at 21°C and medical liquids are prepared at 4°C, according to the Biomedical Laboratory of the Stuttgart Textile Fiber and Research Institute.

11.4.5 Testing of Implantable Materials

In vitro testing of textile implants includes a broad range of chemical, mechanical, and biological tests, and standards have been developed for

TABLE 11.1

International Testing Standards for MEDITECH Products

S. No.	International Standard	Scope/Test Parameter
1.	EN14683	The European market uses this standard to check the requirements and performance of surgical face masks
2.	EN 13795	To determine the requirements for surgical drapes, gowns, and clean air suits used as medical devices
3.	EN 14079	Surgical gauze specification
4.	EN 1372-1	Surgical dressings
5.	EN 862-2	Alcohol repellency
6.	EN 1734	Water resistance
7.	EN 20811	Liquid repellency
8.	EN 556	Sterilization of healthcare products
9.	ASTM F 2100	This specification provides for the classification of medical face mask material performance
10.	ASTM F 2407	This establishes requirements for the performance and labeling of surgical gowns
11.	ASTM F1983	Standard for assessment of compatibility of absorbable/restorable biomaterial for implantable devices
12.	ASTM F2026	Standard for polyetheretherketone (PEEK) polymers for surgically implantable devices
13.	ASTM E 96	Water vapor permeability
14.	ASTM ES-21	Blood repellency
15.	ASTM D 734	Porosity
16.	ASTM D 894	Peel strength
17.	ASTM D 671	Water vapor permeability
18.	ASTM D 357	Air permeability of hospital linen and bandages
19.	ASTMD 4751	Pore size
20.	AS 3789.8	Textiles for recyclable barrier fabrics
21.	AS 3789.6	Textiles for institutional fabric specifications
22.	AS 4369.4	Indicates conditions for the manufacturing of absorbent wadding
23.	NBR13904	Brazilian standard giving specifications of sutures
24.	IST 10.1	Wicking rate
25.	IST 80.0	Absorbency
26.	IST 80.9	Water repellency
27.	IST 90.0	Softness
28.	IST 50.0	Flame retardance
29.	IST 80.6	Alcohol repellency
30.	BS 2823	Water resistance
31.	BS 4745	Thermal resistance
32.	BS 1425	Microbial resistance
33.	ISO 811	Water resistance
34.	ISO 16603	Synthetic blood penetration

(Continued)

TABLE 11.1 (*Continued*)

International Testing Standards for MEDITECH Products

S. No.	International Standard	Scope/Test Parameter
35.	ISO 16604	Viral penetration
36.	ISO 1420 A	Water resistance
37.	ISO 3781	Wet tensile strength
38.	IS0 11193-1	Knot strength
39.	ISO 10993-5	Biocompatibility
40.	ISO 17190-1	pH of polyacrylate
41.	ISO 17190-2	Residual monomers
42.	ISO 17190-3	Residual size distribution
43.	ISO 25539-1	Graft testing
44.	ISO 11137	Sterilization of healthcare products
45.	ISO 9949-2	Defines nine terms used in the field of urine absorbing aids and comprises the vocabulary for products
46.	ISO 22610	Test method to determine the resistance to wet bacterial penetration
47.	ISO 22612	Test methods for resistance to dry microbial penetration

most of these. ISO 10993 [2], titled "Biological evaluation of medical devices," summarizes the in vitro test methods for medical devices such as textile implants. There are strict regulations governing the clinical use of medical implants and regenerative therapies. Depending on the nature of the treatment, these products may be classified as a medical device, such as wound-dressing material that supports the regeneration of skin, a pharmaceutical, such as a bone morphogenetic protein, a cell therapy, such as chondrocyte transplanted into the knee or a combination product.

Other international testing standards of medical textile products and their parameters are shown in Table 11.1.

References

1. T. Sureshram, Specifications/properties required for the meditech products and their testing, The South India Textile Research Association, Coimbatore, India, pp. 1–16.
2. ISO 10993-1, Biological evaluation of medical devices—Part 1: Evaluation and testing within a risk management process, 2009.
3. ASTM F2100-07, Standard specification for performance of materials used in medical face masks, ASTM International, West Conshohocken, PA, 2007.
4. ASTM F2101-14, Standard test method for evaluating the bacterial filtration efficiency (BFE) of medical face mask materials, using a biological aerosol of *Staphylococcus aureus*, ASTM International, West Conshohocken, PA, 2014.

5. ASTM F1862-00a, Standard test method for resistance of medical face masks to penetration by synthetic blood (horizontal projection of fixed volume at a known velocity), ASTM International, West Conshohocken, PA, 2000.
6. AATCC TM42-13, Water resistance: Impact penetration test, 2013.
7. ASTM F1670-08, Standard test method for resistance of materials used in protective clothing to penetration by synthetic blood, ASTM International, West Conshohocken, PA, 2008.
8. ASTM D3217/D3217M-15, Standard test methods for breaking tenacity of manufactured textile fibers in loop or knot configurations, ASTM International, West Conshohocken, PA, 2015.

12

Smart and Electronic Textiles

Iftikhar Ali Sahito and Awais Khatri

CONTENTS

12.1 Introduction

Textiles are the most versatile materials found today with respect to their ease of fabrication, the design to which they can be engineered, their conventional and emerging applications, and their flexibility, lightweight, and low cost (Dastjerdi & Montazer, 2010; Kadolph, 2007). Among modern forms of textiles, electronic textiles or e-textiles have found their way into everyday life in the form of technological equipment within regular clothing to improve the quality of life (Bowman & Mattes, 2005; Carpi & De Rossi, 2005; Curone et al., 2010; Shim, Chen, Doty, Xu, & Kotov, 2008). Textile fabrics are such a flexible class of materials that might replace silicon wafers in the future, in fact many researchers say that fabrics are the new silicon wafers

(Dhawan, Seyam, Ghosh, & Muth, 2004; Post, Orth, Russo, & Gershenfeld, 2000; Yoo, 2013). We are at the beginning of the use of portable and wearable electronic devices, which will drive through a new path for textiles that hitherto have only been considered as something to be worn and used for decorative purposes. Yet, there are a lot of obstacles to be overcome before e-textiles can replace the novel silicon wafer in the future and build their own commercial market with standardized procedures for their production, along with reproducible results. So far e-textiles are in the research phase and it might take a few more years before they reach our doors as ordinary clothing.

Today, the concept of technical textiles has gone far beyond its application to civil engineering, agriculture, aircraft, sportswear, and bulletproof vests. Smart textiles are being successfully used in various medical transplants (Eriksson & Sandsjö, 2015; Van Langenhove, 2007). Many brands have shown a lot of interest in and used e-textiles as wearable technology, including the "Hi-call" Bluetooth-enabled phone glove, the Adidas miCoach, Nike Fit, and Cute Circuit's Galaxy Dress and T-shirt-OS (Jost, Dion, & Gogotsi, 2014).

Smart and e-textiles have emerged as one of the most promising materials to address one of the biggest challenges facing the world, namely, an ever increasing energy crisis (Pu et al., 2016; Zou, Ping, Cai, & Liu, 2015). E-textiles, due to their electrical conductivity together with their flexibility, have attracted many people to engineer various types of textiles, including cotton (Sahito, Sun, Arbab, Qadir, & Jeong, 2015; Xu et al., 2014), lyocell (Mengal et al., 2016), polyester (Arbab, Sun, Sahito, Qadir, & Jeong, 2015), nylon (Yun, Hong, Kim, Jun, & Kim, 2013), and wool (Javed, Galib, Yang, Chen, & Wang, 2014) fabrics, for application in batteries, fuel cells, supercapacitors and solar cells, and sensing and display inventions (Cherenack, Zysset, Kinkeldei, Münzenrieder, & Tröster, 2010).

Scientists hold two different views as to how to define smart textiles: some say that they have the ability to sense changes in the environment, others say that they must not only sense but also take certain predefined actions based on what has been sensed (Stoppa & Chiolerio, 2014). These smart and intelligent materials have found their way into monitoring the health and physiological condition of people, such as the beat rate of the heart (Stoppa & Chiolerio, 2014), heat generated by the body to calculate loss of calories (Harifi & Montazer, 2015), and running speed and various other exercises by athletes, accompanied by flexible displays, integrated keyboards in clothing, printable electronics, and much more (Kelly, Meunier, Cochrane, & Koncar, 2013; H. M. Lee, Choi, Jung, & Ko, 2013). However, one of the main challenges faced is how well the characteristics of the fabric can be maintained during the production of smart or e-textiles. To address this key issue, many researchers have focused on textile fibers, as they are the basic building blocks of the fabric, and dedicated themselves to yarns.

Smart textiles are an entirely new class of fibers, fabrics, and the various products made from them. It is important to understand that this new

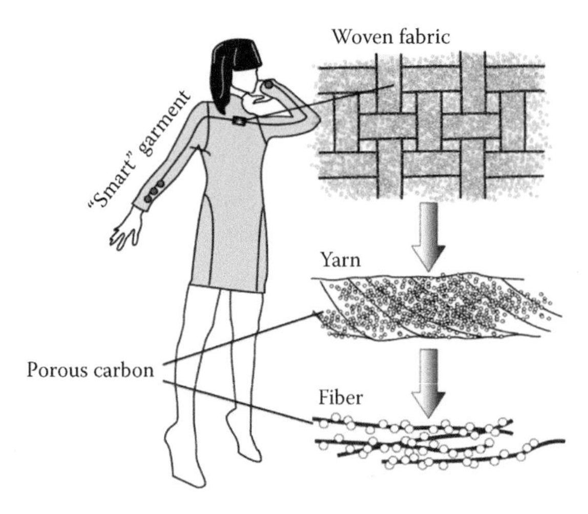

FIGURE 12.1
Design concept of a porous textile integrated into a smart garment, demonstrating porous carbon impregnation from the weave, to the yarn, to the fibers. Carbon-coated textiles for flexible energy storage. (From Jost, K. et al., *Energy Environ. Sci.*, 4(12), 5060, 2011.)

generation of textile materials, apart from their many technical and smart uses, are primarily made for the safety and protection of military personnel (Dalsgaard & Sterrett, 2014; Nayak, Wang, & Padhye, 2015). Figure 12.1 shows how a porous carbon material can be applied to fibers and fabricated yarn so as to form an electrically conductive fabric, which can then be used with electronic devices to work as a smart garment.

It is not unknown for military personnel to be prone to extreme weather conditions, even with the chance of a battle ahead of them. Therefore, the rapid development of smart garments is due to their application to military purposes; however, added fashion and convenience remain only secondary to their precise end use. It is also important to note that the advent of smart textiles with various characteristics in a single material or those that produce a composite and hybrid structure using a range of new classes of materials has brought various disciplines of the engineering sciences together. For example, electrical, mechanical, chemical, electronics, and textile engineers work hand in hand to produce smart and e-textiles for various applications (Ten Bhömer, Tomico, Kleinsmann, Kuusk, & Wensveen, 2013).

12.2 Interactive Fabrics

Communication through various new technologies has taken the place of conventional ones since mobile phones began to use various new software for connecting people to each other. People want these devices to be flexible, small, and truly portable so they can have them around their person all the time without the need to recharge them (Kumar, Rai, Oh, Harbaugh, & Varadan, 2014). We know that portable devices are in an age of rapidly growing innovation and development and represent an enormous market share (Di et al., 2016; Kim, 2015; Özdemir & Kılınç, 2015; Roggen, Tröster, & Bulling, 2013). The incorporation and integration of portable electronic devices with low cost textile materials carry a natural source of interest for technology and scientists.

Fabric-based calculators and keyboards are already available. They were the first invention made on a single layer of fabric using capacitive sensing. In such devices, the point of contact can be an array of embroidered or silk-screened electrodes (Post et al., 2000). The concept was to measure total capacitance, which can vary depending upon the touch of a finger on a capacitive electrode. Furthermore, the capacitive sensor using the sensing array can also indicate whether the garment fits the wearer or not, as in this case pressure signals can be detected (Takamatsu, Yamashita, Imai, & Itoh, 2014; Takamatsu et al., 2015).

A schematic diagram of the preparation and use of a capacitive fabric sensor working as a keyboard is shown in Figure 12.2. Seiichi Takamatsu et al. showed that a nylon fiber coated with a PEDOT:PSS conductive polymer results in an electrically conductive thread, which is then woven into a fabric. The device works on a capacitive sensor and at every event of

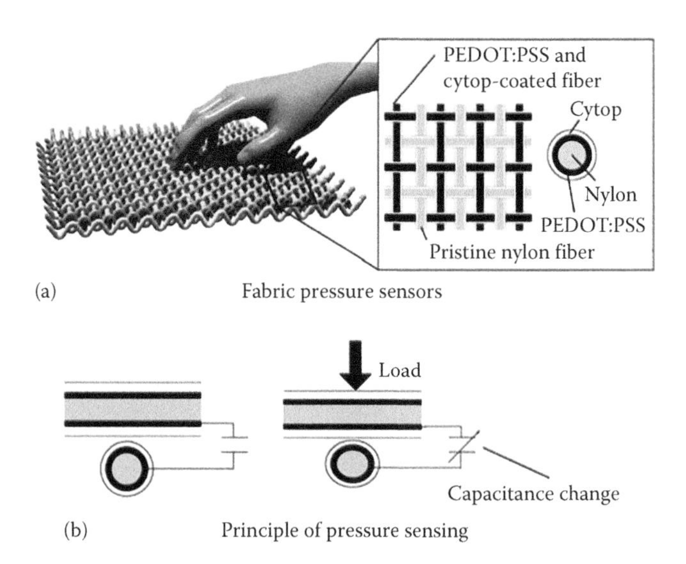

FIGURE 12.2
Concept of fabric pressure sensor array. (a) In the structure of the sensor, conductive polymer and dielectric film-coated fibers are woven as wefts and warps, forming capacitive type of pressure sensors. (b) In the sensing mechanism, applied pressure induces capacitance change. (From Takamatsu, S. et al., *Sensors Actuat. A: Phys.*, 184, 57, 2012.)

pressure exerted by the fingers the capacitance changes, giving off a signal. The theme is simple; however, flexibility and stability of the conductive polymer coating is important. In this way, scalable production of such devices is not a big challenge to realize.

The sensitivity of these e-textiles can be controlled in two ways. First, by the quality of the capacitive sensor itself, and second by the amount of electrical conductivity of the conductive threads used in this product (Guo, Huang, Cai, Liu, & Liu, 2016; Wei, Torah, Yang, Beeby, & Tudor, 2013). A completely rolled keyboard can also be seen in Figure 12.2, which proves that such devices are flexible, durable, and portable. It is important to mention here that such kinds of printed circuit boards are supported by the conventional electronic components necessary to perform the given tasks, such as capacitive sensors and signal processing data streams.

During the early development of these devices, KENPO jackets showed a lot of interest in these e-textiles and prepared 50 jackets equipped with a small MP3 player connected to a keypad. This was a test application; they said they had an excellent response from the people who wore the garment. Not only KENPO jackets, but several other brands have been interested in producing these garments with integrated mobile phones. However, there are two main reasons why these garments have not been truly commercialized: first, it is not easy to wash them, and second, a battery is required, which needs charging. This drives through to building textiles for energy harvesting and storage so that the devices can work by solar energy and charge themselves in future.

12.3 Classification of Smart and E-Textiles

Broadly, the different parameters that can be sensed by smart textiles are thermal, mechanical, chemical, electrical, magnetic, and optical. In this way such textiles are categorized into three types:

1. *Passive smart textile.* These are materials that can only sense a change in the surrounding environment or any stimuli.
2. *Active smart textile.* These materials react upon sensing a change in the surroundings or a perturbation. They are equipped with actuators in conjunction with sensors and take predetermined actions.
3. *Ultra-smart textile.* These materials are also called very smart textiles and they possess the ability to adapt themselves to environmental perturbation or any variation in the signals received. Essentially, they consist of a central processing unit acting as a brain with all the predefined programs in place.

12.4 Fabrication Techniques Used to Produce E-Textiles

There are various methods of imparting the necessary electrical conductivity to textile materials, all of them are different in nature and produce materials with different levels of conductivity and stability of the coated material (Matsuhisa et al., 2015; Yamashita, Takamatsu, Miyake, & Itoh, 2013). Figure 12.3 shows the methods of enabling electrical conductivity before and after the fabric manufacturing process.

12.4.1 Producing E-Textiles at the Fiber Stage by Using Nanofibers and the Core Sheath Mechanism

Yarns made of electrically conductive fibers or fiber supercapacitors can be woven, knitted, or stitched to form textiles or wearable garments. Therefore, many researchers have focused on producing energy harvesting devices based on fibers that can be woven and eventually used as wearable electronic devices. Figure 12.4 shows a few fibrous designs for working as supercapacitors in e-textiles (Cheng et al., 2013; Jost et al., 2013; J. A. Lee et al., 2013; Meng et al., 2013).

12.4.2 Producing E-Textiles through Weaving or Knitting Electrically Conductive Threads in Fabric

Production of electro-conductive textiles using conductive threads and weaving or knitted them into the fabric may be one of the most reliable and stable processes. However, there are certain limitations to the conductive yarns that can be used to produce such e-textiles (de Vries & Peerlings, 2014; Gunnarsson, Karlsteen, Berglin, & Stray, 2015; Huang et al., 2015; Rundqvist, Nilsson, Lund, Sandsjö, & Hagström, 2014). It is well-known that during the weaving process especially, and the knitting process as well, that the

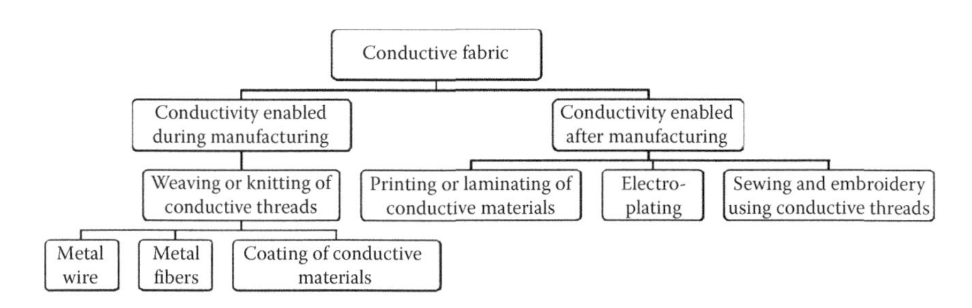

FIGURE 12.3
Techniques to enable conductivity in fabrics. (From Castano, L.M. and Flatau, A.B., *Smart Mater. Struct.*, 23(5), 053001, 2014.)

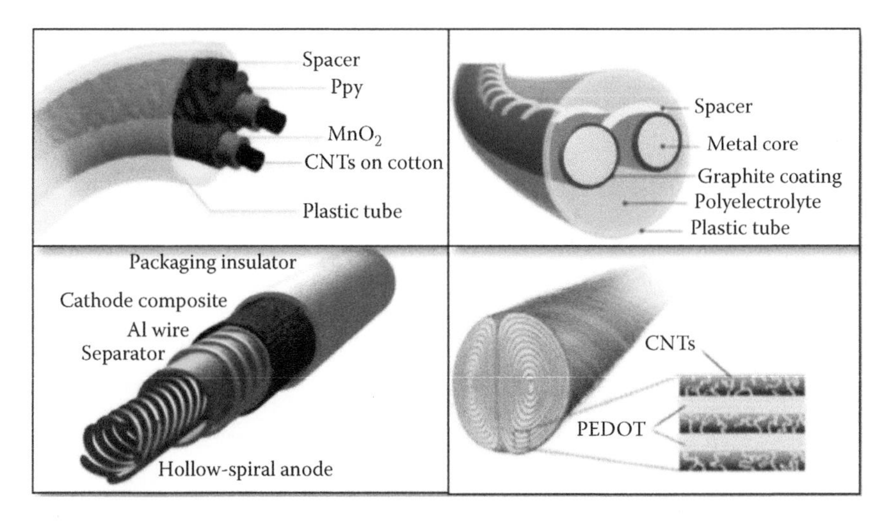

FIGURE 12.4
Various fiber designs for energy harvesting and storage. (From Jost, K. et al., *J. Mater. Chem. A*, 2(28), 10776, 2014.)

yarn is under a lot of tension to form a firm fabric. Therefore, yarn used in warp needs to withstand weaving stresses and must be sufficiently flexible. Conversely, metallic yarn is not flexible, rather it is brittle, therefore it is difficult for a weaver or knitter to produce fabric made of metallic yarn. As can be seen in Figure 12.5a, a metallic filament-made yarn is used to knit the electro-conductive fabric, which is shown in Figure 12.5b to be a localized part of the knitted fabric. On the other hand a copper fiber is seen to be wrapped in some synthetic filament-made yarn and then plain-woven into the fabric, as shown in Figure 12.5c. Now it can become apparent that all three different structures have different levels of electro-conductivity depending upon the amount of conductive yarn used.

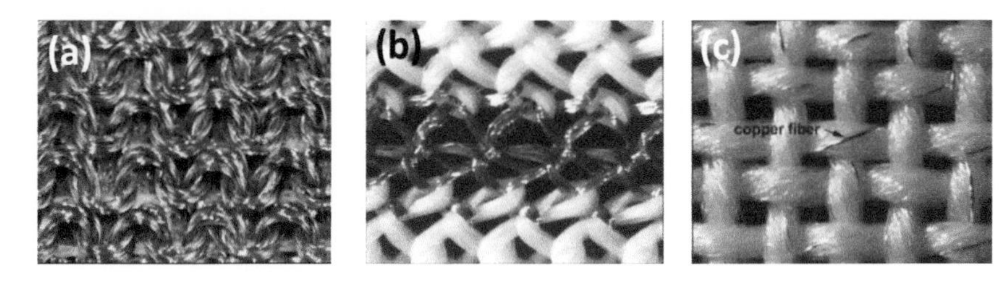

FIGURE 12.5
Electrically conductive knitted fabric (a) produced by knitting electrically conductive yarns, (b) a localized construction of electrically conductive threads knitted with common yarns and (c) copper wire woven in with wool yarns. (From Monfeld, C., Smart textiles textiles with enhanced functionality, www.smarttextile.de, 2015. Accessed September 15, 2017.)

12.4.3 Producing E-Textiles by Coating Electrically Conductive Carbon Nanomaterials on Fibers, Yarns, and Fabrics

Carbon materials are one of the most studied electrically conductive materials today. Materials such as activated carbon, carbon nanotubes, graphene, and their composites are now well established to produce an electrically conductive coating on fibers, yarns, and fabrics (Aboutalebi et al., 2014; Liu et al., 2013). The knitted and woven structures of cotton (Liu, Hu, & Yang, 2016), lyocell (Mengal et al., 2016), polyester (Molina et al., 2013), silk (Lu, Mao, & Zhang, 2015), and wool (Javed et al., 2014) have been made electrically conductive using a simple dip and dry technique. However, cotton fiber, due to its intrinsic hydrophilic characteristic, responds best to being directly coated by these carbon materials. Other fibers, not being hydrophilic, need some binders to enable the proper attachment of carbon materials to them.

E-textiles were first used in an electrochemical system by Yi Cui's group at Stanford University; the applications were supercapacitors and batteries (Hu et al., 2010). They used a nonwoven, cotton, fiber-based structure; an ink-like supersuspension of single-walled carbon nanotubes was coated on the substrate by using a simple and quick dip and dry procedure, as shown in Figure 12.6. The nonwoven structure is porous and can easily absorb an adequate amount of aqueous solution. It also showed the stability of the coated material through various techniques, including Scotch tape taste and in aqueous solution.

Another carbon material, two-dimensional graphene, has an extremely high electrical conductivity and has revolutionized research recently (Liu, Yan, Lang, Peng, & Xue, 2012) with its matchless characteristics coupled with its production in large quantities, which has made it easier to fabricate graphene-coated conductive composites (Sahito et al., 2015). B. Fugetsu in Hokkaido University first reported the application of the oxidized form of graphite, graphene oxide (GO), onto a cotton textile in the same way as dyes are commonly applied to textiles. It is important to note that GO is an electrical insulator and, to bring back electrical conductivity, must be returned to graphene by various reduction methods, including chemical and thermal methods. In this way the coated material is called reduced graphene oxide (rGO). For textile materials only the chemical reduction method is viable, as only it can be realized at temperatures below 100°C. On the other hand, the thermal reduction of GO for better results is carried out at a very high temperature, more than 600°C, which cannot be sustained by common textile fibers. The application of rGO and its composites to textiles has been followed by various other researchers (Molina et al., 2013; Sahito et al., 2016; Samad, Li, Alhassan, & Liao, 2014; Shateri-Khalilabad & Yazdanshenas, 2013; Yu et al., 2011; Yun et al., 2013); however, Sahito et al. have claimed to have produced the lowest surface resistance at only 7 Ω/sq. Figure 12.7 shows a scanning electron microscope (SEM) image of their result, showing a smooth coating of graphene with a layer that can be seen in cross-sectional view and which follows the fabric's contours. The stability of these coatings will be discussed later in this chapter.

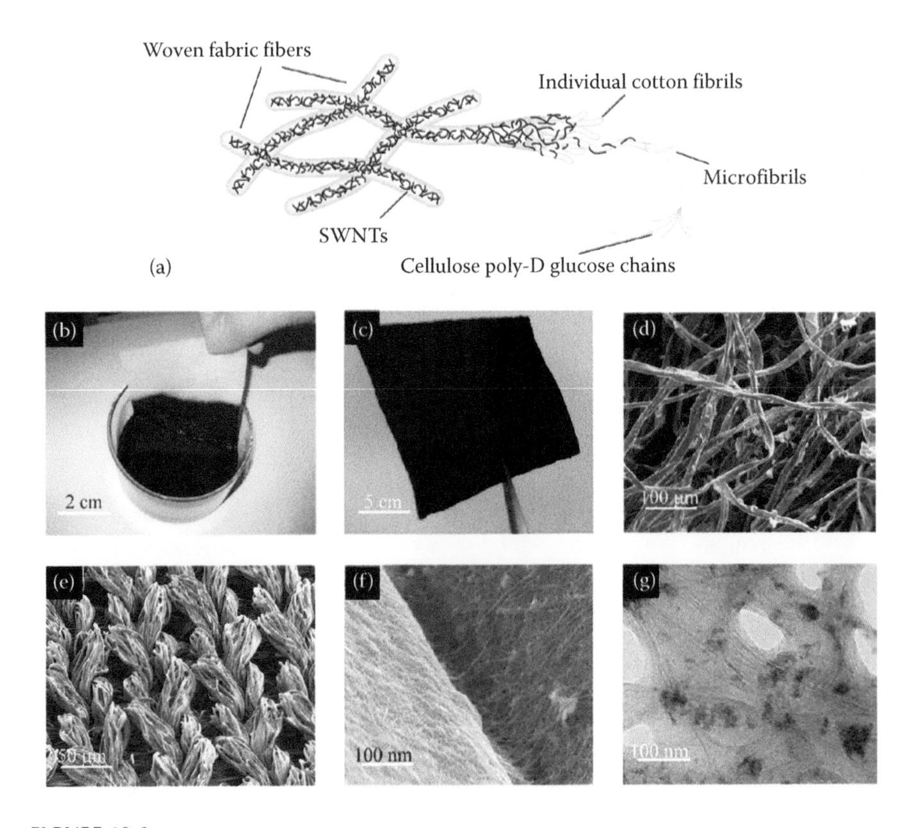

FIGURE 12.6
Porous textile conductor fabrication. (a) Schematic of SWNTs wrapping around cellulose fibers, (b) Dipping textile into an aqueous SWNT ink, (c) A thin, 10 × 10 cm textile conductor, (d) SEM image of coated cotton with SWNTs on the cotton fiber surface, (e) SEM image of fabric sheet coated with SWNTs on the fabric fiber surface, (f) Conformal coating of SWNT on the fabric fibers, (g) TEM image of SWNTs on cotton fibers. (From Hu, L. et al., *Nano Lett.*, 10(2), 708, 2010.)

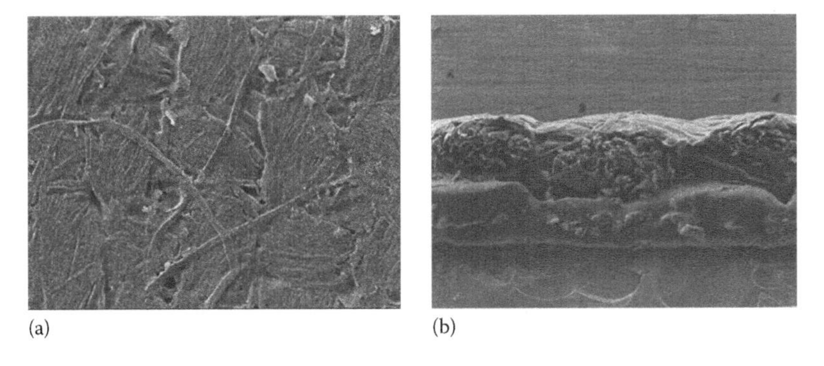

FIGURE 12.7
SEM image of reduced graphene oxide coated on cotton. (a) Top view and (b) cross-sectional view. (From Sahito, I.A. et al., *J. Power Source*, 319, 90, 2016.)

TABLE 12.1

Comparison of the Surface Resistance Measured by Various Researchers at Different Process Parameters and Using Various Chemicals

Material	Reducing Agent	Time (min)	Temperature (°C)	Surface Resistance (Ω/sq.)	References
Cotton fabric	Ascorbic acid	60	95	840	Shateri-Khalilabad and Yazdanshenas (2013)
	Sodium borohydride	720	Room Temperature	560	Xu, Guo, Liu, and Bian (2015)
	Sodium hydrosulfite	60	95	374	Shateri-Khalilabad and Yazdanshenas (2013)
	Thermal reduction	120	300	225	Liu, Yan, Lang, Peng, and Xue (2012)
	Hydrazine hydrate	2440	100	206	Gu and Zhao (2011)
	Sodium hydrosulfite	30	90	100	Fugetsu, Sano, Yu, Mori, and Tanaka (2010)
	Hydrazine hydrate	30	100	7	Sahito et al. (2016)

Table 12.1 compares the results for the surface resistance of various researchers and reveals different times, temperatures, and reducing agents.

12.4.4 Producing E-Textiles by Coating Electrically Conductive Polymers on Textiles

Electro-conductive coatings can be mainly categorized into two types, typically those that may be inherently conductive polymers or those that may be conductive fillers in a polymer; these latter are also called conductive doped polymers. Conductive fillers used in a polymer are metallic particles that bring a very high electro-conductivity into polymeric materials. Carbon materials are also used, which are not only easily controllable for dissolution, but are also low cost materials. Single-walled carbon nanotubes and graphite are typical examples of doping carbon materials in a polymer solution. Carbon materials in the form of nanotubes or nanoparticles possess very high surface area, which results in very high electro-conductivity, sometimes even more than the metallic particles doped conducting polymers. Electrically conducting polymers can also produce e-textiles, either by coating them on textiles or through electrospinning, by preparing a nanofiber web of the conducting polymers, such as polyaniline (PANI), polypyrrole (PPy), polyacetylene, polythiophene, and (poly(3,4-ethylenedioxythiophene)(4-styrenesulfonate)) PEDOT–PSS by chemical and electrochemical deposition methods (Akşit et al., 2009; Babu, Dhandapani, Maruthamuthu, & Kulandainathan, 2012;

Babu, Subramanian, & Kulandainathan, 2013; Dall'Acqua et al., 2006; Onar et al., 2009; Varesano, Aluigi, Florio, & Fabris, 2009). Through a lengthy chemical synthesis process, Yoshihiro Egami et al. coated polypyrrole nanoparticles onto a polyester/nylon blend and obtained very low surface resistance in the range of 3–52 Ω/sq. (Egami et al., 2011). In another study Jie Xu et al. used a 100% cotton plain weave fabric and first applied nickel plating by a multistep process. Their method was time consuming and incurred a number of steps, though it resulted in a very low surface resistance of only 1.5 Ω/sq. initially. Further, when they coated polypyrrole during 2 h of polymerization, the surface resistance value increased and reached 41.5 Ω/sq. (Xu et al., 2014).

Electrospinning is a well-known technique these days requiring no big set-up; it works over a short period of time and requires no complicated steps. Above all, a lot of variation in the product can be controlled with high precision and repeatability, which may not be realized in a dip and dry process. The thickness of the nanofiber sheet, the diameter of the nanofibers, and their porosity can be controlled to produce various effects with the fibers. PPy, PEDOT:PSS, PANI, and other conducting polymers have been successfully electrospun on textile substrates directly and have shown tremendous results (Chen, Miao, & Liu, 2013; Yanılmaz & Sarac, 2014). Titanium carbide (TiC) nano-felt was chlorinated at a temperature between 200°C and 1000°C and converted to carbon derived carbide by Yu Gao et al. and demonstrated promise for high power supercapacitors.

12.5 Electrical Resistance of E-Textiles

In order to assess how good the transportation of electrons in a material is, its electrical conductivity or resistivity is measured. However, for e-textiles the conductivity of the materials is not as high as for metals, especially, because e-textiles serve the purpose of being flexible, which limits their electrical conductivity. Therefore, e-textiles are measured for their surface resistance in Ω/sq., which is reciprocal to electrical conductivity in siemens.

The surface resistance of e-textiles is measured using either a two-probe or four-probe resistivity meter. A four-probe meter gives more precision of results compared to a two-probe one. However, when the resistance is measured using a two-probe meter, silver or any other metallic paste is used and applied to the e-textiles at two points, to provide better current collection; the two probes are then placed on the two points, as shown in Figure 12.8. The four-probe meter uses four electrodes, as shown in Figure 12.9, which are placed over the substrate and measure the passage of electrical current.

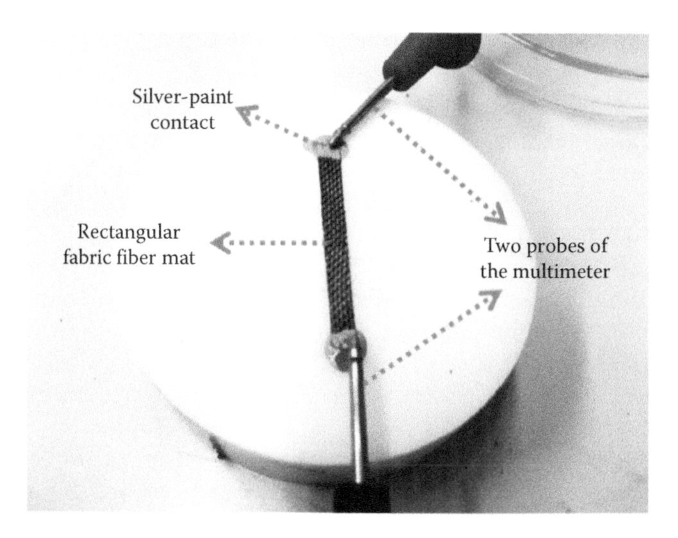

FIGURE 12.8
Measuring electrical resistance at the fabric surface using a two point probe. (From Samad, Y.A. et al., *RSC Adv.*, 4(33), 16935, 2014.)

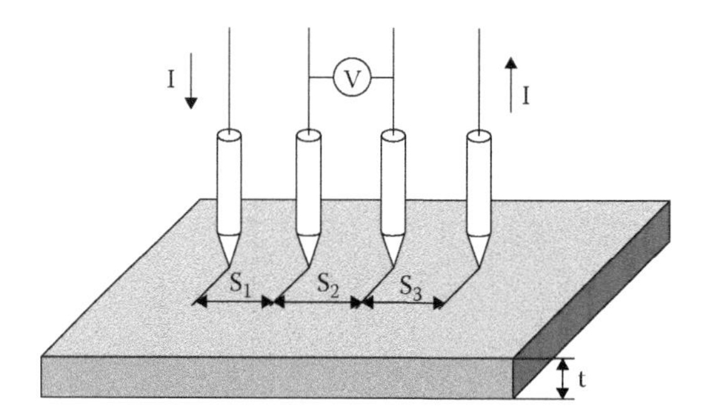

FIGURE 12.9
Schematic diagram of measuring electrical resistance at the fabric surface using a two point probe. (From Shimamoto, A. et al., *J. Pressure Vessel Technol.*, 135(2), 021501, 2013.)

12.6 Current and Voltage Measurement of E-Textiles

Electrochemical impedance spectroscopy (EIS) is one of the most useful techniques to measure the various characteristics of e-textiles. It is important to mention here that the EIS technique measures a variety of tests, including electrical resistance, cyclic voltammetry, current versus time, internal resistance of the e-textile, the flow of electrons in any electrochemical cell consisting of e-textiles, and capacitance. The impedance analyzer works on a four-point probe method, which gives better precision for the results and hence it is the most widely used technique to measure the internal and surface resistance of e-textiles. More than anything else, e-textiles are focused on due to their flexibility. To ensure that these materials perform the same given task at any point of curvature they need to be checked for the flow of current under a fixed voltage. In other words it is the ability of the material to conduct electricity at the same rate when the material is under bending conditions. In order to do this, the stability of e-textiles has been measured by various researchers. For example, Xu et al. (2014) connected the fabric with an electrochemical impedance analyzer and measured the flow of current at a fixed voltage of 2 V. The PPy/nickel-coated fabric was bent at five different positions and the flow of current was recorded. The authors observed that there was no change in the flow of current at any bending angle.

Sahito et al. (2016) also measured the current flow in a graphene-coated cotton fabric; the constant voltage this time was 1 V. As can be seen in Figure 12.10, the current remained unchanged during all bending angles, which can be seen in the insets of the figure. This shows that such e-textiles perform well when they are bent and the garment takes on any body contour without device malfunction.

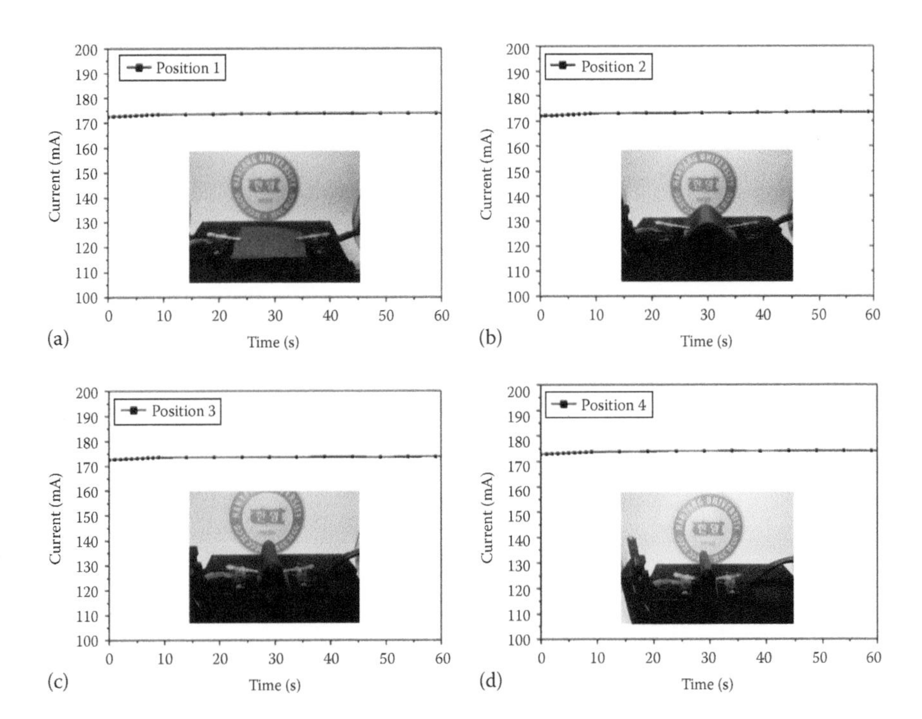

FIGURE 12.10
Current vs. time measurement of e-textile at a constant voltage of 1 V. (a) At position 1, (b) at position 2, (c) at position 3 and (d) at position 4. (From Sahito, I.A. et al., *J. Power Sources*, 319, 90, 2016.)

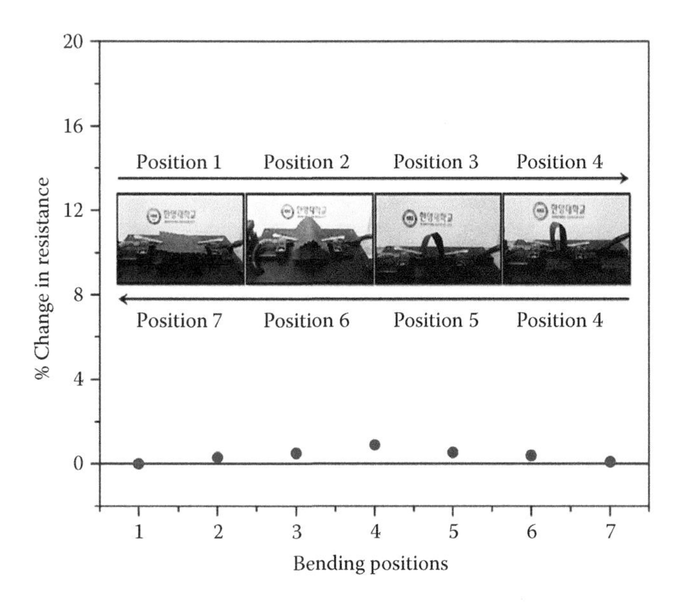

FIGURE 12.11
Change in surface resistance of e-textiles against various bending positions.

Further, to ensure that the surface resistance of e-textiles remains unchanged at various bending cycles, the two opposite ends of the e-textile are connected to the electrochemical impedance analyzer and the change in resistance is measured against different bending positions.

Mengal et al. (2016) measured the change in resistance in percentage against various bending angles. It is more reasonable to believe that at the largest bending angle, when the e-textile is completely folded, the surface resistance may change as the fabric is stretched to an extent that the material may crack at the surface of the fabric. Their results show that a negligible change of 1.5% in the surface resistance was observed and that was at the highest bending angle: position 4 in Figure 12.11.

12.7 Stability of the Coating on Fabric and Compatibility with Electrolyte Solution

For the e-textile to work in an electrochemical cell, including dye sensitized solar cells, liquid electrolyte batteries, fuel cells, and supercapacitors, the stability of the coated e-textile must be assessed. Researchers have done this for a carbon-coated e-textile in water and an iodine-based electrolyte solution (Arbab et al., 2015; Sahito et al., 2016). This test is quite simple and the

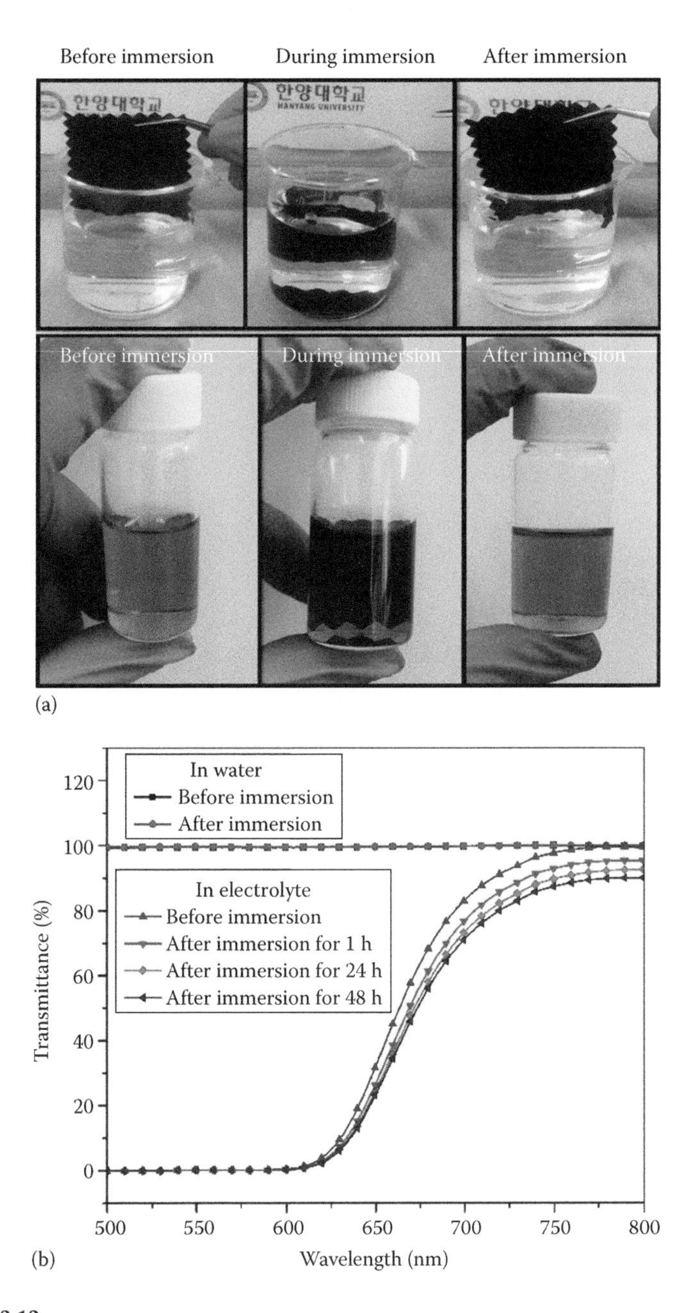

FIGURE 12.12
Transmission spectra of the electrolyte solution and DI water before and after immersion of the electro conductive textile. (From Mengal, N. et al., *Carbohydr. Polym.*, 152, 19, 2016.)

results are very accurate depending upon the operator. In this test, the deionized (DI) water and the electrolyte solution are measured for their transmission spectra using a UV-Vis spectroscopy technique. At the first check both the DI water and the electrolyte solution do show a particular transmission spectra, which can be seen in Figure 12.12. Afterwards, the carbon-coated e-textile is put in water and an electrolyte solution for a certain amount of time, which was 48 h in the report by Mengal et al. (2016). They recorded the transmission spectra in the range of 260–800 nm. Representative samples of DI water and the electrolyte solution were measured for transmission spectra after the fabric had remained in the water and electrolyte solution for 1, 24, and 48 h. They reported that there was no change in the spectrum of water, even after 48 h; however, there was a slight change in the spectrum of the electrolyte solution, which can be seen in Figure 12.12, which they claimed was negligible, though the exact percentage change was not reported.

References

Aboutalebi, S. H., Jalili, R., Esrafilzadeh, D., Salari, M., Gholamvand, Z., Aminorroaya Yamini, S.,... Moulton, S. E. (2014). High-performance multifunctional graphene yarns: Toward wearable all-carbon energy storage textiles. *ACS Nano, 8*(3), 2456–2466.

Akşit, A. C., Onar, N., Ebeoglugil, M. F., Birlik, I., Celik, E., & Ozdemir, I. (2009). Electromagnetic and electrical properties of coated cotton fabric with barium ferrite doped polyaniline film. *Journal of Applied Polymer Science, 113*(1), 358–366.

Arbab, A. A., Sun, K. C., Sahito, I. A., Qadir, M. B., & Jeong, S. H. (2015). Multiwalled carbon nanotube coated polyester fabric as textile based flexible counter electrode for dye sensitized solar cell. *Physical Chemistry Chemical Physics, 17*(19), 12957–12969. doi:10.1039/C5CP00818B.

Babu, K. F., Dhandapani, P., Maruthamuthu, S., & Kulandainathan, M. A. (2012). One pot synthesis of polypyrrole silver nanocomposite on cotton fabrics for multifunctional property. *Carbohydrate Polymers, 90*(4), 1557–1563.

Babu, K. F., Subramanian, S. S., & Kulandainathan, M. A. (2013). Functionalisation of fabrics with conducting polymer for tuning capacitance and fabrication of supercapacitor. *Carbohydrate Polymers, 94*(1), 487–495.

Bowman, D., & Mattes, B. R. (2005). Conductive fibre prepared from ultra-high molecular weight polyaniline for smart fabric and interactive textile applications. *Synthetic Metals, 154*(1), 29–32.

Carpi, F., & De Rossi, D. (2005). Electroactive polymer-based devices for e-textiles in biomedicine. *IEEE Transactions on Information Technology in Biomedicine, 9*(3), 295–318.

Castano, L. M., & Flatau, A. B. (2014). Smart fabric sensors and e-textile technologies: A review. *Smart Materials and Structures, 23*(5), 053001.

Chen, D., Miao, Y.-E., & Liu, T. (2013). Electrically conductive polyaniline/polyimide nanofiber membranes prepared via a combination of electrospinning and subsequent in situ polymerization growth. *ACS Applied Materials & Interfaces, 5*(4), 1206–1212.

Cheng, H., Dong, Z., Hu, C., Zhao, Y., Hu, Y., Qu, L.,... Dai, L. (2013). Textile electrodes woven by carbon nanotube–graphene hybrid fibers for flexible electrochemical capacitors. *Nanoscale, 5*(8), 3428–3434.

Cherenack, K., Zysset, C., Kinkeldei, T., Münzenrieder, N., & Tröster, G. (2010). Woven electronic fibers with sensing and display functions for smart textiles. *Advanced Materials, 22*(45), 5178–5182.

Curone, D., Secco, E. L., Tognetti, A., Loriga, G., Dudnik, G., Risatti, M., ... Magenes, G. (2010). Smart garments for emergency operators: The ProeTEX project. *IEEE Transactions on Information Technology in Biomedicine, 14*(3), 694–701.

Dall'Acqua, L., Tonin, C., Varesano, A., Canetti, M., Porzio, W., & Catellani, M. (2006). Vapour phase polymerisation of pyrrole on cellulose-based textile substrates. *Synthetic Metals, 156*(5–6), 379–386.

Dalsgaard, C., & Sterrett, R. (2014). White paper on smart textile garments and devices: A market overview of smart textile wearable technologies. In *Market opportunities for smart textiles*. Aarhus, Denmark: Ohmatex ApS.

Dastjerdi, R., & Montazer, M. (2010). A review on the application of inorganic nanostructured materials in the modification of textiles: Focus on anti-microbial properties. *Colloids and Surfaces B: Biointerfaces, 79*(1), 5–18.

de Vries, H., & Peerlings, R. (2014). Predicting conducting yarn failure in woven electronic textiles. *Microelectronics Reliability, 54*(12), 2956–2960.

Dhawan, A., Seyam, A. M., Ghosh, T. K., & Muth, J. F. (2004). Woven fabric-based electrical circuits part I: Evaluating interconnect methods. *Textile Research Journal, 74*(10), 913–919.

Di, J., Zhang, X., Yong, Z., Zhang, Y., Li, D., Li, R., & Li, Q. (2016). Carbon-nanotube fibers for wearable devices and smart textiles. *Advanced Materials, 28*(47), 10529–10538.

Egami, Y., Suzuki, K., Tanaka, T., Yasuhara, T., Higuchi, E., & Inoue, H. (2011). Preparation and characterization of conductive fabrics coated uniformly with polypyrrole nanoparticles. *Synthetic Metals, 161*(3), 219–224.

Eriksson, S., & Sandsjö, L. (2015). Three-dimensional fabrics as medical textiles. In X. Chen (Ed.), *Advances in 3D textiles* (pp. 305–340). Woodhead Publishing Limited, A volume in Woodhead Publishing Series in Textiles.

Fugetsu, B., Sano, E., Yu, H., Mori, K., & Tanaka, T. (2010). Graphene oxide as dyestuffs for the creation of electrically conductive fabrics. *Carbon, 48*(12), 3340–3345.

Gu, W. L., & Zhao, Y. N. (2011). Graphene modified cotton textiles. *Advanced Materials Research, 331*, 93–96.

Gunnarsson, E., Karlsteen, M., Berglin, L., & Stray, J. (2015). A novel technique for direct measurements of contact resistance between interlaced conductive yarns in a plain weave. *Textile Research Journal, 85*(5), 499–511.

Guo, X., Huang, Y., Cai, X., Liu, C., & Liu, P. (2016). Capacitive wearable tactile sensor based on smart textile substrate with carbon black/silicone rubber composite dielectric. *Measurement Science and Technology, 27*(4), 045105.

Harifi, T., & Montazer, M. (2015). Application of nanotechnology in sports clothing and flooring for enhanced sport activities, performance, efficiency and comfort: A review. *Journal of Industrial Textiles, 46*(5), 1147–1169.

Hu, L., Pasta, M., Mantia, F. L., Cui, L., Jeong, S., Deshazer, H. D.,... Cui, Y. (2010). Stretchable, porous, and conductive energy textiles. *Nano Letters, 10*(2), 708–714.

Huang, Y., Hu, H., Huang, Y., Zhu, M., Meng, W., Liu, C., ... Zhi, C. (2015). From industrially weavable and knittable highly conductive yarns to large wearable energy storage textiles. *ACS Nano, 9*(5), 4766–4775.

Javed, K., Galib, C., Yang, F., Chen, C.-M., & Wang, C. (2014). A new approach to fabricate graphene electro-conductive networks on natural fibers by ultraviolet curing method. *Synthetic Metals, 193*, 41–47.

Jost, K., Dion, G., & Gogotsi, Y. (2014). Textile energy storage in perspective. *Journal of Materials Chemistry A, 2*(28), 10776–10787.

Jost, K., Perez, C. R., McDonough, J. K., Presser, V., Heon, M., Dion, G., & Gogotsi, Y. (2011). Carbon coated textiles for flexible energy storage. *Energy & Environmental Science, 4*(12), 5060–5067.

Jost, K., Stenger, D., Perez, C. R., McDonough, J. K., Lian, K., Gogotsi, Y., & Dion, G. (2013). Knitted and screen printed carbon-fiber supercapacitors for applications in wearable electronics. *Energy & Environmental Science, 6*(9), 2698–2705.

Kadolph, S., & Langford, A. (eds.). (2007). *Textiles*. Upper Saddle River, NJ: Prentice Hall.

Kelly, F. M., Meunier, L., Cochrane, C., and Koncar, V. (2013). Polyaniline: Application as solid state electrochromic in a flexible textile display. *Displays, 34*(1), 1–7.

Kim, R.-H. (2015). Cure performance and effectiveness of portable smart healthcare wear system using electro-conductive textiles. *Procedia Manufacturing, 3*, 542–549.

Kumar, P., Rai, P., Oh, S., Harbaugh, R. E., & Varadan, V. K. (2014). Smart e-textile-based nanosensors for cardiac monitoring with smart phone and wireless mobile platform. *Micro and smart devices and systems* (pp. 387–401). Springer, New Delhi, India. https://link.springer.com/chapter/10.1007/978-81-322-1913-2_23.

Lee, H. M., Choi, S. Y., Jung, A., & Ko, S. H. (2013). Highly conductive aluminum textile and paper for flexible and wearable electronics. *Angewandte Chemie, 125*(30), 7872–7877.

Lee, J. A., Shin, M. K., Kim, S. H., Cho, H. U., Spinks, G. M., Wallace, G. G., ... Baughman, R. H. (2013). Ultrafast charge and discharge biscrolled yarn supercapacitors for textiles and microdevices. *Nature Communications, 4*, article number 2970, 1–8.

Liu, S., Hu, M., & Yang, J. (2016). A facile way of fabricating a flexible and conductive cotton fabric. *Journal of Materials Chemistry C, 4*(6), 1320–1325.

Liu, B., Wang, X., Chen, H., Wang, Z., Chen, D., Cheng, Y.-B., ... Shen, G. (2013). Hierarchical silicon nanowires-carbon textiles matrix as a binder-free anode for high-performance advanced lithium-ion batteries. *Scientific Reports, 3*, article number 1622, 1–7.

Liu, W.-w., Yan, X.-b., Lang, J.-w., Peng, C., & Xue, Q.-j. (2012). Flexible and conductive nanocomposite electrode based on graphene sheets and cotton cloth for supercapacitor. *Journal of Materials Chemistry, 22*(33), 17245. doi:10.1039/c2jm32659k.

Lu, Z., Mao, C., & Zhang, H. (2015). Highly conductive graphene-coated silk fabricated via a repeated coating-reduction approach. *Journal of Materials Chemistry C, 3*(17), 4265–4268.

Matsuhisa, N., Kaltenbrunner, M., Yokota, T., Jinno, H., Kuribara, K., Sekitani, T., & Someya, T. (2015). Printable elastic conductors with a high conductivity for electronic textile applications. *Nature Communications, 6*, article number 8461, 1–11.

Meng, Y., Zhao, Y., Hu, C., Cheng, H., Hu, Y., Zhang, Z., … Qu, L. (2013). All-graphene core-sheath microfibers for all-solid-state, stretchable fibriform supercapacitors and wearable electronic textiles. *Advanced Materials, 25*(16), 2326–2331.

Mengal, N., Sahito, I. A., Arbab, A. A., Sun, K. C., Qadir, M. B., Memon, A. A., & Jeong, S. H. (2016). Fabrication of a flexible and conductive lyocell fabric decorated with graphene nanosheets as a stable electrode material. *Carbohydrate Polymers, 152*, 19–25.

Molina, J., Fernández, J., Inés, J., Del Río, A., Bonastre, J., & Cases, F. (2013). Electrochemical characterization of reduced graphene oxide-coated polyester fabrics. *Electrochimica Acta, 93*, 44–52.

Monfeld, C. (2015). Smart textiles textiles with enhanced functionality. www.smart-textile.de. Accessed September 15, 2017.

Nayak, R., Wang, L., & Padhye, R. (2015). *Electronic textiles for military personnel* (pp. 239–256). Elsevier, Cambridge, U.K.

Onar, N., Akşit, A. C., Ebeoglugil, M. F., Birlik, I., Celik, E., & Ozdemir, I. (2009). Structural, electrical, and electromagnetic properties of cotton fabrics coated with polyaniline and polypyrrole. *Journal of Applied Polymer Science, 114*(4), 2003–2010.

Özdemir, H., & Kılınç, S. (2015). Smart woven fabrics with portable and wearable vibrating electronics. *Autex Research Journal, 15*(2), 99–103.

Post, E. R., Orth, M., Russo, P., & Gershenfeld, N. (2000). E-broidery: Design and fabrication of textile-based computing. *IBM Systems Journal, 39*(3.4), 840–860.

Pu, X., Song, W., Liu, M., Sun, C., Du, C., Jiang, C., … Wang, Z. L. (2016). Wearable power-textiles by integrating fabric triboelectric nanogenerators and fiber-shaped dye-sensitized solar cells. *Advanced Energy Materials, 6*(20), 1601048.

Roggen, D., Tröster, G., & Bulling, A. (2013). Signal processing technologies for activity-aware smart textiles. *Multidisciplinary know-how for smart-textiles developers* (pp. 329–365). Philadelphia, PA: Elsevier, Woodhead Publishing Series in Textiles.

Rundqvist, K., Nilsson, E., Lund, A., Sandsjö, L., & Hagström, B. (2014). *Piezoelectric textile fibres in woven constructions*, Ambience 1410i3m, Tampere, Finland, September 7–9, 2014.

Sahito, I. A., Sun, K. C., Arbab, A. A., Qadir, M. B., Choi, Y. S., & Jeong, S. H. (2016). Flexible and conductive cotton fabric counter electrode coated with graphene nanosheets for high efficiency dye sensitized solar cell. *Journal of Power Sources, 319*, 90–98.

Sahito, I. A., Sun, K. C., Arbab, A. A., Qadir, M. B., & Jeong, S. H. (2015). Graphene coated cotton fabric as textile structured counter electrode for DSSC. *Electrochimica Acta, 173*, 164–171.

Samad, Y. A., Li, Y., Alhassan, S. M., & Liao, K. (2014). Non-destroyable graphene cladding on a range of textile and other fibers and fiber mats. *RSC Advances, 4*(33), 16935–16938. doi:10.1039/C4RA01373E.

Shateri-Khalilabad, M., & Yazdanshenas, M. E. (2013). Fabricating electroconductive cotton textiles using graphene. *Carbohydrate Polymers, 96*(1), 190–195.

Shim, B. S., Chen, W., Doty, C., Xu, C., & Kotov, N. A. (2008). Smart electronic yarns and wearable fabrics for human biomonitoring made by carbon nanotube coating with polyelectrolytes. *Nano Letters, 8*(12), 4151–4157.

Shimamoto, A., Yamashita, K., Inoue, H., Yang, S.-m., Iwata, M., & Ike, N. (2013). A nondestructive evaluation method: Measuring the fixed strength of spot-welded joint points by surface electrical resistivity. *Journal of Pressure Vessel Technology, 135*(2), 021501–021501-7. doi:10.1115/1.4007957.

Stoppa, M., & Chiolerio, A. (2014). Wearable electronics and smart textiles: A critical review. *Sensors, 14*(7), 11957–11992.

Takamatsu, S., Kobayashi, T., Shibayama, N., Miyake, K., & Itoh, T. (2012). Fabric pressure sensor array fabricated with die-coating and weaving techniques. *Sensors and Actuators A: Physical, 184*, 57–63.

Takamatsu, S., Lonjaret, T., Ismailova, E., Masuda, A., Itoh, T., & Malliaras, G. G. (2015). Wearable keyboard using conducting polymer electrodes on textiles. *Advanced Materials, 28*(22), 4485–4488.

Takamatsu, S., Yamashita, T., Imai, T., & Itoh, T. (2014). Lightweight flexible keyboard with a conductive polymer-based touch sensor fabric. *Sensors and Actuators A: Physical, 220*, 153–158.

Ten Bhömer, M., Tomico, O., Kleinsmann, M., Kuusk, K., & Wensveen, S. (2013, February 8–10). *Designing smart textile services through value networks; team mental models and shared ownership*. Paper presented at the ServDes. 2012 Conference Proceedings Co-Creating Services; The Third Service Design and Service Innovation Conference, Espoo, Finland.

Van Langenhove, L. (2007). *Smart textiles for medicine and healthcare: Materials, systems and applications*. Elsevier: Woodhead Publishing Limited, Cambridge, U.K.

Varesano, A., Aluigi, A., Florio, L., & Fabris, R. (2009). Multifunctional cotton fabrics. *Synthetic Metals, 159*(11), 1082–1089.

Wei, Y., Torah, R., Yang, K., Beeby, S., & Tudor, J. (2013). Screen printing of a capacitive cantilever-based motion sensor on fabric using a novel sacrificial layer process for smart fabric applications. *Measurement Science and Technology, 24*(7), 075104.

Xu, J., Li, M., Wu, L., Sun, Y., Zhu, L., Gu, S., … Xu, W. (2014). A flexible polypyrrole-coated fabric counter electrode for dye-sensitized solar cells. *Journal of Power Sources, 257*, 230–236.

Xu, L.-L., Guo, M.-X., Liu, S., & Bian, S.-W. (2015). Graphene/cotton composite fabrics as flexible electrode materials for electrochemical capacitors. *RSC Advances, 5*(32), 25244–25249.

Yamashita, T., Takamatsu, S., Miyake, K., & Itoh, T. (2013). Fabrication and evaluation of a conductive polymer coated elastomer contact structure for woven electronic textile. *Sensors and Actuators A: Physical, 195*, 213–218.

Yanılmaz, M., & Sarac, A. S. (2014). A review: Effect of conductive polymers on the conductivities of electrospun mats. *Textile Research Journal, 84*(12), 1325–1342.

Yoo, H.-J. (2013). Your heart on your sleeve: Advances in textile-based electronics are weaving computers right into the clothes we wear. *IEEE Solid-State Circuits Magazine, 5*(1), 59–70.

Yu, G., Hu, L., Vosgueritchian, M., Wang, H., Xie, X., McDonough, J. R., & Bao, Z. (2011). Solution-processed graphene/MnO_2 nanostructured textiles for high-performance electrochemical capacitors. *Nano Letters, 11*(7), 2905–2911. doi:10.1021/nl2013828.

Yun, Y. J., Hong, W. G., Kim, W.-J., Jun, Y., & Kim, B. H. (2013). A novel method for applying reduced graphene oxide directly to electronic textiles from yarns to fabrics. *Advanced Materials, 25*(40), 5701–5705. doi:10.1002/adma.201303225.

Zou, D., Ping, M., Cai, X., & Liu, G. (2015). *Efficient energy devices in the fiber shape*. Paper presented at the Photonics for Energy, Wuhan, China.

Index

B

C

D

R

S

For Product Safety Concerns and Information please contact our EU representative GPSR@taylorandfrancis.com Taylor & Francis Verlag GmbH, Kaufingerstraße 24, 80331 München, Germany

T - #0324 - 160425 - C63 - 234/156/18 - PB - 9781138746336 - Gloss Lamination